BIOMASS, BIOFUELS, BIOCHEMICALS

Series Editor

Ashok Pandey

Centre for Innovation and Translational Research,
CSIR-Indian Institute of Toxicology Research,
Lucknow, India

BIOMASS, BIOFUELS, BIOCHEMICALS
Green-Economy: Systems Analysis for Sustainability

Edited by

GANTI S. MURTHY

Biological and Ecological Engineering, Oregon State University, Corvallis, OR, United States; Biosciences and Biomedical Engineering, Indian Institute of Technology, Indore, India

EDGARD GNANSOUNOU

Ecole Polytechnique Fédérale de Lausanne (EPFL), School of Environment, Civil Engineering, and Architecture, Institute of Civil Engineering, Bioenergy and Energy Planning Research Group, Lausanne, Switzerland

SAMIR KUMAR KHANAL

Department of Molecular Biosciences & Bioengineering, University of Hawai'i at Manoa, Honolulu, USA; Department of Biology, Hong Kong Baptist University, Kowloon Tong, Hong Kong

ASHOK PANDEY

Centre for Innovation and Translational Research, CSIR-Indian Institute of Toxicology Research, Lucknow, India

ELSEVIER

Elsevier
Radarweg 29, PO Box 211, 1000 AE Amsterdam, Netherlands
The Boulevard, Langford Lane, Kidlington, Oxford OX5 1GB, United Kingdom
50 Hampshire Street, 5th Floor, Cambridge, MA 02139, United States

Copyright © 2022 Elsevier BV. All rights reserved.

No part of this publication may be reproduced or transmitted in any form or by any means, electronic or mechanical, including photocopying, recording, or any information storage and retrieval system, without permission in writing from the publisher. Details on how to seek permission, further information about the Publisher's permissions policies and our arrangements with organizations such as the Copyright Clearance Center and the Copyright Licensing Agency, can be found at our website: www.elsevier.com/permissions.

This book and the individual contributions contained in it are protected under copyright by the Publisher (other than as may be noted herein).

Notices
Knowledge and best practice in this field are constantly changing. As new research and experience broaden our understanding, changes in research methods, professional practices, or medical treatment may become necessary.

Practitioners and researchers must always rely on their own experience and knowledge in evaluating and using any information, methods, compounds, or experiments described herein. In using such information or methods they should be mindful of their own safety and the safety of others, including parties for whom they have a professional responsibility.

To the fullest extent of the law, neither the Publisher nor the authors, contributors, or editors, assume any liability for any injury and/or damage to persons or property as a matter of products liability, negligence or otherwise, or from any use or operation of any methods, products, instructions, or ideas contained in the material herein.

British Library Cataloguing-in-Publication Data
A catalogue record for this book is available from the British Library

Library of Congress Cataloging-in-Publication Data
A catalog record for this book is available from the Library of Congress

ISBN: 978-0-12-819242-9

For Information on all Elsevier publications visit our website at
https://www.elsevier.com/books-and-journals

Publisher: Susan Dennis
Acquisitions Editor: Kostas Marinakis
Editorial Project Manager: Andrea Dulberger
Production Project Manager: Joy Christel Neumarin Honest Thangiah
Cover Designer: Greg Harris

Typeset by Aptara, New Delhi, India

Contents

Contributors   *xi*
Preface   *xiii*

1. Systems analysis and its relevance for the sustainability transitions   1
Ganti S. Murthy, Ashok Pandey

1.1 Introduction   1
1.2 Importance of systems analysis for sustainable development   2
1.3 Understanding the systems   3
1.4 Structure and behavior of systems   4
1.5 Making sense of data and understanding bias in analyzing systems   12
1.6 Relevance of systems analysis for a transition to bioeconomy   13
1.7 Conclusions and perspectives   15
References   15

2. Techno-economic assessment   17
Ganti S. Murthy

2.1 Introduction   17
2.2 Different methods used in techno-economic analysis/assessment   17
2.3 Basic Steps of techno-economic analysis/assessment   18
2.4 Uncertainty and sensitivity analysis   22
2.5 Real option analysis   24
2.6 Tools, software, and data sources to conduct techno-economic analysis/assessment   25
2.7 Worked example   26
2.8 Conclusions and perspectives   29
References   30

3. Environmental impacts   33
Ganti S. Murthy

3.1 Introduction   33
3.2 Methods used for assessing the environmental impacts   33
3.3 Life cycle assessment   37
3.4 Life cycle assessment/analysis methodology   39
3.5 Life cycle assessment/analysis software and life cycle inventory databases   47
3.6 Worked example   48
3.7 Perspectives   50
3.8 Conclusions and perspectives   50
References   50

4. Environmental risk assessment — 53
Ganti S. Murthy
4.1 Introduction — 53
4.2 What is risk analysis? — 54
4.3 Risk analysis method — 56
4.4 Databases, tools, and software — 65
4.5 Examples — 65
4.6 Perspectives — 69
4.7 Conclusions and perspectives — 73
References — 73

5. Resource assessment — 75
Ganti S. Murthy
5.1 Introduction — 75
5.2 Land resources — 76
5.3 Water resources — 81
5.4 Nutrient resources — 86
5.5 Metals and minerals — 90
5.6 Examples — 91
5.7 Conclusions and perspectives — 96
References — 96

6. Policy, governance, and social aspects — 99
Ganti S. Murthy
6.1 Introduction — 99
6.2 Complexities of policy making — 99
6.3 Commonly used policy making models — 101
6.4 Policy making frameworks — 103
6.5 Social and governance aspects — 105
6.6 Case studies — 106
6.7 Conclusions and perspectives — 109
References — 110

7. Resilience thinking — 113
Ganti S. Murthy
7.1 Introduction — 113
7.2 Understanding and quantifying resilience — 115
7.3 Resilience thinking in systems analysis — 123
7.4 Conclusions and perspectives — 124
References — 126

8. General logic-based method for assessing the greenness of products and systems — 127
Edgard Gnansounou
- 8.1 Introduction — 127
- 8.2 The sustainability value added — 129
- 8.3 The logic-based model — 131
- 8.4 Application for assessing the sustainability of products and systems — 135
- 8.5 Conclusions and perspectives — 145
- References — 145

9. A systems analysis of first- and second-generation ethanol in the United States — 147
Ganti S. Murthy
- 9.1 Introduction — 147
- 9.2 Systems analysis of ethanol technologies — 153
- 9.3 Conclusions and perspectives — 172
- References — 172

10. Solar energy in India — 175
Ganti S. Murthy
- 10.1 Introduction — 175
- 10.2 Development of solar energy in India — 180
- 10.3 Challenges to solar energy in India — 182
- 10.4 Innovative responses to the challenges — 188
- 10.5 Overall scenario — 191
- 10.6 Conclusions and perspectives — 192
- References — 193

11. A systems analysis of solar and wind energy in the United States — 195
Kyle Proctor, Ganti S. Murthy
- 11.1 Introduction — 195
- 11.2 Technical feasibility analysis — 196
- 11.3 Environmental Impact assessment — 199
- 11.4 Resource sustainability analysis — 201
- 11.5 Policy, governance, and social impact analysis — 205
- 11.6 Conclusions and perspectives — 207
- References — 207

12. Biofuels and bioproducts in India — 209
Ganti S. Murthy
- 12.1 Introduction — 209
- 12.2 Systems analysis of biofuel technologies — 209

12.3	Resource assessment for bioethanol from agricultural residues	210
12.4	Techno-economic analysis	216
12.5	Environmental impact assessment	222
12.6	Policy and social aspects of biofuels in India	224
12.7	Conclusions and perspectives	228
	References	229

13. A case study on integrated systems analysis for biomethane use — 231
Sarath C. Gowd, Deepak Kumar, Karthik Rajendran

13.1	Introduction	231
13.2	Dimensions of systems analysis	233
13.3	Case study of Ireland for biomethane use	238
13.4	Conclusions and perspectives	240
	References	241

14. Alternative ammonia production processes and the use of renewables — 243
Gal Hochman, Alan Goldman, Frank A. Felder

14.1	Introduction	243
14.2	Ammonia production via current practices	244
14.3	Haber–Bosch using electrochemical H_2 production (E/H–B)	248
14.4	Direct electrochemical nitrogen reduction	248
14.5	Conclusions and perspectives	255
	References	256

15. Regional strategy of advanced biofuels for transportation in West Africa — 259
Edgard Gnansounou, Bénédicte Nsalambi

15.1	Introduction	259
15.2	Case of West Africa	261
15.3	Conclusions and perspectives	274
	References	277

16. Advanced biofuels for transportation in West Africa: Common referential state-based strategies — 279
Edgard Gnansounou, Elia Ruiz Pachón

16.1	Introduction	279
16.2	Types of feedstock for advanced biofuels	281
16.3	Biofuels for transportation	283
16.4	Cases of West African states	287
16.5	Conclusions and perspectives	309
	References	309

17. Semantic sustainability characterization of biorefineries: A logic-based model 311
Edgard Gnansounou, Catarina M. Alves, Elia Ruiz Pachón, Pavel Vaskan

17.1 Introduction 311
17.2 The problematic of sustainability characterization 313
17.3 Case study 316
17.4 Conclusions and perspectives 337
References 338

18. Solid biofuels 343
Ashish Manandhar, Seyed Hashem Mousavi-Avval, Jaden Tatum, Esha Shrestha, Parisa Nazemi, Ajay Shah

18.1 Introduction 343
18.2 Solid biofuel types 344
18.3 Solid biofuel properties 348
18.4 Chemical properties 350
18.5 Costs of solid biofuels supply 352
18.6 Life-cycle environmental impacts 355
18.7 Solid biofuel policies 361
18.8 Opportunities for using solid biofuels 363
18.9 Challenges for solid biofuels 363
18.10 Conclusions and perspectives 365
References 365

19. Potential value-added products from wineries residues 371
Prasad Mandade, Edgard Gnansounou

19.1 Introduction 371
19.2 A large diversity of wastes/residues of grape 373
19.3 Valorization of the residues and wastes 375
19.4 Proposed biorefinery scenario using zero-waste cascading valorization of wastes and residues 387
19.5 Conclusions and perspectives 390
References 391

Index 397

Contributors

Catarina M. Alves
Ecole Polytechnique Fédérale de Lausanne (EPFL), School of Environment, Civil Engineering, and Architecture, Institute of Civil Engineering, Bioenergy and Energy Planning Research Group, Lausanne, Switzerland

Frank A. Felder
Center for Energy, Economic & Environmental Policy, Edward J. Bloustein School of Planning and Public Policy, Rutgers, The State University of New Jersey, New Brunswick, NJ, United States

Edgard Gnansounou
Ecole Polytechnique Fédérale de Lausanne (EPFL), School of Environment, Civil Engineering, and Architecture, Institute of Civil Engineering, Bioenergy and Energy Planning Research Group, Lausanne, Switzerland

Alan Goldman
Department of Chemistry and Chemical Biology, Rutgers, The State University of New Jersey, New Brunswick, NJ, United States

Sarath C. Gowd
Department of Environmental Science, SRM University-AP, Amaravati, India; School of Engineering and Applied Sciences, SRM University-AP, Amaravati, India

Gal Hochman
Department of Agriculture, Food & Resource Economics, Rutgers, The State University of New Jersey, New Brunswick, NJ, United States

Deepak Kumar
Department of Chemical Engineering, State University of New York College of Environmental Science and Forestry, Syracuse, NY, United States

Ashish Manandhar
Department of Food, Agricultural and Biological Engineering, The Ohio State University, Wooster, OH, United States

Prasad Mandade
Ecole Polytechnique Fédérale de Lausanne (EPFL), School of Environment, Civil Engineering, and Architecture, Institute of Civil Engineering, Bioenergy and Energy Planning Research, Group, Lausanne, Switzerland

Seyed Hashem Mousavi-Avval
Department of Food, Agricultural and Biological Engineering, The Ohio State University, Wooster, OH, United States

Ganti S. Murthy
Biological and Ecological Engineering, Oregon State University, Corvallis, OR, United States; Biosciences and Biomedical Engineering, Indian Institute of Technology, Indore, India

Parisa Nazemi
Department of Food, Agricultural and Biological Engineering, The Ohio State University, Wooster, OH, United States

Bénédicte Nsalambi
Ecole Polytechnique Fédérale de Lausanne (EPFL), School of Environment, Civil Engineering, and Architecture, Institute of Civil Engineering, Bioenergy, and Energy Planning Research Group, Lausanne, Switzerland

Elia Ruiz Pachón
Ecole Polytechnique Fédérale de Lausanne (EPFL), School of Environment, Civil Engineering, and Architecture, Institute of Civil Engineering, Bioenergy and Energy Planning Research Group, Lausanne, Switzerland

Ashok Pandey
Centre for Innovation and Translational Research, CSIR-Indian Institute of Toxicology Research, Lucknow, India

Kyle Proctor
Biological and Ecological Engineering, Oregon State University, Corvallis, OR, United States

Karthik Rajendran
Department of Environmental Science, SRM University-AP, Amaravati, India

Ajay Shah
Department of Food, Agricultural and Biological Engineering, The Ohio State University, Wooster, OH, United States

Esha Shrestha
Department of Food, Agricultural and Biological Engineering, The Ohio State University, Wooster, OH, United States

Jaden Tatum
Department of Food, Agricultural and Biological Engineering, The Ohio State University, Wooster, OH, United States

Pavel Vaskan
Ecole Polytechnique Fédérale de Lausanne (EPFL), School of Environment, Civil Engineering, and Architecture, Institute of Civil Engineering, Bioenergy and Energy Planning Research Group, Lausanne, Switzerland

Preface

Assessing the sustainability of green economy focusing on the emerging bioeconomy is a strong imperative and is an integral part of evaluating new technologies, processes/unit operations, products, and policies. It is increasingly recognized by academia, industry, government, and society that sustainability is multifaceted, and interdisciplinary systems analysis approaches are required to provide a comprehensive understanding. Various elements of sustainability analysis include assessing technical feasibility, economic viability, environmental impacts, resource sustainability, and social aspects of engineered systems. Emerging green, circular economy, and bioeconomy concepts incorporating biofuels and bioproducts, resource recovery from wastes/residues and utilization of technologies must have a long-term sustainability focus right from ideation.

Increasingly, there is a recognition that comprehensive sustainability analysis is necessary for evaluating technologies/processes or policies for the emerging green economy. While the scientific literature has many examples of qualitative analyses, quantitative, and objective analyses are sorely lacking. This is due to a lack of necessary interdisciplinary integration in curriculums across the world. Individual aspects such as techno-economic analysis, sustainability analysis, policy aspects are covered in various disciplines; however, there are currently no books that address the systems analysis for sustainability comprehensively. This book addresses this critical gap in the available books for students and practitioners of sustainability analysis at the systems level.

Systems analysis for sustainability is an emerging discipline where any technologies/processes or policies are evaluated comprehensively for sustainability. Trifold sustainability metrics such as technical feasibility, economic viability, and environmental impacts are commonly used to assess sustainability. In addition to these metrics, it is important to consider resource sustainability, policies, and social aspects for evaluating the sustainability of any proposed alternative. This approach results in identifying possible pitfalls early in the process so that truly sustainable alternative technologies, products, and policies can be formulated. This book provides a theoretical background to perform such analyses. It demonstrates the application of these principles in practice with case studies carefully chosen from around the world to emerging technologies in the green economy.

The book is written with a focus on enabling interdisciplinary understanding and cross-fertilization of ideas and methods from various disciplines. The book could be adopted as a reference book for interdisciplinary courses in environmental, ecological, chemical, and mechanical engineering disciplines. It is organized into two main sections. The first nine chapters of the book lay a comprehensive foundation for systems analysis of bioeconomy, and various methods used for performing systems analysis of biofuels. The chapters focus on techno-economic analysis, environmental impact assessments, environmental risk analysis, resource sustainability analysis, policy, and social aspects, and general logic-based methods. The last chapter in the first part of the book discusses the emerging concept of resilience and resilience thinking in the context of adaptations to climate change and bioeconomy. All the chapters in the first part of the book present theoretical foundations for each technique, quantitative, and qualitative frameworks, and worked examples to aid the student in a solid understanding of these techniques. The second part of the book presents case studies from around the world starting with chapters focused on the liquid biofuels and solar energy potential for the United States and India. Two chapters are devoted to assessing the potential for various biofuels production in the African continent. The integrated analysis of biomethane in Ireland, solid biofuels production in the United States, utilization of winery residues is other case studies included in the second part of the book. With the increasing fraction of grid electricity being produced from the renewable sources, there is a need to utilize the excess electricity produced. One chapter specifically focuses on the decentralized ammonia production using excess renewable electricity thus potentially displacing the fossil fuel-dependent Haber–Bosch process. It is hoped that the reader will find these case studies provide concrete examples of the application of system analysis techniques for bioeconomy.

Preparing this book was indeed a challenge. As with any challenging project, this is a product of a wonderful team of authors with excellent teamwork. We would like to thank all the authors for their participation in the preparation of the chapters. We would like to acknowledge the reviewers for their constructive comments. We thank the entire Elsevier team led by Dr. Marianakis Kostas, Senior Book Acquisition Editor and Andrea Dulberger, Editorial Project Manager for their patience, understanding, and help during this long and arduous process of completing this book. We sincerely hope this book will be useful to senior undergraduate students, beginning graduate students, and researchers.

Editors
Ganti S. Murthy
Edgard Gnansounou
Samir Kumar Khanal
Ashok Pandey

CHAPTER ONE

Systems analysis and its relevance for the sustainability transitions

Ganti S. Murthy[a,b], Ashok Pandey[c]
[a]Biological and Ecological Engineering, Oregon State University, Corvallis, OR, United States, [b]Biosciences and Biomedical Engineering, Indian Institute of Technology, Indore, India, [c]Centre for Innovation and Translational Research, CSIR-Indian Institute of Toxicology Research, Lucknow, India

1.1 Introduction

We are living in Anthropocene, an era in which human activity is having a significant impact on the Earth's ecosystems and geology. Realizing this simple fact requires a deep appreciation and acceptance of the tectonic changes in the biosphere and the planetary system that are taking place due to human activity [1]. Remarkable 1972 study, "limits to growth" pointed to the unsustainable trajectories of humanity [2]. Unfortunately, the trajectories remained largely in line with the business-as-usual case even after 30 years [3]. How does humanity extricate itself from the unsustainable trajectories of population growth, resource consumption, and over-exploitation of the natural resources from a bleak future and possible collapse? The thinking and action patterns that have resulted in the current situation must be changed. But how? As Einstein said, "We cannot solve our problems with the same thinking we used when we created them." We cannot continue to use paradigms of unending growth in a world of inexhaustible resources. The reductionist and linear causal thinking that led to vast improvements in human conditions but put humanity on an unsustainable path cannot be the future. The thought and action patterns that worked when the population of the world was 500 million will not work in a world of 7.5 billion that uses several times the per capita natural resources to sustain the way of life.

The human systems for producing food, managing our societies were built and have evolved over a long time. While much of the study of these human systems are mostly focused on the recent events, understanding how these systems were built, how they evolved over time is important for visualizing and planning for possible future sustainable scenarios. How the human societal, technological, economic, and political systems evolved contains important information regarding the elements, functions, and their relationships which is critical to understanding the various characteristics of the system. Simply observing the past and current events and using them to visualize the future is bound to fail as the systems we have evolved and their interactions with the natural systems are too complex to be modeled and require systems thinking.

Richmond defined systems thinking as "Systems Thinking is the art and science of linking structure to performance, and performance to structure—often for purposes of changing structure (relationships) so as to improve performance" [4]. Systems thinking paradigm recognizes the common adage 'sum is greater than the parts [5]. Systems thinking looks at the systems as a whole to understand the components, their functions, and their interactions. Systems thinking helps in developing a greater appreciation and understanding of the individual parts and their interactions. Such a paradigm is needed to understand the complex sub-systems for the extraction of resources from nature, produce adequate food and feed, maintain our national and international trading, and preserving our societal and cultural identities.

1.2 Importance of systems analysis for sustainable development

Systems thinking uses systems analysis tools for conceptualization of the problems and possible scenarios [6]. Systems analysis consists of multiple aspects including (1) mapping the structures and flows of the systems including the elements, functions, and relationships; (2) drawing of the causal loops with clear identification of stocks, flows, feedback, and delays; (3) system dynamics. Systems analysis also uses back of envelope calculations to quantify the system flows and stocks, and use of statistics, modeling software, spreadsheets to quantify various aspects of the systems [6]. Systems analysis tools in the engineering and military domain were developed from the operations research after the second world war [7]. System analysis tools can help us develop various perspectives of the system that helps us understand why a system behaves in the way it does and how it will behave if it is perturbed in a particular way. It is important to realize that systems analysis is not a prediction tool. Systems analysis tools will only help in generating multiple "what if" scenarios, i.e., possible future scenarios. Scenario that will play out in the future is dependent on the actions of the individuals, society, and the systems response to those actions.

Many solutions have been proposed for moving humanity on a more sustainable path. Since the socio-economic-ecological system is extremely complex with so many feedback loops that are not understood well or not even identified at all, it is impossible to plan with certainty. It has been well recognized that we cannot consume our way out to a sustainable path [8]. This is often seen in cases where the individual actions however well-intentioned may move the system deeper into unsustainable paths if not carefully thought out. All these challenges point to the need for a careful analysis from a systems perspective before we embark on the sustainability adaptations and transformations of systems toward sustainability. It is very important to recognize here, that the call to systems analysis is not a call to abandon reductionist analysis of individual systems. The reductionist approaches can teach us a lot about the components of the systems while the systems analysis will help us in understanding their behavior in a system. Combining these two approaches will help in developing truly sustainable solutions to the global existential challenges we face today.

1.3 Understanding the systems

All systems are composed of elements, interrelationships, and functions/purpose. All systems have a purpose which is property of the whole system and not of the individual parts. While all three are essential for a functioning system, the function/purpose is the heart of gives the system its character. If the function of the purpose of the system is changed, the system no longer remains the original system. The individual parts of the system function together through their various interconnections to fulfill the purpose of the system [5]. For example, a football team consists of players (elements) who function together through various interdependencies and interactions in the field as per the rules of the game to win a match. The purpose of the football team as a system is to win a match as per the rules of the game. All parts of the system help in fulfilling the purpose but cannot change the purpose of the system. Note that the individual players can be replaced with other players and the functions and interdependencies changed but the system will still retain its purpose. As long as the system retains its purpose, the system can be said to have retained its character and is the same system. The structures of the system form the heart of a system which results in the patterns of system behavior. Patterns of behavior give rise to the events that are mostly observed in the real world. It is very difficult to discern the structures of the system just from the events. It requires long-term observation of the trends, patterns to understand the underlying structures of the system [9]. Interconnectedness, circular feedback loops instead of linear causality, emergent behaviors, a whole system outlook, and identifying relationships between various components of the systems are six tools to identify the underlying system [10].

The linear event-based thinking (Fig. 1.1) that dominates the conventional problem-solving processes focuses can perceive the events and their timelines effectively [11]. It is difficult or impossible for such analysis to capture fundamental behavior patterns and the system response to various inputs that culminate in a manifested event for a complex system [5]. For example, the financial systems are complex, and the prices of commodities vary over time. Econometrics can help in fitting the time series data using various techniques and can "predict" the future prices as long as the inputs do not change. If the underlying inputs to the systems change resulting in a fundamentally different behavior of the system, the regression-based predictions will not be able to capture the changes [5]. It is in such situations that systems thinking can provide a valuable alternative.

Fig. 1.1 Linear event-based thinking (adapted from [11]).

1.4 Structure and behavior of systems

System dynamics is used to understand the system's responses to various changes in the system. Since the system structure determines the response of the system to various changes, the systems thinking places great emphasis on understanding the underlying structure of the systems. Identifying the structures of the systems is one of the most important aspects of systems thinking. A few fundamental motifs, called archetypes in systems thinking literature, are found to be repeating in various systems (Fig. 1.2). All fundamental archetypes/motifs are made of stocks and flows. The stocks are the quantifiable and

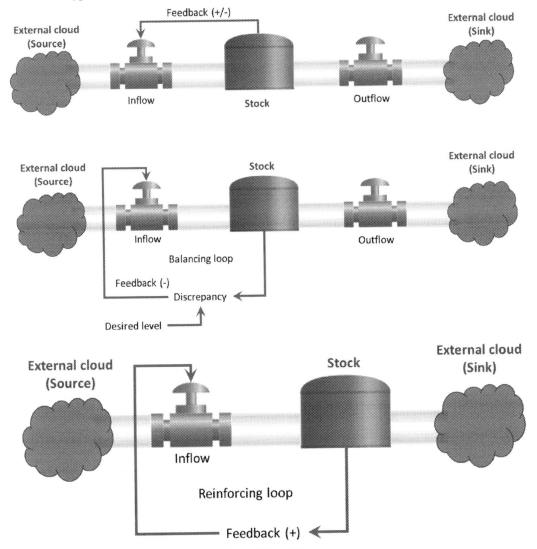

Fig. 1.2 Basic components of systems (adapted from [5]).

tangible aspects of the system and contain integrates the information on changing flows within the system. The flows depict the exchange of material/information flows between the stocks. The stocks and flows together form two types of loops: reinforcing and balancing loops. Reinforcing loops have a self-strengthening feedback that results in exponential runaway growths and ultimately a collapse. A nuclear fission or bacterial growth on the unlimited substrate are examples of a reinforcing loop. The balancing loops on the other hand have an inbuilt mechanism to moderate the responses and achieve an equilibrium over time. The balancing loops can be a source of stability and inertia in the system. Earth-orbiting the Sun in a stable orbit and circulation patterns in the Earth's oceans are examples of balancing loop. In addition to the reinforcement and balancing loops made of stocks and flows, the systems also have delay elements. Delay in a reinforcement loop slows the response of the system while the delays in balancing loop can lead to oscillatory behavior. It is important to note that the strength of the various inputs and the loops vary with time and different parts of the system may dominate over one another thus governing the system outputs. The systems with similar structure will have similar behaviors [5]. Just as few motifs found in the biological networks are responsible for a wider range of behaviors, the archetypes that arise from these simple components give rise to a wide range of real-world behaviors that are difficult to capture using the linear thinking models.

All archetypes are made of reinforcing and balancing loops with delays. There are many types of archetypes described in the literature. Some of which are listed below.
1. Fixes that fail: this archetype occurs when the problems occurring in the system are fixed using a short-term quick-fix solution that does not necessarily solve the underlying issues (Fig. 1.3). Unfortunately, the system structure and dynamics mean

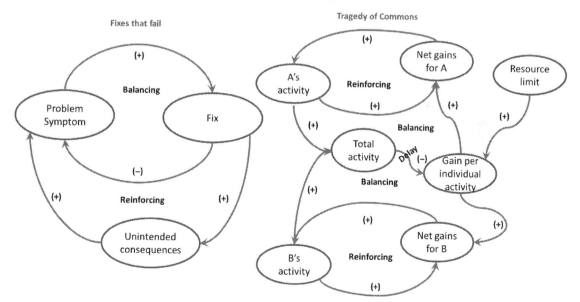

Fig. 1.3 System archetypes: fixes that fail and tragedy of commons (adapted from [5,9]).

that there are unintended consequences that appear with a delay, counteracting the quick-fix solutions, and building destabilizing pressures in the original system. The underlying cause of the system resistance to new policy/fix is the nonalignment of the goals of various actors in the system. The best way to come out of the negative consequences of such an archetype is to abandon the ineffective policies or fixes and implement a solution that addresses the fundamental causes and aligns the goals of the actors with the system goals. Expending any more energy on a failing policy/fix will only build more resentment among the actors and require an increasing amount of energy to maintain the system functions. Examples of the fixes that fail archetype is taking a loan to make the interest payments on a previous loan. While taking a new loan temporarily solves the immediate need for additional finances, it creates a bigger problem in the near future. The best course of action for such a situation is to divest, sell off some of the assets, or declare a bankruptcy and rationalize the business structure.

2. The tragedy of commons: when a common resource is used by many actors and the condition of the common resource does not influence the decisions of the actors, the system will always be overexploited and result in a systemic collapse (Fig. 1.3). This happens because the feedback about the condition of the common resource is absent, weak, or has a very long delay such that the immediate needs of the actors take precedence over the long-term concerns regarding the overexploitation of the common resource. One of the important characteristics of such an archetype is that the benefits of the use of common resource are owned by the individual actor but the costs are shared across the system. Over-fishing in the world's oceans is a current example of the tragedy of the commons. No individual fisherman has an incentive to reduce the fish catch and is in fact incentivized to exploit the diminishing fish supplies using even more advanced technologies to obtain more profit in the near term. Ultimately, the overexploitation of the limited resource due to the actions of all actors has resulted in the collapse of the ocean fisheries. Similarly, the costs of the global greenhouse gas emissions are shared by the world through climate change impacts but the benefits of cheap but polluting manufacturing processes are shared by an individual actor. The costs of GHG emissions will be paid in the long term and the feedback mechanisms are very weak while the incentives from the emissions are immediate. Therefore, we can observe huge disagreements among countries over mandatory commitments to reduce greenhouse gas emissions. The best way to break out of a tragedy of commons situation is to strengthen the feedback, educate, and sensitize the actors about the long-term consequences, and regulating the actors by imposing short-term costs. Native American Chief Seattle said, "We do not inherit the earth, we borrow it from our children."

3. Drift to low performance: in this archetype, when there is a gap in between the actual and perceived performance goal or the desired system state, the actors in the

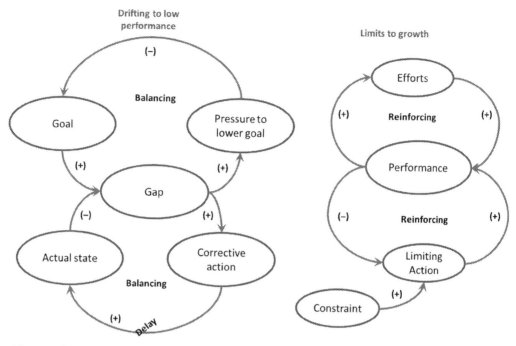

Fig. 1.4 System archetypes: drifting to low performance and limits to growth (adapted from [5,9]).

system react by lowering the performance goal (Fig. 1.4). This leads to a reinforcing loop of lower performance and a further lowering of the performance goals. The lowering of the performance goals or lowering the expectations from the system can happen slowly over the long term that the actors in the system do not perceive the lowering of the goals, but it nevertheless manifests in time. A common example of such an archetype in action is the death of a frog in slowly boiling water despite the opportunities for it to leave the vessel. The only way to extricate out of the negative effects of the reinforcing loops is to resist the urge to lower standards and keep the standards high regardless of the performance. A second approach is to make the goals more sensitive to the best performance than the worst performance. Using this strategy, the same feedback loops can be used to improve the performance of the system over time by reinforcing the improvements in performance over time.

4. Limits to growth: this is a classic archetype that was also used as the title of the report by the pioneers of the systems thinking field [2]. This archetype occurs when the growth in the system is limited by external resource constraints (Fig. 1.4). Initially, the system grows in a reinforcing loop, but the increasing growth/performance strengthens a balancing loop which gains strength and reduces the performance/growth. This combination of a reinforcing and balancing loops results in a limit to the maximum possible performance/growth in the systems. The ultimate state of

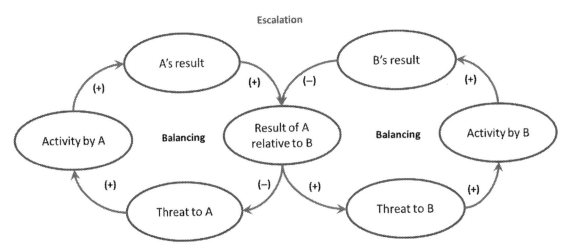

Fig. 1.5 System archetypes: escalation (adapted from [5,9]).

this archetype is the system collapse unless there is the input of new resources into the system. Growth of bacteria in a Petri dish with a limited substrate is a classic example of limits to growth archetype. Unfortunately, human society may also be on a similar dynamic of a limits to growth. As described earlier, the only way to come out of the undesirable end state is to shift to more of the renewable resources and move to a steady-state utilizing the renewable resources.

5. Escalation: this archetype occurs when there are at least two stocks/actors is trying to surpass the state of another stock/actor, it leads to a mutually reinforcing loops leading to exponential escalation of the responses culminating in the collapse of one of the stocks/actors (Fig. 1.5). The cold war nuclear arms race was an example escalation that ended with the dissolution of the USSR. The only way to avoid an escalating response archetype is to avoid getting caught in an escalating system response. Unilateral disengagement/refusing to compete in unproductive escalatory ladder are the best ways to manage if caught in an escalatory archetype response.

6. Success to the successful: when the shared/common resources are allocated in proportion to the success of an actor, the availability of additional resources to that actor increases the chances of success and future allocation of even more resources thus building a positive reinforcing loop for the successful actor while building a negative reinforcement loop for the actors who are not successful (Fig. 1.6). Rising income inequalities in the society where the rich tend to get richer over time while the poor are getting poorer is an example of this archetype. The cycle if continued indefinitely can lead to the complete exclusion of the loosing actors leading to a build-up of resentment, non-cooperation/participation, and revolution from the masses who can no longer participate in the system. Many political and social revolutions that occurred historically can be traced to the unfortunate end of this archetype. Limiting

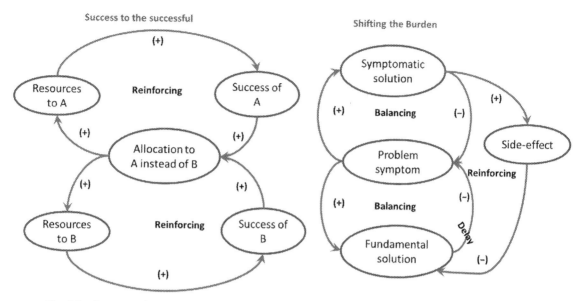

Fig. 1.6 System archetypes: success to the successful and Shifting the burden (adapted from [5,9]).

the gains by the successful beyond a certain point and leveling the field by removing some of the advantages of the successful or strengthening the weaker players and diversification are some of the strategies that can be used to mitigate the negative effects of this archetype. It must be noted that completely removing success as a basis for allocation of resources is also counterproductive as it leads to the elimination of incentives for any actor to compete and the mediocracy will rule. The system will then transition to drift to low-performance archetype in a race to the bottom.

7. Shifting the burden/addiction: this type of archetype occurs when a fundamental solution to a problem takes effect over a long time due to delays, but the symptomatic solution has no delay in improving problem symptom temporarily (Fig. 1.6). The symptomatic solution has a reinforcing loop of side effects which increase the system's dependence on the symptomatic solution at the cost of the more sustainable fundamental solution. A very common example of such an archetype is the consumption of caffeine to combat low energy levels without paying any attention to longer-term more viable solutions such as more exercise and rest. This leads to an increasing dependence on caffeine to address increasingly severe symptoms of lower energy levels. Another example is the formulation biodiesel policies of the EU that lead to increasing biodiesel production in Malaysia and Indonesia resulting in the shifting of the environmental burden to these countries. The best way to avoid getting into this archetype is by strengthening the capacity of the system to handle the stresses. If already in the archetype, a good way to handle it is through a renewed focus on the fundamental solutions while slowly disincentivizing the symptomatic solutions.

8. Rule beating: this archetype refers to the evasive actions taken by mostly lower levels of system hierarchy to subvert the intent of the system rules while appearing to follow it in a letter. This happens in the lower level of system hierarchy as a response to the overly rigid, unworkable, and ill-defined rules formulated in a top-down manner. Examples of rule beating are the shift in the passenger behavior to carry larger cabin luggage in response to the changes in the airlines policies to charge for the check-in luggage. Implementing the rule more strictly will only serve to build greater resentment and subversion of the intent of the rule. The rules must be redesigned to actually achieve the purpose/intent of the rules than the mere appearance of adherence by the participants through a consultative rule-making process.

9. Seeking the wrong goal: the function/purpose of the system is one of the fundamental aspects of the system behavior. If the purpose of the system is not defined carefully to be in sync with intent of the system's purpose, the system's behavior will produce a results that align with the stated system purpose but not necessarily the intent of the systems function. For example, if the higher gross national product (GNP) of a country is the stated goal with policies will encourage higher GNP without necessarily increasing the comfort or happiness of people. It has been shown in many surveys that solely pursuing the higher GDP does not necessarily increase the happiness of the citizens. In a commendable move, the Government of Bhutan has switched to Gross National Happiness as the measure of development. Identifying goals that align with their underlying intent and being careful to ensure that means are not confused with results (higher GNP is one of the means to increase welfare and happiness of citizens) is critical to overcome the negative consequences of this archetype.

10. Accidental adversaries: the parts of the system start off in a reinforcing synergistic loop that improves the success of both parts of the system. However, some of the unintended activities of one of the parties can lead to reactions from the other party that turns the negative reinforcing loop of increasing animosity turning former partners into adversaries (Fig. 1.7). An example of the archetype in action is the tension between the customer-facing marketing section and the design engineering teams through the scope creep. Providing new information late into the design cycle by the marketing team causes additional efforts and deadline slippages for the engineering team which responds by putting design freezes. In future, the marketing team will front-load the design goals to beat the design freeze strategy of the engineering team leading to avoidable tension between collaborators. To avoid the negative consequences of such an archetype, clear communication of intent and actions without delays, revisiting the shared vision periodically, and strengthening the collaborations are some of the actions that can be taken by the actors in the system.

11. Growth and underinvestment: this archetype occurs when the successful growth is limited by the capacity limitations which could be eliminated/reduced if investments are made (Fig. 1.8). The process loop to augment the capacity have a delay in

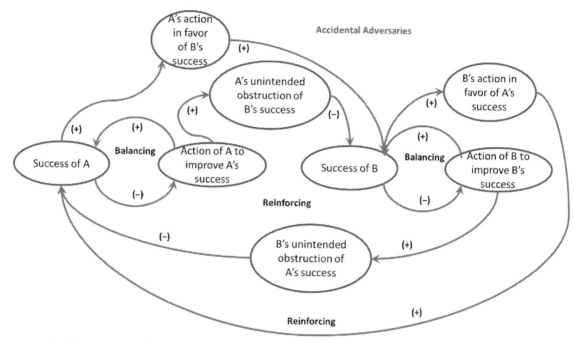

Fig. 1.7 System archetypes: accidental adversaries (adapted from [5,9]).

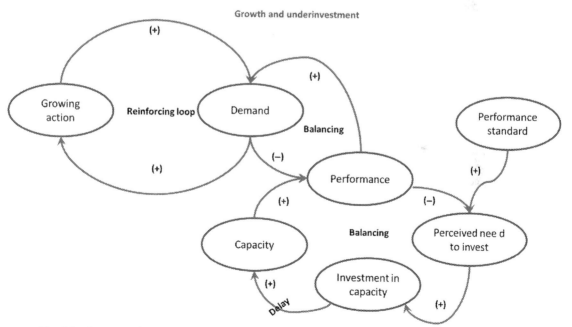

Fig. 1.8 System archetypes: growth and underinvestment (adapted from [5,9]).

perceiving the need for investment, making a policy decision to invest, committing resources to increase the capacity, and finally the increase in capacity. The growth is restricted due to lack of capacity before the process loop to increase in the capacity is completed leading to the withholding of the investment and further constricting the growth and reducing the overall performance. This archetype is commonly observed in vary large and older companies that have highly hierarchical and bureaucratic setups with long chains for decisions. The negative effects of this archetype can be reduced by looking into the criteria for capacity investments, reducing the delays in the decision processes, and building capacity for future growth proactively.

1.5 Making sense of data and understanding bias in analyzing systems

Understanding the system structure requires one to observe the system responses and the events for a reasonable amount of time and discern the trends and discard the noise from the signal. This is a particularly difficult task when looking at complex systems as the responses are not linear, data are noisy, and often incomplete. In a pioneering work [12], Hans Rosling et al. identify ten dramatic instincts that we must be aware of when interpreting data trends (Fig. 1.4). These instincts discuses below are critical when studying the systems as the systemic events that are most often observed are presented as the data and the systems analysts must make inferences about not only the system's events but also the underlying patterns of behavior, structures of the systems and most importantly the purpose of the system. This cannot be efficiently done without addressing some of the inherent biases of human mind when analyzing data.

1. The gap instinct: the gap instinct points to the common misconception of simply looking at the gap between the two extremes in the data. Looking at the complete distribution of the data where the majority of the data points lie, may show that the extreme gap is only at the continuous distribution of the observations more likely will show that the vast majority of the data lie somewhere in the middle of the perceived gap.
2. Negativity instinct: to control the negativity instinct, we must be aware that negative and bad news are reported and grabs our attention much more frequently than the good news and positive trends. Correcting for this reporting bias by expecting bad news and looking at the long-term trends rather than the isolated downward trends in a long-term positive trajectory can help in controlling this observational bias.
3. The straight-line instinct: linear extrapolation of complex systems far into the future and expecting the linear trends to continue is a straight-line instinct. This instinct can be controlled by asking the question about the validity of linear projections far into the future and imagining saturating responses.
4. Fear instinct: similar to the gap instinct, fear gets the attention and hence the dramatic attention to fearful things is the cause of a very common bias. Understanding

the real risk of the event rather than the biased fears and remembering that risk is a product of the probability of occurrence of an event and the consequences arising from it helps mitigate this instinct.
5. Size instinct: numbers reported without the context often seem large and hide the real magnitude of their relative importance. Comparing the sizes relative to the populations, understanding the rates, and identifying the most significant contributors are some of the ways to mitigate the bias associate with the size instinct.
6. Generalization instinct: this instinct happens when the data are categorized and used in explanations. The variabilities within the categories are important as are the similarities among various groups. Differences across groups and the size of the majority are some of the other aspects that should be considered when looking at data to mitigate the generalization instinct bias.
7. Destiny instinct: many systemic changes happen over long timelines. It is important to avoid instincts to assume constancy when looking at slowly changing systems. Keeping track of small improvements, continuous update of the data, and observing the differences in the system's values and behaviors at points separated by long timelines can help in mitigating this bias.
8. Single perspective instinct: observing a complex system from a single perspective can provide only a limited understanding of the system's behavior. Challenging ideas about the system, expanding the expertise and using multiple tools to analyze the problem, and being aware of simple solutions but completely wrong for complex problems are approaches to combat this bias.
9. Blame instinct: holding one single person or cause responsible for the system's behavior and outcomes (both positive and negative) is an unhealthy bias that prevents exploration of other possible explanations and causes. Understanding the causes by studying the underlying system dynamics is the best way to avoid this unfortunately too common bias.
10. Urgency instinct: calls for urgent action in response to perceived crisis must be examined critically. Often such urgent responses may only lead to the failing fixes or shifting the burden type archetypes discussed above. Taking time to understand the best course of long-term solutions, insisting on data, and being careful about drastic actions are some of the methods to combat the urgency instinct (Fig 1.9).

1.6 Relevance of systems analysis for a transition to bioeconomy

The world we live in is a complex world. The industrial, agricultural, financial, and resource extraction systems span across the globe. These systems were built in the last 300 years since the industrial revolution on a global scale. These systems are also tightly integrated and interact with the biosphere in inseparable complex ways. As has been discussed earlier, we are living Anthropocene and humanity is on an unsustainable trajectory. The need for transition to a more sustainability has been recognized widely

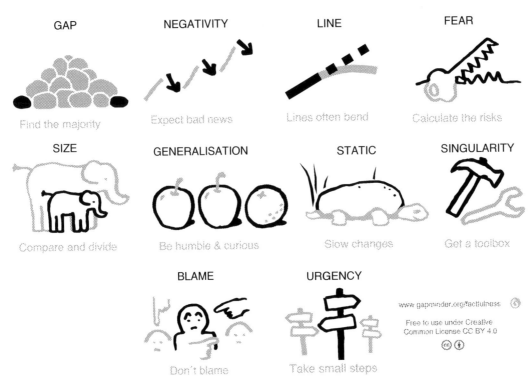

Fig. 1.9 Factfulness tools to identify the trends in data (figure from [13]).

and 15,364 scientists from 187 countries published "the worlds scientists' warning to humanity: A second notice" on the eve of 25 years of the first notice in 2017 [14]. The scientists note with concern that the current trends of ozone depletion, freshwater availability, marine life depletion, ocean dead zones, forest loss, biodiversity destruction, climate change, and continues human population growth are pushing the earth's ecosystems beyond its capacities to support life and we are approaching limits of causing and substantial and irreversible harm. The scientists note with alarm that these unsustainable trajectories have become continued since the publication of the "Worlds scientists' waning to humanity" in 1992 [14]. They identified few strategies (Table 1.1) that could put the humanity sustainable transitions to a more sustainable future.

The proposed sustainable bioeconomy involves shifts from the conventional fossil fuel based economy to renewable energy and biobased products. For the world that is addicted to fossil fuels and is dependent on extractive use of resources, this change represents a tectonic and truly historical shift. Such transitions have never been done in the past and hopefully, humanity will never need such massive transitions in future. These shifts represent a massive rewiring of the global financial, energy, food, and trading systems. To navigate the sustainable transition to a green economy requires a deep understanding of the human societal, technological, economic, and political systems.

Table 1.1 Effective steps for humanity for a transition to a sustainable future [14].
Effective Strategies for a sustainable transition to a sustainable future (not in the order of importance or urgency).

1. Prioritizing the enactment of connected well-funded and well-managed reserves for a significant proportion of the world's terrestrial, marine, freshwater, and aerial habitats.
2. Maintaining nature's ecosystem services by halting the conversion of forests, grasslands, and other native habitats.
3. Restoring native plant communities at large scales, particularly forest landscapes;
4. Rewilding regions with native species, especially apex predators, to restore ecological processes and dynamics.
5. Developing and adopting adequate policy instruments to remedy defaunation, the poaching crisis, and the exploitation and trade of threatened species.

Systems analysis can provide tools for developing such an understanding of the systems from a technical feasibility, economic viability, resource sustainability, environmental impacts, environmental risk, policy, and societal acceptability perspectives.

Combining the quantitative tools used such as technoeconomic analysis, life cycle assessment, environmental risk assessment, resource assessment, policy analysis, and societal acceptance with the systems dynamics can provide an unmatched comprehensive perspective of the various proposed technologies, policies, and strategies for the sustainability transition. Developing such a comprehensive view will help us transition into a sustainable and resilient future.

1.7 Conclusions and perspectives

The need for systems analysis for a transition to a more sustainable future from the fossil fuel based unsustainable world was discussed comprehensively. The development of systems thinking paradigm and various core concepts such as the importance of the structure in determining its behavior, the fundamental building blocks of the systems structures, eleven archetypes, their structure, and dynamics were discussed. Factfulness approaches for addressing inherent ten common biases and the methods to overcome them when analyzing complex data were discussed. Finally, the importance of the systems analysis tools for the green economy is discussed to stress the importance of understanding from multiple perspectives before the course of actions are determined for a sustainable and resilient future.

References

[1] N.G. Society, Anthropocene, National Geographic Society, 2019. http://www.nationalgeographic.org/encyclopedia/anthropocene/ [accessed February 13, 2021].
[2] D. Meadows, D. Meadows, J. Randers, J.W. Behrens III, Limits to Growth, Potomac Associates-Universe Books, New York, (1972). ISBN 0876631650. DOI: http:dx.doi.org/10.1349/ddlp.
[3] G. Turner, A comparison of *The Limits to Growth* with 30 years of reality, Global Environ. Change 18 (2008) 397–411.

[4] Richmond, B., 1994. Systems thinking/system dynamics: Let's just get on with it. System Dynamics Review 10, 135–157.
[5] D.H. Meadows, D. Wright, Thinking in Systems: a Primer (2009).
[6] Sverdrup, H. Systems thinking, systems analysis and systems dynamics in science. 2018. 81: https://sds.memberclicks.net/assets/conferences/2018/Systems%20thinking%2C%20systems%20analysis.pdf [accessed February 13, 2021].
[7] Hoag, M.W. An introduction to systems analysis. 1956. RM-1678. Project RAND. The RAND corporation, Santa Monica, CA, USA.
[8] Grant, L.K. Can we consume our way out of climate change? A call for analysis. Behav. Anal. 34:245-266.
[9] Kim, D. Introduction to systems thinking. The systems thinker. 2016. https://thesystemsthinker.com/introduction-to-systems-thinking/: [accessed February 13, 2021].
[10] Acaroglu, L. Tools for systems thinkers: the 6 fundamental concepts of systems thinking. 2021. https://medium.com/disruptive-design/tools-for-systems-thinkers-the-6-fundamental-concepts-of-systems-thinking-379cdac3dc6a [accessed February 13, 2021].
[11] P. Hjorth, A. Bagheri, Navigating towards sustainable development: a system dynamics approach, Futures 38 (2006) 74–92.
[12] H. Rosling, O. Rosling, A.R. Rönnlund, Factfulness: Ten Reasons We're Wrong About the World and Why Things are Better Than You Think, Flatiron Books, New York, NY, 2018.
[13] Gapminder. Factfulness poster. 2021. https://www.gapminder.org/factfulness/ [accessed 13 February 2021].
[14] W.J. Ripple, C. Wolf, T.M. Newsome, M. Galetti, M. Alamgir, E. Crist, M.I. Mahmoud, W.F. Laurance, 15, 364 scientist signatories from 184 countries, World scientists' warning to humanity: a second notice, BioScience 67 (2017) 1026–1028.

CHAPTER TWO

Techno-economic assessment

Ganti S. Murthy[a,b]
[a]Biological and Ecological Engineering, Oregon State University, Corvallis, OR, United States, [b]Biosciences and Biomedical Engineering, Indian Institute of Technology, Indore, India

2.1 Introduction

One of the most important components of systems analysis is the techno-economic analysis/assessment (TEA). TEA is used for assessing the economic performance of a process. TEA deals with the question of the most suitable combination of technologies that will result in the best economic performance. It is important to note that the most suitable technology is not always the best available technology/technologies but depends on the goals of the designer and the economic performance. Most of us perform such analysis very intuitively in our daily life. For example, when selecting a home appliance, many would balance the available features with the budgets available for the purchase. While making such decisions for such mundane situations poses a low economic and technical risk and hence can be made with little analysis. There is a need for systematic approaches when dealing with more complex projects and situations. Economic feasibility of the project can be assessed using TEA. Techno-economic analysis is a systematic way of making decisions about the choice of the combination of technologies and the economic analysis to arrive at optimal solutions.

2.2 Different methods used in techno-economic analysis/assessment

Techno-economic analysis is formally defined as the process to evaluate the technical and economic performance of a process, product or systems. As is the case with systems analysis, performing a good TEA is an interdisciplinary effort and requires sex-cellent engineering skills, very good knowledge of economics, a good understanding of uncertainties and statistical analysis, and deep domain knowledge of the technologies/systems being investigated. Often, good process modeling skills are invaluable when evaluating complex processes.

Once the candidate technologies and processes are identified, the goals of the TEA process are defined before beginning the analysis and will determine the tools required for performing the TEA analysis. The TEA process begins with the identification of the candidate technologies and processes. Then the economic analysis for process configuration is performed. Different levels of details/types of methods are

Biomass, Biofuels, Biochemicals
DOI: https://doi.org/10.1016/B978-0-12-819242-9.00019-1

often used for conducting the economic analysis. Commonly recognized methods for TEA analysis are [1]:

1. Methods for zero-order estimates: performing TEA analysis requires resources and often a preliminary economic analysis is warranted to determine if a detailed economic analysis is actually needed. Many quick methods are used in practice which is used only to obtain a preliminary idea of the cost-benefits. Static cost-benefit assessment is performed using the data for an average year to compare the benefits and costs. Annuity methods are a modification of the static cost-benefit assessment method where the initial capital investment and interest are distributed equally over the life of the project. Net cash flow methods are a modification of the annuity method and use actual costs and benefits rather than the average values. It is emphasized that these methods are only intended to be used to commit resources for more detailed economic analysis using the methods discussed below.

2. Net present value and internal rate of return methods: this method specifically incorporates the time value of money and is often also called the discounted cash flow method. The net present value (NPV) method recognizes the time value of money (i.e., discount rate is negligible) and hence $100 in the current time is worth more than a promise to pay $100 at a future date. Thus, future money is discounted at a discount rate which reflects the investor perception of the risk for the project. Economically viable projects must have a positive NPV.

The Internal rate of return is defined as the discount rate for which the project has a zero NPV and is calculated iteratively by varying the discount rate until the NPV of the project is zero. Just as NPV represents the investor perspective of the project, the IRR represents the project developer's view of the project. Therefore, a project developer will not go forward with the project unless the IRR > discount rate for the project. An IRR that is higher than the discount rate implies that the project developer expects the process to be profitable even after meeting the capital and operational costs, paying the interest, and other miscellaneous expenses. A positive NPV on the other hand implies that the investment is profitable for the investor and the discount rate accurately reflects the risk associated with this project for the investor. Major challenge with the NPV method is the accurate determination of the discount rate. Additionally, future cash flows are dependent on the inputs costs and revenues which are uncertain and therefore cannot be determined a priori with certainty.

2.3 Basic Steps of techno-economic analysis/assessment

As noted earlier, Tea can be performed at many levels and most commonly classified as the following when simulation processing plants based on the scope of analysis;

1. Order of magnitude estimate: ±10 to 50% accuracy.
2. Study estimate (factored estimate): ±30.

3. Preliminary estimate (budget authorization estimate): ±20.
4. Definitive estimate (project control estimate): ±10.
5. Detailed estimate (firm or contractor's estimate): ±5.

As can be seen from this classification, the accuracy and precision of the estimates are higher for the detailed estimate compared to the order of the magnitude estimate but also require a much higher amount of resources to perform. Typically, an order of magnitude analysis costs $3000–$13,000 while the detailed estimate costs could be more than $1,000,000 for complex plants [2].

Before embarking on a TEA, a preliminary evaluation of the project to assess the overall technical feasibility, necessary intellectual property resources, safety, availability of raw materials, and other inputs such as infrastructure and utilities, current, and future market size and competition must be assessed. Most of the TEA process uses the following steps:

1. Base design: all unit operations and major equipment used in the process are specified without necessarily specifying the size of the equipment. A process flow diagram connecting all the inputs, outputs, and utilities used in each unit operation is drawn along with the specifications for the process conditions which usually include the process flow rates, temperatures, pressures, and residence times.
2. Material and energy balance: the process flow diagram specified in the previous step is used to perform the mass and energy balances for the process conditions. The solution process is often iterative especially if the process incorporates any recycle streams. Design iterations for incorporating additional processes, investigating the combination of various processes, minimizing the utilities, maximizing the process recycle streams are performed in this step. While simpler processes can be solved in spreadsheet software, complex processes involving recycle streams thermodynamic calculations are most commonly solved using process simulation software.
3. Fixed capital expenses: after the performance of the mass and energy balances, the fixed capital expenses are estimated as the sum of the direct costs and indirect costs. Often the fixed capital costs (called CAPEX) are expressed as the cost per unit capacity of the plant to enable comparison of the CAPEX with other plants with different sizes and technologies. The CAPEX is dependent on the plant size and therefore care must be taken to compare CAPEX of plants that are of similar size to avoid errors. The direct costs are estimated based on the total installed equipment needed for the plant. The cost for the purchase of major equipment is called the purchase equipment cost (PEC) and is a very important cost that needs to be calculated based on the process equipment size. Often the size and pricing information available for the equipment does not match with the required sizes of the equipment and needs to be estimated from the available data. The new cost of equipment for a

new size based on the original size and cost information is typically calculated using the following power law equation:

$$\text{New cost} = \text{Original cost} \left(\frac{\text{New size}^*}{\text{Original size}^*} \right)^{\exp}$$

*or characteristic linearly related to the size.

The exponent (exp) ranges from 0.6 to 0.7 and can be found for specific equipment in one of the many engineering handbooks such as Perry's Chemical Engineering Handbook [3] and RS Means Handbook [4]. Once the new cost of equipment is obtained using the above equation, the costs need to be adjusted for age of the data as inflation and other factors lead to increase in equipment prices. Chemical Engineering Plant Cost Index, Marshall–Swift Index (MSI), and consumer price index are some of the indices used to adjust the equipment costs for inflation (Table 2.1). The CPECI is specific to chemical engineering equipment while the MSI is a composite index for the industrial chemical equipment that includes inflation. The consumer price index on the other hand is a generic consumer index that accounts for the price increases of general goods, results in overestimation of the equipment costs, and must the used as the last option.

Total cost of the plant includes costs such as infrastructure development, utilities, buildings, electrical equipment, instrumentation, laboratories, and the installation costs in addition to the capital equipment costs estimated above. A common approach to estimate these costs which can often exceed the capital equipment costs is to use industry-specific factors for their estimation. A more detailed estimate of factors for each category such as buildings, installation, instrumentation can be estimated based on factors obtained from handbooks and expert advice. Typical factors for estimating the fixed capital expenses from the total equipment cost are provided in Table 2.2. In the case of limited resources to perform TEA analysis, these factors can be lumped into a single factor ("lang factor") which usually varies between 3–5 for most types of processing plants and used as a multiplier for the PEC to estimate the CAPEX.

Table 2.1 Comparison of various cost indices.

Year	MSI (1926 = 100)	CEPCI (1958–1959 = 100)	CPI (1982 = 100)
1990	915.1	357.6	135.44
1995	1027.5	381.1	157.93
2000	1089.0	394.1	178.45
2005	1260.9	468.2	202.38
2010	1457.4	550.8	225.96
2015	1598.1	556.8	237.017
2018	1638.2	603.1	251.07
2019	–	607.5	255.657

CEPCI, Chemical engineering plant cost index; CPI, consumer price index; MSI, Marshall–Swift index.

Table 2.2 Factors for estimating the fixed capital expenses [2].

Direct costs	
Purchased equipment costs (PEC)	**15%–40% of the fixed capital expenses**
Installation	25%–55% of PEC
Instrumentation	6%–30% of PEC
Piping	10%–80% of PEC
Electrical	10%–40% of PEC
Land and buildings (new site)	10%–80% of PEC
Site preparation and yard improvement	8%–20% of PEC
Utilities and other services	30%–80% of PEC
Indirect costs	
Engineering and supervision	15%–30% of PEC
Contractors fee	2%–8% of PEC
Contingency	5%–15% of PEC
Startup expenses	8%–10% of PEC

In addition to the direct capital costs, the plant construction also has additional expenses such as contractor fees, engineering costs, and contingency costs which are classified as indirect costs. Indirect costs are often specified as a percentage of the direct costs or purchased equipment costs.

Purchases Equipment cost, PEC

Direct capital costs, DC = PEC × (combined lang factor or sum of individual lang factors for each type of costs such as installation).

Indirect Capital Costs, IDC = direct capital costs × indirect cost factor.

Fixed capital expenses, FCE = DC + I DC.

1. Operational costs: operational costs of a plant are classified into variable and fixed operational costs. The fixed operational costs include costs that are incurred regardless of the state of operation of the plant and are also called facility-dependent costs. Facility-dependent costs include the interest on loans, local taxes, insurance, and depreciation. The variable operational costs are costs that vary depending on the state of operation of the plant and include costs such as the raw material and other input costs, utilities, labor costs, consumables, waste product disposal, royalties, regular maintenance, and product advertising. The sum of facility-dependent costs and the variable operational costs is called the operational expenses and is typically specified on the unit product bases. The CAPEX and operational expenses are two very important costs that are very commonly used in the TEA analysis.

2. Cash flow analysis: after the completion of the CAPEX and OPE estimation, the expected production and the sales data are used to setup a flowsheet for conducting the cash flow analysis. Cash flow analysis includes both revenues and expenses and this data is used to calculate the NPV, internal rate of return, and payback period. The cash flow analysis is very important for financial planning.

Net cash flow (for a given year) = Revenues − Costs

The payback period refers to the time required to recover the investment from the project. Therefore, payback period must always be less than the design project life for viable projects. The payback period is only indicative of how quickly the initial investment is realized without any information about the accumulation of profits from the project.

$$\text{Payback period} = n^- + \frac{S_{n^-}}{S_{n^-+1}}$$

where n^- is the last period with a negative cumulative cash flow, S_{n^-} is the cumulative cash flow at the end of the time period n^-, and S_{n^-+1} is the cumulative cash flow for the time period $n^- + 1$.

Once the discount rate (i) also called the opportunity cost of investment is specified, the NPV can be calculated from the cash flow analysis data. The discount rate reflects the cost of capital which is the expected returns from a profitable venture elsewhere with a similar financial risk profile. The NPV must be greater than zero for a profitable project and projects with higher NPV are selected. The NPV is sensitive to the distribution of the cash flows. The NPV for the project is calculated using the following equation:

$$\text{NPV} = \sum_{n=0}^{N} \frac{C_n}{(1+i)^n}$$

The internal rate of return (IRR, r) is defined as the value of the discount rate in the above equation when the NPV is equal to zero. The IRR indicates the ability of the project to return profits and therefore must be greater than the discount rate. Keeping all other factors (such as revenue flows) constant, projects with higher upfront capital costs will have lower IRR compared to projects with a more uniform distribution of the capital costs. The IRR can be calculated using the following equation:

$$\text{NPV} = \sum_{n=0}^{N} \frac{C_n}{(1+r)^n} = 0$$

where C_n refers to the cash flow during the year n, for a total of N years.

2.4 Uncertainty and sensitivity analysis

Uncertainty which can be defined as the lack of perfect knowledge can be traced to three sources: aleatory, epistemic, and linguistic uncertainty [5]. Aleatory uncertainty pertains the inherent randomness in the process and therefore cannot be reduced by acquiring more data/knowledge about the process. Epistemic uncertainty refers to a lack of knowledge about the system while linguistic uncertainty arises due to imprecise descriptions of the system. It is possible to reduce and quantify the linguistic uncertainty using precise scientific language to describe the system. On the other hand, epistemic and aleatory uncertainties can be further classified into six categories [6]:

- Inherent randomness: refers to the uncertainties due arising due to the nature of the process itself. No matter how well the input conditions are known, the evolution of the process is uncertain. Given sufficient observations of the process, this type of randomness can be specified to be within bounds. This type of uncertainty is the hallmark of natural processes such as weather. This type of uncertainty can be addressed by describing using probabilistic modeling methods.
- Measurement error: measurement errors give rise to uncertainties in the measured quantities. This type of error can be relatively easily addressed through careful data collection, better training, and using probabilistic methods to bound the uncertainties.
- Systematic error: systematic errors are difficult to notice/quantify and arise from measurement bias. A thorough review of the models and experimental protocols by independent experts is a way to address such errors.
- Natural variation: natural variation arises due to changes in the systems due to Spatio-temporal variations and parameters. For example, there is uncertainty in the crop response due to different soil, weather, and management practices. This type of uncertainty can be quantified by carefully considering the range of factors that influence the system. For example, the uncertainty in the crop yield for a particular variety can be specified by considering the varietal response to soils, management practices, and weather. However, due to many possible combinations and uncertainty in weather, there is a natural variation in the crop yield.
- Model uncertainty: model uncertainties arise since the model is only an abstract representation of a complex system and does not account for all factors that influence the real process. The uncertainty due to model parameter and model structure are two subclasses of model uncertainty. While the model parameter uncertainty can be quantified using statistical methods and sensitivity analysis, the model structure uncertainty is more difficult to quantify. One approach to quantify the model structure uncertainty is to compare the outputs of different models developed for the process using independent processes and comparing their outputs. If the model outputs from models created from various criteria and real observations result in similar outputs, then one can be reasonably certain that none of the important underlying processes have been neglected in the model, thus reducing the model structure uncertainty.
- Subjective judgment: use of appropriate input data, interpretation of output data especially in limited data contexts, often requires subjective inputs of the practitioners. Under such conditions, it is possible that the decision to include/exclude a data

set is based on subjective criteria used by the practitioner and leads to subjective judgment uncertainty.

The complexities of real-world mean that engineers and analysts often operate under resources, data, and knowledge deficit environment, and therefore adequate attention must be paid to the sources of uncertainty in the analysis and clear strategies developed to address them. When operating under resource-limited environment, it is often necessary to prioritize which aspects/parameters of the model require attention first to improve the model fidelity. Sensitivity analysis is performed for TEA models to address this issue.

Sensitivity analysis is performed typically by changing the model input and parameter values, one at a time, and observing the changes in the output. After normalization of the results, the inputs or parameters are ranked in the order of their impact on outputs. This process results in the identification of the most important model parameters and inputs and any efforts to resolve the uncertainties in their values must be prioritized over other inputs/parameters.

After completion of sensitivity analysis, a comprehensive model uncertainty is evaluated by varying all the inputs and parameters together using Monte Carlo analysis. The inputs and parameter uncertainties are described by probability distributions and, these distributions are sampled, and the model is simulated for each Monte Carlo run. By running many such simulations by sampling the input and parameter distributions, a comprehensive data set of output variations is developed. These outputs are used to construct output distributions and analyzed to quantify the uncertainty in the model outputs due to uncertainties in the model inputs and parameters.

2.5 Real option analysis

Traditional investment strategies involve fully committing to a technology/process strategy by investing in the selected technology completely. The investment is made with the hope that the future market and economic conditions are favorable, and the risks of failure are all incorporated into the discount rate used in the NPV calculation. The traditional NPV based investment decisions are typically used to optimize decisions with respect to the available capital and investment alternatives that exist at the beginning of the project. They completely ignore the ability and value of delaying investment for the future. They do not recognize the ability to raise capital in the future and reevaluate the investment options in the future (Fig. 2.1).

A real options strategy is defined as "the right but not the obligation to acquire, expand, contract, abandon or switch some or all of an economic asset on fixed terms on or before the time the opportunity ceases to be available" [7]. The real options concept has been attributed to Meyers [8]. For example, an oil exploration company will company will commit some resources for preliminary exploration of potential oil field

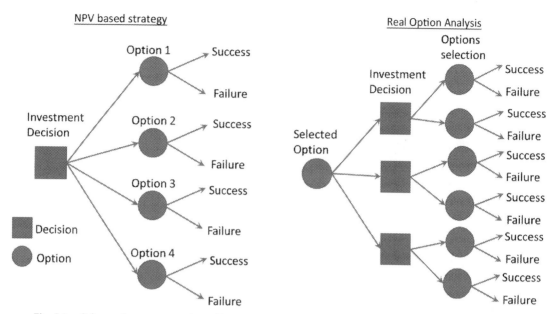

Fig. 2.1 Schematic representation of NPV and real option analysis based strategies. *NPV*, net present value.

sites and invest in further detailed testing in some of the promising sites. Only after confirmation through detailed testing will the company invest in the infrastructure for drilling the actual production wells (Fig. 2.2). This strategy of investing in stages in physical or economic assets which provides flexibility in the time-sensitive management decisions to optimize the opportunities is called real options and the analysis to identify such options at the initial stages of the project is called the real option analysis (ROA).

It is important to recognize that the traditional strategies are adequate when there is less uncertainty about future scenarios and the available options are all similar with no large differences in returns on the complexity of technology. The ROA is very useful when the uncertainties in the market are large, the investment plans involve huge capital outlays, the technologies/processes have very long project life and cannot be reversed easily without incurring large losses and there are non-linear responses to the investment decisions. Two of the common analytical approaches for performing the ROA are contingent claims analysis and dynamic programming [9] and a detailed discussion of these methods is out of scope for this chapter and the reader is referred to [7,10].

2.6 Tools, software, and data sources to conduct techno-economic analysis/assessment

The types of tools used for performing TEA can vary from spreadsheets to dedicated software. The process modeling and economic analysis can be performed using software such as Aspen Plus and SuperPro Designer. This software enable process

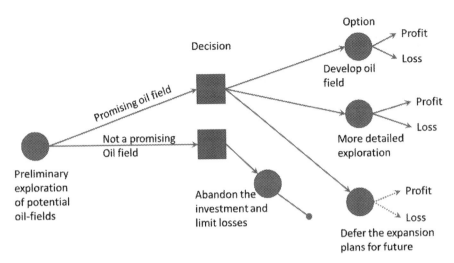

Fig. 2.2 Application of real option analysis to oil field development.

schematic designs, mass and energy balances, specification of equipment data and models, and usually have inbuilt thermodynamic property databases, economic, and utility databases for many equipment and capabilities for performing optimization, simulations, and generating reports. Financial modeling is usually performed using spreadsheet-based software such as @Risk, Crystal Ball, and open-source software packages such as Python and R. Obtaining reliable and accurate data for building the TEA models is critical for obtaining reliable results. The Engineers's cost handbook [2], Perry's Chemical Engineering handbook [3], RS Means [4], and online databases such as Matche [11] are some of the authoritative sources to obtain the data pertaining the equipment, scaling factors, cost factors, and other sizing data. Economic data for the analysis can be obtained from the published Govt. sources in each country.

2.7 Worked example

A bioethanol plant using 2 million tons/year sugarcane purchased at $30/ton and producing 84 L ethanol per ton sugarcane has a fixed equipment cost of $25 million. The plant obtains all the utility needs through sugarcane bagasse combustion. The excess energy is used to produce electricity (175 kWh/ton sugarcane) which is sold to a power company at $0.08/kWh. Materials required for the fermentation of the sugarcane juice are estimated to cost $0.02/ton sugarcane. The plant requires 10 operators who have a yearly benefit of $30,000 and the overhead costs (insurance, supervision, laboratory costs) are estimated at 85% of the yearly salary for an operator. Other costs (miscellaneous materials, maintenance, capital charges, insurance, taxes, and others) are estimated to be 25% of the fixed capital expenses. Assume general overheads as 5% of the direct costs.

Techno-economic assessment

Various factors for direct costs are piping = 0.4, instrumentation = 0.3, insulation = 0.1, electrical facilities = 0.05, buildings = 0.2, yard improvement = 0.1, and other auxiliary facilities = 0.2 and unlisted equipment to be 0.075 of fixed equipment cost. Assume indirect costs factors as engineering = 0.25, construction = 0.35, contractor's fee = 0.05, and contingency = 0.1 of direct costs.

The projected cash flow is provided in the first three columns of the table below.

Year	Expenses	Revenues	Year	Expenses	Revenues
0	70	0	11	69	91.56
1	40	0	12	65	99.96
2	17.5	0	13	69	94.92
3	72	96.6	14	73	99.96
4	68	99.96	15	65	88.2
5	67	86.52	16	72	88.2
6	72	99.96	17	73	91.56
7	70	96.6	18	66	86.52
8	65	99.96	19	73	98.28
9	65	88.2	20	69	94.92
10	66	96.6			

1. Calculate the fixed capital expenses for this plant.
2. Calculate the variable costs, fixed costs, direct costs, coproduct revenue, and the unit production cost for the plant.
3. Calculate the payback period, IRR, and NPV assuming a discount rate of 10%.

   Given Data:
   - Capacity, Design life: 2 million tons/year sugarcane purchased at $30 ton and producing 84 L ethanol /ton sugarcane.
   - Fixed equipment cost: $30 million.
   - Capital cost factors: piping = 0.4, instrumentation = 0.3, insulation = 0.1, electrical facilities = 0.05, buildings = 0.2, yard improvement = 0.1 and other auxiliary facilities = 0.2.
     Unlisted equipment to be 0.075 of fixed equipment cost.
   - Indirect cost factors: engineering = 0.25, construction = 0.35, contractor's fee = 0.05, and contingency = 0.1.
   - Operating cost factors:
     * Electricity is produced @ 175 kWh/ton sugarcane which is sold to a power company at $0.08/kWh.
     * Materials costs are $0.02/ton sugarcane
     * Labor costs are $30,000/year-person with an 85% overhead and 10 operators.
     * Miscellaneous costs (miscellaneous materials, maintenance, capital charges, insurance, taxes and others) are 25% of fixed capital expenses
     * General overhead is 5% of direct costs.

Assumptions: Plant operates 330 days a year

Calculate: variable costs, fixed costs, direct costs, coproduct revenue, and unit production cost.

Solution:

Direct costs = Fixed equipment cost × (1.0 + piping + instrumentation + insulation + electrical facilities + buildings + yard improvement + other auxiliary facilities + unlisted equipment)

= $30 million × (1.0 + 0.4 + 0.3 + 0.1 + 0.05 + 0.2 + 0.1 + 0.2 + 0.075) = $72.75 million ~$73 million/year

Indirect costs = Direct costs × (engineering + construction + contractor's fee + contingency)

= $72.75 million × (0.25 + 0.35 + 0.05 + 0.1) = $54.56 million

Fixed capital expenses (CAPEX) = Direct costs + Indirect costs = $72.75 + $54.56 = $127.31 million ~$127.5 million.

Operational costs (per year):
- Feedstock costs = $30/ton sugarcane × 2 million tons/year = $60 million.
- Labor costs = 10 operators × $30000/operator (1 + 0.85) = $555,000.
- Material costs = $0.02 × 2 million tons/year = $40,000.
- Miscellaneous costs = 0.25 × $127.31 million = $31.83 million.
- General overhead = 0.05 × 72.755 million = $3.64 million.

Total operational costs = feedstock costs + labor costs + material costs + miscellaneous costs + general overhead = $96.06 million/year ~$96 million/year.

Ethanol produced = 84 L/ton sugarcane × 2 million tons/year sugarcane = 168 million L/year.

Electricity produced = 175 kWh/ton sugarcane × 2 million tons/year sugarcane = 350 million kWh/year.

Coproduct revenue = Electricity produced × Electricity price = $28 million/year.

Production cost = Total operational cost−coproduct revenue = $68.06 million ~$68 million/year.

Unit production cost = Production cost/Ethanol production = $68.06/168 L = $0.405/L ethanol produced.

Answer: *Direct costs, fixed capital costs (CAPEX), operational costs, coproduct revenue, and unit production cost are $73 million, $ 127.5million, $96 million/year, $28 million/year, and $0.405/L ethanol, respectively.*

Cash flow analysis for ethanol plant (all numbers in millions $)

	Given data			Calculated data		
					IRR calculation	NPV calculation
				Cumulative	11.3%	10.0%
Year	Expenses	Revenues	Net cash flow	cash flow		
0	70	0	−70	−70	−70.00	−70.00
1	40	0	−40	−110	−35.95	−36.36
2	17.5	0	−17.5	−127.5	−14.14	−14.46

(continued)

Techno-economic assessment

	Given data			Calculated data		
				Cumulative	IRR calculation	NPV calculation
Year	Expenses	Revenues	Net cash flow	cash flow	11.3%	10.0%
3	72	96.6	24.6	−102.9	17.86	18.48
4	68	99.96	31.96	−70.94	20.86	21.83
5	67	86.52	19.52	−51.42	11.45	12.12
6	72	99.96	27.96	−23.46	14.74	15.78
7	70	96.6	26.6	3.14	12.61	13.65
8	65	99.96	34.96	38.1	14.89	16.31
9	65	88.2	23.2	61.3	8.88	9.84
10	66	96.6	30.6	91.9	10.53	11.80
11	69	91.56	22.56	114.46	6.98	7.91
12	65	99.96	34.96	149.42	9.72	11.14
13	69	94.92	25.92	175.34	6.48	7.51
14	73	99.96	26.96	202.3	6.06	7.10
15	65	88.2	23.2	225.5	4.68	5.55
16	72	88.2	16.2	241.7	2.94	3.53
17	73	91.56	18.56	260.26	3.03	3.67
18	66	86.52	20.52	280.78	3.01	3.69
19	73	98.28	25.28	306.06	3.33	4.13
20	69	94.92	25.92	331.98	3.07	3.85
				Total NPV	0.00	57.07

Payback period, P = 6 + (|−23|/26.6) = 6.9 years.
IRR = 0.113 corresponding to the value of discount rate when the total NPV of 0.
Total NPV with a discount rate of 10% is $57 million.

These numbers indicate that the plant will pay itself in 6.9 years after the start of construction and will have an internal rate of return of 11.3%. The investment of $127.5 million in this ethanol plant venture (barring any unforeseen financial risks and costs) would result in returns that will be equal to $57 million in current dollars over its lifetime. Since the IRR of 11.3% is better than the discount rate (10%) used in NPV calculation, it indicates that this is a good investment opportunity as returns are higher compared to other investment opportunities. However, since this difference between the NPV and IRR, the economic performance will be very dependent on the uncertainties in the operational costs and the revenues. Therefore, the discount rate used in the NPV should be evaluated carefully to reflect this financial risk.

2.8 Conclusions and perspectives

Basic questions about the technology choice and economic performance are answered through TEA. TEA is used to determine the overall profitability of the projects. For example, TEA is often used to determine the plant and equipment sizing, profitability, cash flow analysis, and is an integral part of engineering practice. However, TEA is not limited to these options and often extends to include identifying the tradeoffs, identifying the optimal choice of technologies under external uncertainties, the optimization of the process parameters, identifying the optimal size of the processing plants as detailed in the literature (for some examples, pleae see [13-22]). It is important to note that this chapter only provides a high-level overview of this topic, which is adequate for the first introduction to this topic. For a detailed study, the reader is referred to several excellent textbooks [2,12] that cover this topic more extensively.

This chapter introduced to the techno-economic analysis which is a very important and critical tool for assessing the techno-economic feasibility of a project. Establishing TEA feasibility through study of the process energy and mass balances, economic analysis including payback period, NPV, cash flow analysis and uncertainty and sensitivity analysis is a necessary condition for the success of any commercially viable projects. The next few chapters will deal with the environmental impact, resource sustainability, and social perspectives of systems sustainability analysis.

References

[1] Lauer, M. Methodology guideline on techno economic assessment (TEA): generate in the framework of ThermalNet WP3B Economics. 2008. https://ec.europa.eu/energy/intelligent/projects/sites/iee-projects/files/projects/documents/thermalnet_methodology_guideline_on_techno_economic_assessment.pdf [accessed April 25, 2020].
[2] M.S. Peters, K.D. Timmerhaus, Plant Design and Economics, 4th edn, McGraw-Hill, New York, NY, 1991.
[3] Process economics: fixed capital cost estimation, in: R.H. Perry, D.W. Green (Eds.), Perry's Chemical Engineers' Handbook7th edn., McGraw-Hill, New York, NY, 1998 Section.
[4] RS Means (Ed.), RSMeans Estimating Handbook3rd edn., John Wiley & Sons Ltd, Chichester, 2009.
[5] L. Uusitalo, A. Lehikoinen, I. Helle, K. Myrberg, An overview of methods to evaluate uncertainty of deterministic models in decision support, Env. Model. Soft. 63 (2015) 24–31.
[6] H.M. Regan, M. Colyvan, M.A. Burgman, A taxonomy and treatment of uncertainty in the environmental modelling process-a framework and guidance, Environ. Model. Softw. 22 (2002) 1543–1556.
[7] Ian Robinson Tim Kyng IAAust Biennial Convention 2003. Real option analysis: the challenge and the opportunity https://actuaries.asn.au/Library/Events/Conventions/2003/6e-conv03presrobinson.pdf.
[8] S. Myers, Determinants of corporate borrowing, J. Fin. Econ. 5 (1977) 147–175.
[9] A. Borison, Real options analysis: where are the emperor's clothes? Presented at Real Options Conference, Washington,DC, 2003. www.realoptions.org/abstracts/abstracts03.html [accessed April 25, 2020].
[10] D. Lander, G. Pinches, Challenges to the practical implementation of modelling and valuing real options, Q. Review of Econ. Finance 38 (1998) 537–567.
[11] Matches. Matches' process equipment cost estimates. (2013). http://www.matche.com/equipcost/Default.html, [accessed April 2016].
[12] R.E. Westney, The Engineer's Cost Handbook: Tools for Managing Project Costs, CRC Press, Boca Raton, FL, 1997.
[13] T. Eggeman, R.T. Elander, Process and economic analysis of pretreatment technologies, Bioresour. Technol. 96 (2005) 2019–2025.
[14] E. Gnansounou, A. Dauriat, Techno- economic analysis of lignocellulosic ethanol: a review, Bioresour. Technol. 101 (2010) 4980–4991.
[15] F.K. Kazi, J.A. Fortman, R.P. Anex, et al., Techno-economic comparison of process technologies for biochemical ethanol production from corn stover, Fuel 89 (2010) S20–S28.

[16] A. Juneja, D. Kumar, G.S. Murthy, Economic feasibility and environmental life cycle assessment of ethanol production from lignocellulosic feedstocks in Pacific Northwest U.S., J. Ren. Sust. Energy 5 (2013) 023142.

[17] H.J. Kadhum, K. Rajendran, G.S. Murthy, Optimization of surfactant addition in cellulosic ethanol process using integrated techno-economic and life cycle assessment for bioprocess design, ACS Sust. Chem. Eng. 6 (2018) 13687–13695.

[18] K. Rajendran, G.S. Murthy, Techno-economic and life cycle assessments of anaerobic digestion—a review, Biocatl. Agric. Biotechnol. 20 (2019) 101207.

[19] X. Zhao, T.R. Brown, W.E. Tyner, Stochastic techno-economic evaluation of cellulosic biofuel pathways, Biores. Technol. 198 (2015) 755–763.

[20] J.D. Kern, A.M. Hise, G.W. Characklis, R. Gerlach, S. Viamjala, R.D. Gardner, Using life cycle assessment and techno-economic analysis in a real options framework to inform the design of algal biofuel production facilities, Biores. Technol. 225 (2017) 418–428.

[21] M.J. Walsh, L.G.V. Doren, N. Shete, A. Prakash, U. Salim, Financial tradeoffs of energy and food uses of algal biomass under stochastic conditions, Appl. Energy. 210 (2018) 591–603.

[22] A. Pivoriene, Flexibility valuation under uncertain economic conditions, Procedia Social Behavi. Sc. 213 (2015) 436–441.

CHAPTER THREE

Environmental impacts

Ganti S. Murthy[a,b]

[a]Biological and Ecological Engineering, Oregon State University, Corvallis, OR, United States, [b]Biosciences and Biomedical Engineering, Indian Institute of Technology, Indore, India

3.1 Introduction

Environmental impacts due to human actions have long been recognized by human society. Themes related to environmental protection can be found in almost all societies and are generally much stronger in the oriental religions. World Scientists have recently released a second notice to humanity cautioning that the very survival of humans (not life!) on earth is at stake if we do not urgently address the anthropogenic climate change, deforestation, agricultural production especially for ruminant meat production, unleashing of the sixth mass extinction event in 540 million years, and many other issues of environmental protection [1]. In recent times, minimizing the environmental impacts and assuring sustainability have taken centerstage again with increasing recognition of the dangers posed by unchecked exploitation of natural resources for the sole benefit of humans. In this context, understanding environmental impacts of a project are very important and critical aspect of the systems analysis. The environmental impacts are assessed from various perspectives and several qualitative and quantitative methods have been developed to meet the various needs for environmental impact assessment.

3.2 Methods used for assessing the environmental impacts

As can be recognized, the environmental impacts of a project can be widely distributed spatio-temporally. For example, some of the impacts of water pollution or solid waste disposal may only impact the surrounding areas whereas the emission of toxic gases into the atmosphere could have regional impacts. Some of the environmental releases such as local emissions of high BOD effluents, small quantities of ammonia releases during fertilization may pose a threat for a few days–months before completely disappearing. On the other hand, the release of persistent pollutants such as dioxins, lead, and radioactive waste into the local environment will continue to pose a threat and have adverse environmental impacts for decades or even centuries. Export of toxic and hazardous wastes to less developed countries for "recycling" is an example of environmental impacts that are spatially separated from the point of origin. An example of a threat that is globally distributed and evolved over long times is the global anthropogenic CO_2 emissions which have accumulated over the last two centuries in the atmosphere and now pose a global threat through global climate change. Since the environmental impacts are diverse and varied, it should be no surprise that there are various methods for environmental impacts. Some of the common environmental impact methods are described below:

1. Ecological footprint: ecological footprint was introduced by Mathis Wackernagel and William Rees in 1990 [2]. The ecological footprint measures the natural resources required to sustain a particular lifestyle in the number of hectares of land on the demand side. While on the supply side, a similar metric called biocapacity is used to quantify the available natural resources to meet the demands. The ecological footprint must be lower than the biocapacity for a sustainable lifestyle. The question often asked in the ecological footprint analysis is "how many earths" implying, the number of earth that are needed to supply the natural resources to support the lifestyle if every human being lived according to the given lifestyle choices. The results from the ecological footprint are intuitive and therefore commonly adopted by multiple stakeholders to communicate policy, governance, sustainability goals. A global map of the recently published ecological footprints is presented in Fig. 3.1. Online ecological footprint calculators available [4] can be used to determine the ecological footprint of a particular lifestyle.

    One of the criticisms [5,6] of this method is that it does not offer any meaningful information to formulate public policy. Additionally, this method is anthropocentric that considers the other forms of life on earth and the natural resources as simply the stocks to meet the anthropogenic demands and lifestyles.

2. Water footprint: water footprint measures the consumption of the water to produce a unit product in the product lifecycle (Fig. 3.2). The water footprint concept was first introduced by Hoekstra in 2002 [7]. In this methodology, the water is classified into blue, green, and grey waters. Bluewater sources are surface and groundwater sources while green water refers to rainwater that does not become run-off. Greywater is defined as the volume of freshwater required to dilute a polluted stream of water to reduce the pollutant load to background concentrations and existing water quality standards. Bluewater footprint is important for irrigated agriculture, industry, and domestic water use while the green water footprint is relevant for rainfed agriculture, horticultural, and forestry products. Consumption of blue and green water sources in the supply chain refers to the loss of water through evaporation, return to another catchment area or sea, and incorporation into the product. The water footprints for some of the common items are shown in Table 3.1. Water footprint assessment methodology has provisions for the spatio-temporal differentiation of the water footprints accounting. Detailed methodology for the water footprint evaluation can be found in the Water Footprint Assessment Manual [8]. Various

Environmental impacts

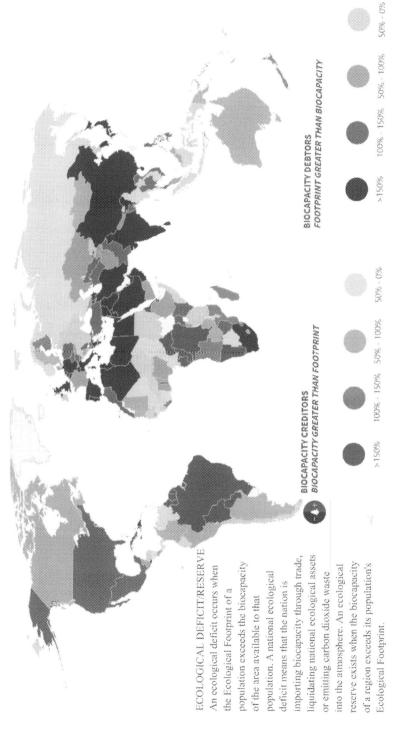

Fig. 3.1 Global Footprint Network [3].

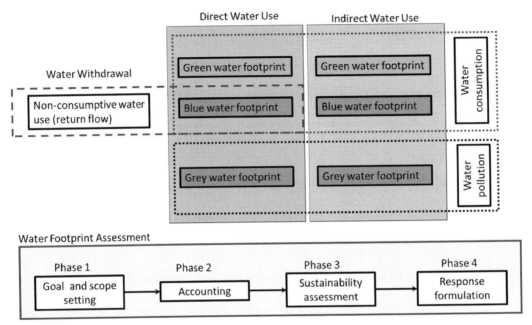

Fig. 3.2 Water footprint concept and methodology (adapted from [8]).

Table 3.1 Water footprints of some common consumer goods.

Product	Water footprint (L)	Remarks
Barley/wheat/Toast	650	for 500 g
Cane Sugar	750	for 500 g
Sorghum	1400	for 500 g
Millet	2500	for 500 g
Tea	90	One pot of 750 mL
Coffee	840	One pot of 750 mL
Milk	1000	1000 mL
Cheese	2500	for 500 g
Cooked rice	250	For 100 g
Chapati/Naan	110	One piece
Tomato	25 L	One piece
Apple	125	One piece
Orange	80 L	One piece
Egg	196	One piece
Chicken	2165	500 g
Lamb	3000	500 g
Beef	7700	500 g
Cotton	10,000	1000 g

interactive water footprint assessment tools have been published and can be found on the water footprint website [9].
3. Carbon footprint: carbon footprint is a concept similar to the ecological and water footprint and measures the carbon intensity of a product/process throughout its supply chain. Carbon footprint is very popular for quantitatively estimating the CO_2 emissions of a product and with the increasing importance of greenhouse gas (GHG) emissions, the carbon footprint has become increasingly important in the last two decades. Many versions of carbon footprint calculations are available including the ones developed by the Global Footprint Network [10] and Carbon Footprint [11]. Carbon footprint calculations are at the core of the California Low Carbon Fuel Standard [12] which are calculated using the GREET software developed by Argonne national laboratory [13]. Similar regulations at the national level in the US mandate reductions of carbon emissions in fuels by Renewable Fuels Standards 2 [14]. These regulations depend on assessing the carbon footprint of the fuels to create the carbon credits that can be traded on the carbon credit markets.
4. Substance flow analysis: substance flow analysis is used to quantify flows of a particular substance(s) of concern. Such analysis is performed when there is a particular concern from a substance such as mercury, lead, pesticides, or radioactive wastes.
5. Material flow analysis: the material flow analysis is similar to substance flow analysis and is usually used to track the flow of materials through the economy. For example, the flows of papers, glass, fertilizers, steel through different sectors of the economy can be tracked using material flow analysis.
6. Environmental risk assessment: environmental risk analysis is performed to quantify the risk associated with a project. The ERA will be discussed in detail in chapter 4.
7. Environmental impact assessment: this is one of the most commonly used methods to assess the environmental impacts of various projects. The emphasis of the EIA is on performing site-specific assessments of environmental impacts. The EIA method involves identification of the main environmental issues and metrics to evaluate impacts on these issues, assessment of the environmental performance of the proposed project against the environmental impact metrics, identification of significant positive and negative impacts, evaluation of overall environmental impacts in comparison to alternatives, and communication and discussion with stakeholders [15]. The standard EIA methodology of the US EPA is shown in Fig. 3.3. A detailed discussion of the EIA process is out of scope for the current chapter and the reader is referred to the detailed and comprehensive manuals for performing the EIA [16].

3.3 Life cycle assessment

The life cycle assessment/analysis (LCA) is one of the most widely used methods to assess and compare the impacts of products and processes with their alternatives. The LCA has roots in the materials flow analysis and has been first introduced in the 1970s

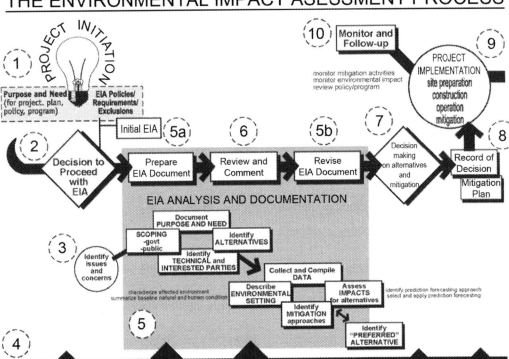

Fig. 3.3 The environmental impact assessment process (figure from [16]).

and was called the resource environmental profile analysis [17]. LCA belongs to the class of methods that use process-oriented metrics and is different from the methods such as sustainability process index and ecological footprint which focus on the depletion of the resources. It is important to note that LCA does not provide any information about the cost/economic performance or the risk associated with the process/product. The TEA (which was discussed in Chapter 2) is used for the former while the latter issues are addressed through the environmental risk assessment methods (discussed in Chapter 4).

The potential environmental impacts and resources used throughout a product's entire life cycle from raw material acquisition to waste management including the intermedia production, use, and reuse stages are assessed through LCA [18]. Over the last few decades, rapid advances in the field have resulted in the formulation of several ISO Standards, practices, tools, and software. The standards for conducting LCA are described primarily in two ISO standards: ISO 14040 Environmental Management-Life cycle assessment-Principles and framework (2006) [18] and ISO 14044 Environmental Management-Life cycle assessment-Requirements and guidelines (2006) [19]. These standards did not have an explicit accounting for water usage and it was increasingly recognized that the water footprint of products and processes are equally important as other environmental impacts. Therefore, a new standard was formulated in 2014, the ISO 14046 Environmental Management-Water footprint-Principles, requirements and guidelines (2014) [20]. The ISO 14046 (2014) describes the methodology and principles for conducting a water footprint analysis.

There are two variants of LCA called attributional and consequential LCA. The attributional LCA (aLCA) is focused on the total emissions from a process/project while the consequential LCA (cLCA) is focused on the changes in the emissions due to the introduction of a process/products [21]. The difference in the focus is reflected in the methodological choices in the use of average data for the aLCA and the marginal data for the cLCA.

3.4 Life cycle assessment/analysis methodology

The LCA as per the ISO standards (ISO 14040-14044) consists of four distinct steps: goal and scope definition, inventory analysis, impact assessment, and interpretation (Fig. 3.4). These four steps are interlinked and are performed iteratively.

3.4.1 Goal definition and scoping

The goal definition and scoping is a critical step of the LCA and is used to define the following questions:
1. What is the purpose of the LCA and what are the uses of this analysis? Who will use these results?
2. What are the product/processes and their alternatives that will be compared?
3. What is the functional unit and how will it be reported?

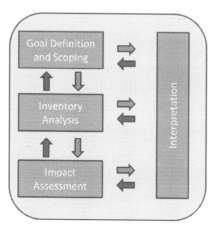

Fig. 3.4 Four steps of the LCA. *LCA*, life cycle assessment/analysis.

4. How will the systems boundary be drawn and what are the criteria for including/excluding a unit process from the systems boundary.
5. How will the results of the LCA be reported?
6. What are the resources (data time, money, technical skills) available to conduct this LCA?
7. What are the characterization methods and impacts that will be estimated?
8. How will the review of the results be carried out? where the functional unit, system boundaries, and the purpose and scope of the LCA are clearly defined.

The functional unit is one of the critical parameters that determine the course of analysis and is defined in the ISO 14040:2006 standard as: "Functional unit defines the quantification of the identified functions of the product. The primary purpose of a functional unit is to provide a reference to which inputs and outputs are related." Additionally, satisfactory answers to the above questions often involve extensive interactions with the stakeholders and the LCA analysts to clearly determine the goal and scope of the analysis.

3.4.2 Life cycle inventory

After the completion of the goal definition and scoping, the life cycle inventory (LCI) is performed using one of the many methods such as the matrix, input–output and hybrid methods. The LCI is the most time-consuming step in conducting an LCA (Fig. 3.5) and consists of developing process flow diagrams, quantification of process flows, coproduct allocation, quantifying the upstream flows through various databases, and finally performing the LCI calculations to obtain a comprehensive output consisting of the following:
- Energy, raw material, and other flows.
- All elementary, intermedia, and product flows to the system.
- All releases to the air, water, and soil.

While performing the LCI, some of the key considerations include the data sources, data quality, age of data, choices of technologies, and technology mix, precision, completeness, representativeness, consistency, reproducibility, source of uncertainty in the data [22]. There are many commercial and free databases that are available for obtaining the required data for the LCI and some of the important ones are listed in Table 3.2.

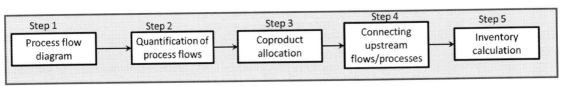

Fig. 3.5 Life cycle inventory steps.

Table 3.2 Life cycle inventory databases.

Name of database	Description	Type
US LCI	Comprehensive database for US data	Free
Digital Commons	Developed by USDA for the LCI of agricultural products with a US focus	Free
ELCD European reference Life Cycle Database	Specific for Europe	Free
Federal LCA commons	US LCI Data	Open and Free. There is also a commercial version developed by GreenDelta GmbH using the free version
Environmental Footprints	Europe Specific Database	Free
Arvi Material Value Chains	Database for wood-polymer composite production	Free
NEEDS (New Energy Externalities Developments for Sustainability)	LCI data for future electricity supply in Europe	Free
EXIOBASE	Global multi-regional environmentally extended supply and use IO database	Free
BioenergieDAT	Bioenergy supply chains with German Background	Free
UVEK LCI Data	Swiss database	Requires Ecoinvent 2.2 license
Agri Balyse	French Agricultural product database	Free but must be connected to Ecoinvent 3.1 in the background.
EcoInvent	Most comprehensive database	Commercial
Gabi	Similar to EcoInvent and one of the most popular databases	Commercial
ESU World Food	Worldwide LCA food database developed by ESU-services	Commercial
LC-Invetories.ch	Updated ecoinvent data 2.2 by ESU services	Commercial
ProBas+	A German LCI database	Commercial
The Evah Pigments Database	Pigments and colorants	Commercial

   a. Matrix method

   The matrix method of LCI also called the process-based LCI was first introduced by Heijungs [23] and described in detail in a foundational text of the subject [24]. In this method, the product system is represented by a system of linear equations that are solved to calculate the LCI. In this method, each process is defined to have a unique main product and the products are used in various processes. The matrix constructed with the rows representing the products and the columns representing the processes is called the process or technology matrix. Since the number of main products is always equal to the number of processes, the technology matrix is always square. Additionally, due to the uniqueness of the main product for a process, all rows and columns of the technology matrix are independent and therefore the technology matrix has the maximal rank and is invertible.

The demand for the products/services is represented by the vector $f_{n \times 1}$ and corresponds to the functional unit of the LCA. The demand/scaling vector $x_{n \times 1}$ determined the activity necessary to generate the functional unit vector $f_{n \times 1}$. This relationship can be represented as follows

$$A_{n \times n} \cdot x_{n \times 1} = f_{n \times 1} \tag{3.1}$$

Since the functional unit vector and the technology matrix are known, the unknown demand/scaling vector can be determined by the following equation:

$$x = A^{-1} f \tag{3.2}$$

The above equation always has a solution since the technology matrix is invertible. It is important to note that the key assumption that makes this possible is that each process has only one main product associated with it and the processes are all independent. This assumption has implications for the way in which the coproducts are considered in the system. The technology matrix is usually a very sparse matrix and the matrix inversion step of the very large technology matrix is the most computationally intensive step. Each process in the technology matrix has a set of emissions that are specified in the environmental intervention/emission matrix B. In the emission matrix, each column corresponds to a process and the rows represent the emissions from each process. The product of the demand/scaling vector (x) and the emission matrix results in the emissions for associated with the demand vector, y.

$$g_{p \times 1} = B_{p \times n} x_{n \times 1} = B_{p \times n} A_{n \times n}^{-1} f_{n \times 1} \tag{3.3}$$

The determination of the emission vector marks the completion of the LCI for the process. The life cycle impact assessment (LCIA) step consists of determining the environmental impacts by considering the emissions and their environmental impacts which are specified in the characterization matrix, Q. The characterization matrix Q consists of characterization factors consists of p mid-point level indicators in each row for each of the m emissions in the columns. Various LCIA methods basically have different characterization matrices including the number of indicators and represent the underlying models used to develop the relationships between the emissions and the indicators. Often, the LICA is calculated in some of the methods (such as ReCiPE) and has what is called mid-point and end-point indicators. The mid-point level indicator vector $h_{m \times 1}$ can be calculated as shown below:

$$h_{m \times 1} = Q_{m \times p} g_{p \times 1} \tag{3.4}$$

The end-point indicator can be calculated from the mid-point indicators by multiplying the $h_{m \times 1}$ with the weighing vector $w_{m \times 1}$ as noted below.

$$ES_{e \times 1} = w_{e \times m} h_{m \times 1} \qquad (3.5)$$

where

Number of products/processes: n
Number of emissions: p
Number of mid-point indicators: m
Number of end-point indicators: e

Technology Matrix, $A_{n \times n}$: rows in the technology matrix represent products from the corresponding column process, i.e, ith row represents the product from the ith column process.

Environmental intervention/emission matrix, $B_{p \times n}$: the matrix represents the p emissions in the rows and the n processes in the columns.

Characterization matrix, $Q_{m \times p}$: the matrix represents weighing factors for each of the m mid-point indicators for the p emissions.

Weighing vector, $w_{m \times e}$: the weighing factors are used to convert the mid-point indicators to end-point environmental indicators.

Functional unit vector, $f_{n \times 1}$: the functional unit for the LCA.

Scaling/demand vector, $x_{n \times 1}$: amounts of various products needed to generate the functional unit.

Emission vector, $g_{p \times 1}$: the amount of emissions associated with the production of one functional unit.

Mid-point level indicators, $h_{m \times 1}$: the mid-point indicators for producing one functional unit.

End-point indicator, $ES_{e \times 1}$: The end-point indicators for producing one functional unit.

b. Input–Output Method

The matrix method described above is one of the most popular methods for conducting the LCI. However, the method leads to unavoidable truncation errors in the analysis as it is physically impossible to include all processes in an economy which theoretically are all connected to the production process associated with the functional unit. The input–output tables describing the activities of the entire economies have been in extensive use since their introduction by Leontief in 1937 [25]. The input–output model of an economy assumes that each industry consumes products from all other sectors of the economy in fixed ratios of the total economic output to generate the outputs required by the economy [26]. Note that the relationships are all based on unit monetary output and are specified in terms of the unit economic value, i.e., the matrix $\tilde{A}$ represents the intermediate consumption among various sectors in the economy. The total equation for the total economy can be written as follows

$$x_{n \times 1} = \tilde{A}_{n \times n} \cdot x_{n \times 1} + y_{n \times 1} \tag{3.6}$$

Since the demand unit vector (y) and the input–output matrix are known, the unknown total demand vector (x) can be determined by the following equation:

$$x = (I - \tilde{A})^{-1} y \tag{3.7}$$

It is important to note the differences between the technology matrix ($A_{n \times n}$) and the IO matrix ($\tilde{A}_{n \times n}$) described above. The total output of the sector in the IO method consists of both intermediate and the final demand (functional unit) for each sector. Due to this difference, the equation for determining the scaling vector is different in both methods. The technology matrix accounts for all the processes needed to generate the functional unit/output vector while the IO matrix describes the inputs to various sectors as a fraction of the total output of each sector. Conceptually, this implies that the IO method accounts for all the processes and does not suffer from the truncation errors as is the case for the matrix method. The system boundary in the IO method is also more complete compared to the matrix methods. On the other hand, the IO tables are prepared for aggregated economic sectors (which vary from country to country) and do not have the granularity associated with the matrix methods. Therefore, IO method cannot differentiate between technology/product options as finely as the matrix methods. A well-constructed technology matrix can have low truncation error while the IO matrix methods give rise to errors resulting from the aggregation of economic sectors. On the other hand, the chief strength of the IO method is its ability to incorporate the service sectors which can be a significant component of the overall emissions for some of the service-intensive sectors such as tourism. The service sector is usually ignored in the process-based matrix methods. Additionally, the systems boundary of the IO methods only includes the domestic sectors for a country and are accurate only for products that have low imported content. Therefore, IO methods are more useful when the resources and time are limited and when the environmental impacts are primarily due to services and have low import content. The only difference in the computation of the LCI between the matrix and the IO methods is in the specification of the technology and IO matrix and the step needed to obtain the demand vector (Eqs. 3.2 and 3.7). All the other steps for calculating the emissions from the emission matrix; mid-point or end-point indicators (Eqs. 3.3–3.5) are exactly similar to the methods described in the matrix method.

c. Hybrid Method

As discussed above, both the matrix and the IO methods have their advantages and disadvantages. There are several hybrid methods developed to integrate the IO and the matrix-based methods, such as the tiered hybrid analysis, IO-based hybrid analysis, and the integrated hybrid analysis [25]. In the tiered hybrid analysis, the process/matrix-based analysis is used for use, disposal, and important upstream process while the remaining inputs and outputs are obtained from the IO analysis. This hybrid approach is fast and can provide reliable results if the differences between the matrix and IO system components are carefully selected to ensure that the data is obtained from the best possible data source (Process or IO database). Integrated hybrid method suggested by Suh and Huppes [26] integrates the technology and IO matrix into a combined model as follows

$$M = \begin{bmatrix} B & 0 \\ 0 & \tilde{B} \end{bmatrix} \begin{bmatrix} A & Y \\ X & I - \tilde{A} \end{bmatrix}^{-1} \begin{bmatrix} f_{n \times 1} \\ 0 \end{bmatrix} \qquad (3.8)$$

where M represents the net emission matrix; X represents the upstream cut-off flows to the LCA system linked with the IO matrix. The elements in X have a unit of monetary value/operation time (scaling factor); Y represents the downstream cut-off flows to the LCA system linked with the IO matrix. The unit of each element of Y is physical unit/monetary value.

Remaining steps of analysis for determining the environmental impacts are similar to other methods discussed above. The relationships between the three methods are shown in Fig. 3.6, and the reader is referred to Suh and Huppes [26] and Suh et al. [27] for detailed discussion of the hybrid methods.

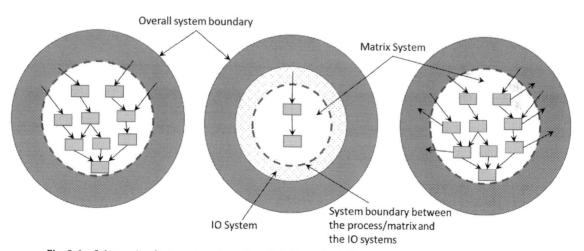

Fig. 3.6 Schematic of interactions in various hybrid methods (figure from [27]).

Real-life processes have multiple coproducts streams that are generated along with the main product. Partitioning the emissions and allocating them to the coproducts is one of the important design decisions that a LCA analyst must make for a robust and reliable LCA. The ISO 14044 standard suggests use of three strategies for coproduct allocation.

1. System partition or system expansion: system expansion which is also known as the displacement/substitution method is the recommended strategy. In this method, the coproduct allocation is completely avoided by expanding the system boundaries or dividing the unit processes into two or more subprocesses.
2. Physical allocation: if the system partition/expansion method cannot be used to avoid the coproduct allocation, then the share of the coproduct allocation is determined based on a physical property such as the mass and energy contents of the main and coproducts. The ratio of emissions for the coproduct is determined as the ratio of the mass/energy of the coproduct to the total mass/energy of the coproduct and the main product in the system.
3. Economic value-based allocation: if any of the approaches described above do not work for the coproduct allocation, then the allocations can be allocated based on the relative economic value of the coproducts.

3.4.3 Life cycle impact assessment

The LCI of the process generates an emission matrix that contains many hundreds of emissions produced from all the upstream and downstream processes. The purpose of the LCIA step is to take these numerous emissions and calculate indicators that can be used by the end-users to meet the purpose of the LCA as defined in the goal and scope definition step. The reduction of the emissions to meaningful impact categories/indicators is accomplished using additional models that relate the emissions to various mid-point/end-point indicators. These steps can be mathematically described by using Eqs 3.4 and 3.5. Since the translation of the emissions to indicators is done through characterization factors that have various errors and uncertainties, the errors in the impact factors accumulate with the introduction of additional models and assumptions (Fig. 3.7). The ISO 14044 standards describe the following optional elements of the LCIA:

- Normalization with respect to a reference scenario.
- Grouping to rank the impact categories.
- Weighting to convert the impact categories into a single score by assigning weights to the impact categories.
- Data quality analysis to assess the reliability of the LCIA results.

3.4.4 Life cycle interpretation

After the completion of the LCIA, the results are evaluated for completeness, sensitivity, and consistency. The results are analyzing for the completeness of the analysis and

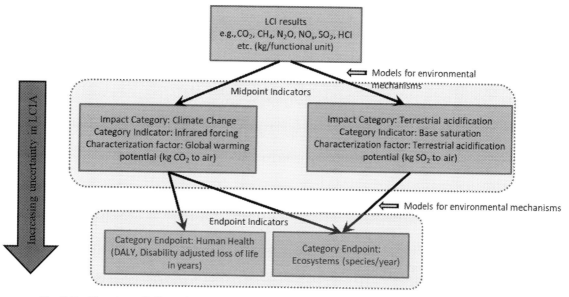

Fig. 3.7 Mapping of LCI results to mid-point and end-point indicators in the ReCiPe-2008 LCIA method. (figure from [22]). *LCI*, life cycle inventory.

identifying any missing/incomplete data that were used to conduct analysis. Special attention must be paid to the technology and spatio-temporal specificity of the data used in the analysis to ensure consistency. The system boundaries are one of the biggest sources of variability in the LCA and it must be ensured that the system boundaries identified in the goal and scope definition are adhered. The confirmation to the system boundaries becomes even more important when using hybrid methods for LCI. Typically, sensitivity and uncertainty analysis are conducted at this stage and the results are incorporated into the interpretation of the results. For details of the techniques used for the uncertainty and sensitivity analysis, the reader is referred to chapter 2. After completion of the interpretation, the goal and scope definition may have to be revised and a second iteration of the LCA performed until the results are consistent. The results of the peer review are also incorporated and the conclusions limitations and the recommendations of the LCA are presented in the step after completion of the iterations.

3.5 Life cycle assessment/analysis software and life cycle inventory databases

The LCA is very data-intensive and calculation-intensive analysis. Therefore having a good database is a prerequisite for conducting a reliable analysis. A list of the common LCA databases was provided in Table 3.2 and can be downloaded from the openLCA website (https://nexus.openlca.org/databases). A list of the software to perform LCA is provided in Table 3.3.

Table 3.3 Life cycle assessment/analysis software.

Software	Website
The Greenhouse gases, Regulated Emissions and Energy use in Transportation (GREET), Argonne National Laboratory	https://greet.es.anl.gov/
Open LCA	http://www.openlca.org/
EIO-LCA	http://www.eiolca.net/
Eco-LCA	http://resilience.eng.ohio-state.edu/eco-lca/index.htm
SimaPro	https://simapro.com/
GaBi	http://www.gabi-software.com/international/index/

LCA, life cycle assessment/analysis.

3.6 Worked example

A bioethanol plant using agriculture residue has the technology matrix, as shown below. The emissions and characterization are also shown below. Calculate the emissions and environmental impacts associated with a functional unit of 23,562 MJ (1000 L) of ethanol.

Technology matrix, A:

Process → Product ↓	Farming	CSE process	Electricity production
Feedstock	1	−2.66878	0
Ethanol	0	1	0
Electricity	0	0.217143	1

Unit Emission Matrix, B:

Emissions↓	1 kg corn *Farming*	1 L ethanol *CSE process*	10000 MJ electricity *Electricity production*
VOC	0.074	0.386	0.148635
CO	0.142	1.278	0.327729
NOx	0.365	1.837	2.235624
PM10	0.037	0.438	2.431774
PM2.5	24.691	145.142	731.7836
SOx	0.193	0.070	5.287524
CH4	0.261	0.070	4.234893
N2O	0.103	0.182	0.023257
CO2	0.069	1.743	1.645955
CO2Biogenic	0.000	−1.740	−0.01789
Greenhouse Gas	0.106	0.058	1.740984

Environmental impacts

Characterization Matrix, Q:

	VOC	CO	NOx	PM10	PM2.5	SOx	CH4	N2O	CO2	CO2Biogenic	Greenhouse Gas
climate change (kg CO2 eq), GHG	0	0	233	0.5	0.5	100	25	233	1	1	1
freshwater eutrophication (kg P eq)	0	0	2.6	0	0	0	0	3	0	0	0
terrestrial acidification (kg SO2 eq)	0	0	0.5	0	0	1	0	0.5	0	0	0

Weighing matrix, w:

GHG	Eutrophication	Acidification
0.05	0.2	0.3

Solution:

Functional unit, f = 23,562 MJ (1000 L) of ethanol.

The functional unit vector can be constructed as: $f = \begin{bmatrix} 0 \\ 1000 \\ 0 \end{bmatrix}$

The inverse of the technology matrix is calculated as:

$$A^{-1} = \begin{bmatrix} 1 & 2.669 & 0 \\ 0 & 1 & 0 \\ 0 & -0.217 & 1 \end{bmatrix}$$

⇨ Scaling/Demand vector, s = A − 1. F

$$s = \begin{bmatrix} 2668.78 \\ 1000 \\ -217.143 \end{bmatrix}$$

⇨ Emission Vector, g = B.s

$g^T =$

| 550.95 | 1585.75 | 2325.11 | 8.83 | 52135.94 | −562.82 | −152.62 | 451.78 | 1568.66 | −1736.85 | −37.34 |

⇨ Environmental impacts vector, h = Q.g

⇨ $h = \begin{bmatrix} 612784.8 \\ 7400.6 \\ 825.63 \end{bmatrix}$

The rows in the above matrix represent the midpoint indicators for climate change (kg CO2 eq), freshwater eutrophication (kg P eq), and terrestrial acidification (kg SO2 eq), respectively.

⇨ Environmental impact score, EI= w.h

EI = 32367.1

Answer:

The environmental impact score of 23,562 MJ (1000 L) of bioethanol is 32,367.1.

This example was limited to demonstration of the basic mathematical operations behind the LCA calculation. For detailed instructions on using software, the reader is referred to the websites of individual software in Table 3.3.

3.7 Perspectives

Most commonly used environmental impact assessment tools were the focus in this chapter. These tools are very useful to understand the environmental impacts of a process or a product. However, all technologies operate in a social, environmental, historical, and policy context. Large-scale adoption of any technology can have various rebound effects that may be unanticipated and are not adequately addressed by the tools discussed here. There is an increasing recognition that the attributional LCA is not suitable for all stakeholders. The rise in the use of consequential LCA to understand the indirect effect of the technologies is a pointer to the acknowledgment of the importance of indirect impacts. As can be intuitively observed, this change in focus requires larger data sets that often have high uncertainties and are subject to much variability especially in the case of economic datasets. The research into methods to address these issues along with the system boundary definition, data sources, and methodological development will be continuing in the future.

3.8 Conclusions and perspectives

This chapter introduced various methods of assessing environmental impacts with a particular focus on the Life-Cycle Assessment. The four-step process described in the ISO standards for conducting the LCA was discussed in detail with a particular focus on the Life-Cycle Inventory step.

References

[1] W.J. Ripple, C. Wolf, M. Galetti, T.M. Newsome, M. Alamgir, E. Crist, M.I. Mahmoud, W.F. Laurance, BioScience 70 (2020) 8–12. https://doi.org/10.1093/biosci/biz088.

[2] Ecological Footprint. https://www.footprintnetwork.org/our-work/ecological-footprint/?_ga=2.131979463.911567551.1588158237-1686574095.1588158237 [Accessed April 25, 2020].

[3] D. Lin, L. Hanscom, A. Murthy, A. Galli, M. Evans, E. Neill, M.S. Mancini, J. Martindill, F.-Z. Medouar, S. Huang, M. Wackernagel, Ecological footprint accounting for countries: updates and results of the national footprint accounts, 2012–2018. Resources 7 (2018) 58 https://data.footprintnetwork.org/#/ [Accessed May 16, 2021].

[4] Ecological Footprint Calculator. https://www.footprintnetwork.org/resources/footprint-calculator/ Accessed: [Accessed April 25, 2020].

[5] J.C.J.M. Van deb Bergh, F. Grazi, Ecological footprint policy? Land use as an environmental indicator, J. Indust. Ecol. 18 (2013). https://doi.org/10.1111/jiec.12045 [Accessed 16 May, 2021].

[6] L. Blomqvist, B.W. Brook, E.C. Ellis, P.M. Kareiva, T. Nordhaus, M. Shellenberger, Does the shoe fit? Real versus imagined ecological footprints, PLoS Biol 11 (2013) e1001700. https://doi.org/10.1371/journal.pbio.1001700.

[7] A.Y. Hoekstra, Virtual water trade, Proceedings of the International Expert Meeting on Virtual Water Trade, 12–13 December 2002, Value of Water Research Report Series No 12, UNESCO-IHE, Delft, 2003. www.waterfootprint.org/Reports/Report12.pdf [Accessed April 25, 2020].

[8] A.Y. Hoekstra, A.K. Chapagain, M.M. Aldaya, M.M. Mekonnen, The water Footprint Assessment Manual: Setting the Global Standard, Earthscan, 2011. https://waterfootprint.org/media/downloads/TheWaterFootprintAssessmentManual_2.pdf [Accessed April 25, 2020].

[9] Interactive tools for water footprint calculation. https://waterfootprint.org/en/resources/interactive-tools/Accessed: [Accessed April 25, 2020].

[10] Global Footprint Network. https://www.footprintnetwork.org/Accessed: [Accessed April 25, 2020].

[11] Carbon Footprint. https://www.carbonfootprint.com/ [Accessed April 25, 2020].

[12] The California low carbon fuel standard. https://ww2.arb.ca.gov/our-work/programs/low-carbon-fuel-standard [Accessed April 25, 2020].

[13] The greenhouse and regulated emissions, and energy use in transportation model. Argonne National Laboratory. https://greet.es.anl.gov/index.php [Accessed April 25, 2020].

[14] Renewable fuel standard 2: final rule. The US Environmental Protection Agency. https://www.epa.gov/renewable-fuel-standard-program/renewable-fuel-standard-rfs2-final-rule [Accessed April 25, 2020].

[15] UNESCO Environmental impact assessment and environmental impact statement guidelines. http://www.unesco.org/new/fileadmin/MULTIMEDIA/HQ/CLT/pdf/ucha_Environmental_Assessment_Method_Southampton.pdf [Accessed April 25, 2020].

[16] US EPA EIA Technical review guidelines for the energy sector. https://www.epa.gov/international-cooperation/eia-technical-review-guidelines-energy-sector [Accessed April 25, 2020].

[17] R.G. Hunt, W.E. Franklin, LCA—how it came about: personal reflections on the origin and development of LCA in the USA, Intl. J. LCA. 1 (1996) 4–7.

[18] ISO (International Standards Organization). 2006. Environmental management—life cycle assessment—principles and framework. ISO 14040:2006. http://www.iso.org [Accessed April 25, 2020].

[19] ISO (International Standards Organization). 2006. Environmental management—life cycle assessment—Requirements and guidelines. ISO 14044:2006. http://www.iso.org [Accessed April 25, 2020].

[20] ISO (International Standards Organization). 2014. Environmental management—water footprint—principles, requirements and guidelines. ISO 14046:2014. http://www.iso.org [Accessed April 25, 2020].

[21] A. Zamagni, J. Guinée, R. Heijungs, P. Masoni, A. Raggi, Lights and shadows in consequential LCA, Int. J. LCA. 17 (2012) 904–918.

[22] G.S. Murthy. Life Cycle Assessment. In book: Bioenergy: Principles and Applications. Eds. Li, Y. and Khanal, S.K. John-Wiley and Sons Inc. Hoboken, NJ, USA, 2016. ISBN: 978-1-118-56831-6. https://www.wiley.com/en-us/Bioenergy%3A+Principles+and+Applications-p-9781118568316

[23] R. Heijungs, A generic method for the identification of options for cleaner products, Ecol. Econ. 10 (1994) 69–81.

[24] R. Heijungs, S. Suh, The Computational Structure of Life Cycle Assessment, Kluwer Academic Publisher, Drodrecht, 2002.

[25] W.W. Leontief, Quantitative input and output relations in the economic systems of the United States, Rev. Econ. Stat. 18 (1936) 105–125.

[26] S. Suh, G. Huppes, Methods foo life cycle inventory of a product, J. Clea. Prod. 13 (2005) 687–697.

[27] S. Suh, M. Lenzen, G.J. Treloar, H. Hondo, A. Horvath, G. Huppes, O. Jolliet, U. Klann, W. Krewitt, Y. Moriguchi, J. Munksgaard, G. Norris, System boundary selection in life-cycle Inventories using hybrid approaches, Env. Sc. Techol. 38 (2004) 657–664.

CHAPTER FOUR

Environmental risk assessment

Ganti S. Murthy[a,b]
[a]Biological and Ecological Engineering, Oregon State University, Corvallis, OR, United States, [b]Biosciences and Biomedical Engineering, Indian Institute of Technology, Indore, India

4.1 Introduction

Assessing, understanding, and mitigating risks is part of decision-making in a complex world where the information is deficient. Decision-makers need to take decisions under uncertainty and risk analysis had been practiced by farmers, businessmen, political and military leaders, and natural resource managers throughout human history. In the past when the natural limits of the earth were not yet breached by humans, the risks were mostly localized and had a limited impact on the overall environment. Adverse impacts were mostly limited to local impacts. However, with the rapid industrialization and increasing complexity of a globalized world, the risk analysis has seen a revival and many approaches have been made to formalize the process. Risk assessment process is different from the safety assessment and is often more objective whereas safety assessment which is a more subjective. The results of safety assessment are often expressed as a simple yes/no question (e.g., is the situation safe? yes/no) versus risk analysis (situation is more/less risky compared to baseline scenarios). Risk although always associated with uncertainty is different from uncertainty in the sense that uncertainty denotes lack of knowledge about the system while the risk is conceptually defined as the product of the probability and the consequence of an uncertain event.

$$Risk = Probability \times Consequence$$

The above formula is only a conceptual formula of risks that are multidimensional, and it is important to note that different stakeholders (technical managers vs. general public) will perceive the same total risk differently depending on the distribution of the probability and the consequences [1]. Elements of risk analysis are present in the Techno-economic analysis and the environmental impacts assessments discussed in earlier chapters. For example, the investment risk is techno-economic analysis is incorporated in the discount rate used for the calculation of the net present value. The risk of exposure to toxic substances is part of the ecotoxicity impact metric used in some of the life cycle impact assessment methods. While the components of the risk analysis are present in some of the calculations, a comprehensive treatment of the topic is needed. This chapter will provide a basic understanding of the risk analysis and the methodologies used to perform risk analysis with a focus on the environmental risk assessment.

4.2 What is risk analysis?

The objective of the risk analysis is to forecast the probability of negative/positive consequences, understand the impact of the consequences, and develop a management plan to mitigate the risks. The risk analysis has broad applicability and is widely used in the technological, environmental, economic, political, and strategic domains.

The risk analysis process is used to answer, "who fears what from where and why?" [2]. In this conceptual model, there are four components:

Who: The stakeholders such as the organizations, policymakers, operators, and the general public.
What: The potential consequences that need to be predicted and managed.
Where: The localization aspects of the risks (macro-level or micro-level risks).
Why: The underlying causal factors for the risk.

Various types of risk analysis are needed depending on the uncertainty and the consequences of being wrong. Risks arise at the macro-level due to the increasing complexity of the global systems, increasing rate of change, and global financial systems. Risks also arise at the micro-level due to uncertainty and natural variability of the systems and typically the impacts due to the risks at the microlevel are contained within a limited spatial and/or temporal frame. The risks can be classified into different types as follows [1]:

- Existing risk: these are risks that exist in the baseline scenario and are currently active. For example, potential location of a biofuel factory in a seismically active zone has an inherent risk due to earthquakes.
- Future risk: these are risks that can arise in the future. For example, a risk of lower biofuel prices due to economic recession.
- Historical risk: this is a risk that is present due to historical reasons. For example, the location for the new biofuel plant may already be contaminated by toxic chemicals released from a plant that was previously operational at the site but has been demolished.
- New risk: a risk that did not exist earlier and is a result of the planned activity is a new risk. Continuing the previous example, establishing a new biofuel plant a new fire risk due to ethanol storage on the plant site.
- Transferred risk: the risks that are transferred to a different stakeholder, location, or time are called transferred risks. Example of a transferred risk is the purchase of fire hazard insurance by biofuel plant to mitigate their risk of losses due to fire hazards which transfers the risk to the insurance company. In return the biofuel plant management pays an insurance premium to the insurance company to take on the risk of fire hazard losses at the biofuel plant. Climate change is another classic example of the risk that was transferred to the entire world due to the industrial activity in the past from the burning of fossil fuels by few countries around the world.

- Transformed risk: when the original risk transforms the nature of the risk it is called transformed risk. For example, the use of flame retardants in the construction of the biofuel plant to mitigate the fire hazard risks may lead to a new risk of cancers in the workers of the biofuel plant.
- Residual risk: the residual risk refers to the remaining risks after the risk management measures have been implemented. For example, the remaining risk of loss from fires after the purchase of fire hazard insurance and measures to mitigate the occurrence of accidents in the biofuel plant is termed residual risk. The residual risk of fire hazards must be acceptable to all stakeholders before the biofuel plant can be built.
- Risk reductions: the management options to reduce the risks are terms as the risk reductions.

While these risks are characterized based on the type of risk, another classification can be based on the source of risks as:

- Life/health/safety risks: risks that have consequences related to life/health/safety of the stakeholders.
- Regulatory risks: risks of not meeting the regulations in force.
- Financial/investment risk: risks of financial nature. These risks can be further subdivided into other specific risks such as interest risk, liquidity risk, inflation risk, credit risk, etc.
- Political risks: risks that arise due to the political systems. For example, operating in a country with dictatorship has different sets of political risks than a democratic country with defined laws and regulations.
- Social risks: risks that arise due to society at large. Often these risks are critical for customer-facing parts of a business and these risks are growing in importance in the recent decades with the advent of social media.
- Strategic risk: risks that are strategic in nature such as security, liabilities, competition, disruptive innovations, program, and project risks.

All risks contain epistemic and aleatory uncertainties. As discussed in detail in Chapter 2, the aleatory uncertainty is related to the natural variation in the process while the epistemic uncertainty is related to the lack of knowledge. As can be discerned, the epistemic uncertainty can be reduced by acquiring more data about the system such as scenarios, system observations, model structure, underlying assumptions used to construct models, and the model parameters. Reducing the epistemic uncertainty is a function of available resources. However, regardless of the available resources and the data collected, the natural variability (i.e., the Aleatory uncertainty) of the system cannot be reduced beyond a limit and is solely governed by the system under study. It is critical to identify the source of uncertainty in the risk analysis as it provides a measure of what can be known and what cannot be known *a priori* and is critical for risk management.

4.3 Risk analysis method

The goals of the risk analysis for a decision or a problem are to identify, and describe the risks, manage the risks, and communicate the risks to stakeholders. Routine risk analysis is adequate when the uncertainty is low, and the consequence of a wrong decision is serious (First Quadrant) while extensive risk analysis with adaptive management is needed when the uncertainty of the event is high, and the consequences of a wrong decision are serious (Second Quadrant) (Fig. 4.1). When the consequences of wrong decision are minor or easily reversed and the uncertainty is high, a modest level of risk analysis is needed (Third Quadrant) while no risk analysis is generally required when the consequences are minor, and uncertainty is low (Fourth Quadrant). There are different frameworks developed by different agencies for conducting the risk analysis but all of them have three essential components of the risk analysis method: risk assessment, risk management, and risk communication (Fig. 4.2). The ISO framework for risk analysis on the other hand has risk assessment as part of the risk management step (Fig. 4.3).

4.3.1 Risk management

This step consists of defining the context of the problem, identifying the problem, identifying the data needs, evaluating the risks, and actions to mitigate/reduce/eliminate the risks so that the levels of risks are acceptable/tolerable. Risk management, therefore, involves identification of technically feasible, economically viable, and socially acceptable actions/plans. It is important to note that the risks although are often talked with a negative connotation, there can be upsides to the risks. The systems called "antifragile"

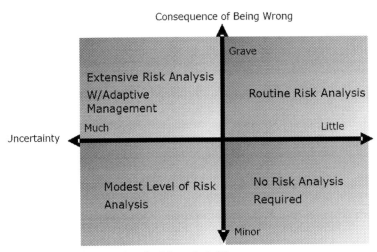

Fig. 4.1 Consequences and uncertainty in risk analysis [1].

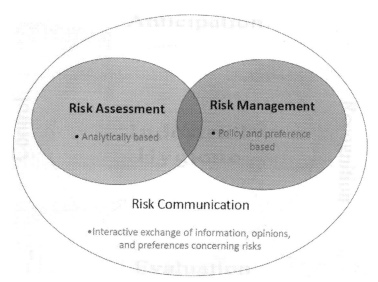

Fig. 4.2 Three steps in risk analysis [1].

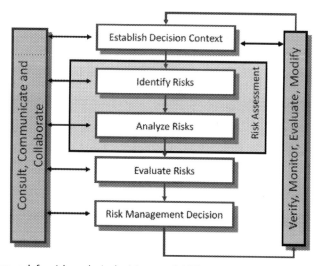

Fig. 4.3 ISO framework for risk analysis decision model [1].

thrive when exposed to uncertainty and are the characteristic of systems that have survived for a long time such as the evolution, forests, bacterial resistance, technological innovation, culture [3]. Therefore, it is important to recognize that such risks with positive upsides must be taken and situations created to exploit such positive risks. The risk reduction and risk-taking strategies for risk management, shown in Table 4.1, illustrate various strategies available to manage the risks.

Table 4.1 Risk management strategies (adapted from [1]).

Risk reduction strategies	Risk taking strategies
Avoidance: this strategy emphasizes the elimination of the probability of a negative consequence event or reducing the impact of the negative consequence event to eliminate the risk	Creation: strategies that increase the probability of a positive consequence from 0 to a positive value
Prevention: this strategy emphasizes the reducing the probability of a negative consequence event	Enhancement: strategies that increase the probability from an existing nonzero value to a higher nonzero value
Mitigation: this strategy reduces risk by reducing the impact of negative consequences through management actions	Exploitation: this strategy involves increasing the impact of the consequence of the event but does not increase the probability of the event
Transfer: this strategy involves transfer of the risk to a different stakeholder willing to bear the risk	Sharing: this strategy maximizes both the probability and the desirable consequences
Retention: when no means for reducing the probability or the consequence of the negative event exists, and the residual risk is still unacceptable even after all mitigation efforts, the only viable strategy is to accept the risk and actively monitor the risks	Ignoring: this strategy involves taking no action to either increase the probability or strengthen the consequences of the desirable event

4.3.2 Risk assessment

As mentioned earlier, the risk can be measured through the probability and consequence of uncertain future events (Fig. 4.1). Risk assessment in this context refers to the estimation of the probabilities and the consequences of risks. The estimation of the probabilities and consequences of risk are two distinct steps and the need to be clearly distinguished for arriving at unbiased risk management options. The estimation can be performed using qualitative and/or quantitative methods. The purpose of risk assessment is to provide an unbiased and objective scientific estimation of the probability and consequences of the risk to help the risk managers chose a risk management option. The risk assessment step is neutral to the course of action and is limited to providing information about the sources, probabilities, consequences of various risks, and available options under different risk management strategies, as discussed in Table 4.1. The decisions about the course of action to manage the risks are made in the risk management step. It is important to distinguish between the risk assessment and the risk management steps to avoid unknown biases from creeping into the analysis.

The risk assessment consists of four steps and involves significant interactions with stakeholders and experts [1].
1. Identifying the risks: in this step, various risks described earlier, and their sources are identified. Such identification and classification of risks provide a clear picture of the origin of risks.

2. Assessing the consequences: after classifying the risks, the next step is to assess the consequences of each type of risk. The uncertainties associated with the consequences are also characterized in this step. Typically, a ranking of risks based on their consequences is also performed in this step.
3. Likelihood assessment: in this step, the probability of various consequences is assessed using various methods such as expert advice and mathematical models.
4. Risk characterization: after estimating the probabilities of various consequences, the risk is characterized using the conceptual formula (Risk = Probability × consequence) described earlier. This step can be performed using qualitative or quantitative methods as the risks are multidimensional and often cannot be reduced to this simple formula.

The risk assessment steps differ based on organizations depending on the type of risks they address and the scope of the risk analysis for their organization. The US EPA for example uses a similar but slightly different four-step risk assessment method (Fig. 4.4). The EPA method reflects the regulatory and toxicity focus of the EPA in contrast to the generic method described above.

There are several qualitative and quantitative risk assessment techniques discussed in the literature. A list of the techniques is presented in Table 4.2. A discussion of few important methods is presented here but the reader is referred to Yoe [1] for a comprehensive discussion.

Risk indices: this is a semiquantitative method that classifies various risks into categories represented by scoring indices on ordinal scales.

Risk index = Hazard probability × probability of contact with hazard × consequence of exposure.

An example of scales for each of the three components is provided in Table 4.3.

After establishing the ordinal scales, the various risks in the project are presented to experts, stakeholders and the hazards, probability of exposure, and the consequences are rated for each type of risk. The results are collated to provide a comprehensive picture of the risks and their uncertainties. The risk indices are only indicative of the risks and cannot be numerically interpolated/extrapolated. For example, it can only be concluded that the risk index of 90 has a higher risk compared to risk index of 45 and not that the risk index of 90 carries a 100% higher risk compared to a scenario with risk index 45 [1].

Hazard analysis and critical control points (HACCP): The HACCP is a risk analysis technique commonly used in food industries to ensure that the food chains are safe from the risks due to physical, chemical, and biological contaminants in the food chain starting from the raw material production to the consumption of the finished product [8]. These techniques were originally developed to ensure the safety of the food for astronauts in the US space program and are currently regulated by the USD food and Drug Administration (FDA). US FDA defines HACCP as, "a systematic approach to the

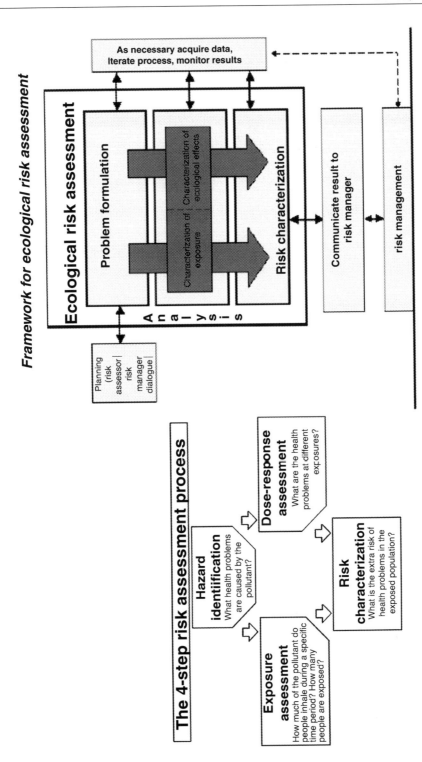

Fig. 4.4 EPA risk assessment method and the ecological risk assessment framework [1].

Table 4.2 Methods of risk characterization [1].

Qualitative methods	Quantitative methods
• Brainstorming • Delhi techniques • Interviews • Checklists • Expert Elicitation • Increase or decrease risk • Risk narratives • Screening • Ratings • Rankings • Risk Matrix • Hazard Analysis and Critical Control Points (HACCP) • Preliminary Hazard Analysis (PHA) • Hazard Operability Study (HAZOP) • Structured What-if (SWIFT) • Go/No-go approach	• Event tree • Fault tree • MCDA • Monte Carlo • Sensitivity analysis • Scenario analysis • Uncertainty decision rules • Subjective probability function • Safety assessment • Fragility curves • Root Cause analysis • Ecological risk assessment

Table 4.3 Possible ordinal scales for evaluating risk indices.

Probability hazard	Probability of exposure	Consequence
1 = Very unlikely	1 = Rare/never	1 = Negligible
2 = Unlikely	2 = Infrequent	2 = Minor
3 = Likely	3 = Frequent	3 = Moderate
4 = Very likely	4 = High	4 = Serious
5 = Inevitable	5 = Constant	5 = Fatal

identification, evaluation, and control of food safety hazards" based on the following seven principles [8]:
1. Conduct a hazard analysis.
2. Determine the critical control points (CCPs).
3. Establish critical limits.
4. Establish monitoring procedures.
5. Establish corrective actions.
6. Establish verification procedures.
7. Establish record-keeping and documentation procedures.

The CCPs are steps in the process chain where the control can be exercised through corrective actions to prevent/eliminate the hazards by bringing the biological, chemical, or physical parameters to within the critical limits. The CCPs are determined by decision trees for each step of the process (Fig. 4.5).

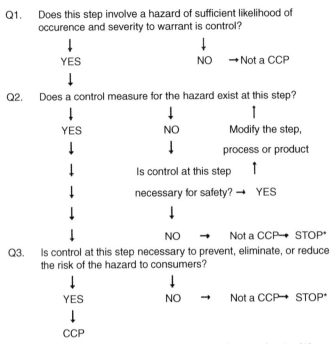

Fig. 4.5 Decision tree to determine if a process step is a critical control point [8].

Event tree: The event tree analysis is a qualitative/quantitative analysis which answers a question, "what happens if an initiating event happens." The event tree is a sequence of nodes and branches that describe the consequences of an event. The root of the event tree is an event that can initiate multiple cascade of the events. All nodes in the event diagram represent the events and the branches describe the possible outcomes. There are no decision points along with the event trees and the events branches are solely determined by the event probabilities, therefore the event trees are also sometimes called probability trees. The event trees are qualitative descriptions of the possible range of outcomes but can be used as quantitative tools if the conditional probabilities for each branch are known. Constructing an event tree requires explicit understanding of the process and a new event tree must be created for each distinct initiating event. The event trees can account for the visually depict the dependence, domino, and timing effects in events sequence that are often difficult to convey verbally. Event tree analysis requires identification of all relevant event sequences *a priori* thus making it difficult to represent events with delays. Fault tree analysis is exact mirror images of the event trees. While the event trees begin with an event and progress forward with the chain of events, the fault trees progress from an outcome to potential initiating events. Fault trees are very commonly used to study events *post-facto* analysis of events such as epidemics and accidents.

Ecological risk assessment (ERA): The ERA was developed by US EPA to address the risks to natural systems and humans as a result of exposures to chemicals and microorganisms, and anthropogenic activity. The US EPA defines the ERA as follows [6]: "An ecological risk assessment is a process for evaluating how likely it is that the environment may be impacted as a result of exposure to one or more environmental stressors such as chemicals, land change, disease, invasive species and climate change."

There are four elements in an ERA which are:
1. Receptors: they are the components of the environment such as humans, plants, and animals as individuals, populations, communities, or ecosystems that are adversely affected by the risk.
2. Exposure: is the co-occurrence of the stressor with the receptors. The quantitative and temporal exposures are considered.
3. Hazard: the type and magnitude of the effect caused by the stressor in the receptors due to exposure.
4. Risk: evaluation of the adverse effects (hazards) in the receptors due to exposure to the stressors.

The ERA (Fig. 4.4) can be broken down into the following phases:
- Planning: problem formulation phase: in this step the information about the receptors that are likely to be impacted, the stressors.
  * Assessment endpoints that reflect management goals and the ecosystem they represent.
  * Conceptual model(s) that represents predicted key relationships between stressor(s) and assessment endpoint(s).
  * Plan for analyzing the risk.
- Analysis phase: in this step, the exposure of the receptors to hazards is determined.
  * Exposure characterization (exposure profile based on Environmental fate and transport data).
  * Ecological effects characterization (stressor-response profile).
  * Uncertainty analysis is also performed here.
  * Risk assessors and risk managers communicate extensively during this phase.
- Risk characterization phase: in this step, the exposure profiles and the hazards are combined to arrive at the risk estimation and interpreted through risk description.
  * The integrated risk characterization includes the assumptions, uncertainties, and strengths and limitations of the analyses. It makes a judgment about the nature of and existence of risks.
- Guidelines:
  - Transparency, clarity, consistency and reasonableness.
  - EPA risk characterization handbook [7].

The ERA can provide a detailed understanding of the nature of risks and the factors that contribute to the environmental risk. It can be used to clearly identify the critical points in the chain of events and identify the appropriate management strategies. However, detailed data are required for this method and without extensive data, the ERA can have a high level of uncertainty.

4.3.3 Risk communication

The principal aim of the risk communication step is to communicate the results of risk assessment, the management strategies, residual risks, and the costs of risk management to stakeholders. The risk communication process must account for the emotional responses of stakeholders and aim to reinforce positive responses and prevent any negative behaviors. There are four risk communication strategies [4] based on how stakeholders perceive the risk. There are typically two dimensions to the risk perceptions: hazard and outrage focus. The focus of stakeholders on either of these two elements results in very different perceptions of the same risk and therefore requires different approaches to risk communication (Fig. 4.6).

Message maps: Message maps are one of the tools used to organize the answers to anticipated questions and concerns from stakeholders [9]. The message maps can provide guidance and to risk managers to convey their message clearly and accurately. The goals of a message map are to educate and inform stakeholders, build trust and maintain credibility and create an informed dialogue and decision-making among stakeholders. The message map organizes the information into three tiers: the first tier identifies the stakeholder along with their questions and concerns. The second tier consists of no more than three key messages to be communicated and the third tier contains the

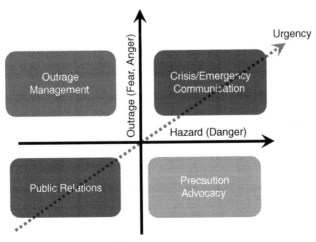

Fig. 4.6 Risk communication strategies [4].

Environmental risk assessment

Stakeholder: Public, health care workers		
Question or Concern: What does the public most need to know about the influenza epidemic?		
Key Message 1	**Key Message 2**	**Key Message 3**
Vaccination a top priority for:	*Symptoms*	*Highly contagious*
Supporting Information 1-1	**Supporting Information 2-1**	**Supporting Information 3-1**
Elderly	Fever	Avoid direct contact with others
Supporting Information 1-2	**Supporting Information 2-2**	**Supporting Information 3-2**
Health care workers	Congestion (cough, sore throat)	Avoid sharing food, drinks
Supporting Information 1-3	**Supporting Information 2-3**	**Supporting Information 3-3**
Immuno-compromised individuals	Muscle aches and pain	Keep bedding (sheets, linens) clean

Fig. 4.7 Example of an overarching message map [9].

additional information to support the three key messages with additional facts and details. An example of a message map to convey the overarching concerns during an influenza outbreak is shown in Fig. 4.7. Such message maps must be prepared for each of the potential questions and used to communicate the risks to the stakeholders. Risk communication is not a publicity spin on technical facts for public relations or damage control but an honest communication with the stakeholders to aid them in making an informed decision about the risks.

4.4 Databases, tools, and software

The US EPA provides comprehensive datasets and tools to conduct the ERA. Various databases containing the toxicity information, technical, and methodological guidance documents, and modeling tools through their EPA EcoBox website [10]. Some of the relevant tools are listed in Table 4.4. The US FDA provides information on the HACCP [8] for the risk assessment in food chains. The Canadian [11] and European Union [12] provide additional regional specific guidance to conduct the risk assessments in various contexts. There are hundreds of risk assessment frameworks that have been tailored for specific domains and the examples provided here are only a few that are focused on the ERAs.

4.5 Examples

Example 1: The dairy industry faces numerous challenges due to contamination from biological, chemical, and physical contaminants throughout the process chains. Sandrou and Aryanitoyannis [13] discuss the implementation of HACCP in the dairy industry. The pasteurized milk is the largest product of dairy industry and is used to reduce the risks due to milk-borne pathogens. Similarly, the spray milk powder is one of the most cost-effective ways to preserve the milk. The flow diagram for the pasteurization of milk and the spray milk powder production process along with the CCPs is

Table 4.4 Tools, databases, and technical documents to conduct ERA (Adapted from [10]).

Authors	Title	Description	URL
U.S. Geological Survey	Acute Toxicity Database	This database summarizes the results from aquatic acute toxicity tests and provides a relative starting point for hazard assessment of contaminants	https://www.cerc.usgs.gov/data/acute/acute.html
U.S. EPA	Aggregated Computational Toxicology Resource (ACToR)	ACToR is an online warehouse of chemical toxicity data that can be searched using chemical name or cas. no. It provides chemical structure, chemical formula, molecular weight, synonyms, physico-chemical values, in vitro assay data, and in vivo toxicology data. Other EPA chemical toxicity databases that can be searched from this website include ToxRefDB (animal toxicity studies) and DSSTox (chemical structures and annotations)	https://actor.epa.gov/actor/home.xhtml
U.S. EPA	Air Emissions Factors and Quantification	This webpage provides information and links for emissions factors tools used for building emissions inventories, guiding air quality management decisions, and developing emissions control strategies	https://www.epa.gov/air-emissions-factors-and-quantification
OECD	eChemPortal	eChemPortal provides information on physical chemical properties, ecotoxicity, environmental fate and behavior, and toxicity of chemicals. Users can search reports and datasets by chemical name, number, or property and also provides chemical exposure and use information	https://www.echemportal.org/echemportal/index?pageID=0&request_locale=en
U.S. EPA	ECOTOXicology Knowledgebase (ECOTOX)	ECOTOX allows the user to query single chemical toxicity data for aquatic and terrestrial plants and wildlife. The database combines information from three databases: AQUIRE, PHYTOTOX, and TERRETOX	https://cfpub.epa.gov/ecotox/
National Library of Medicine (NLM)	Toxicology Data Network (TOXNET)	Toxicology Data Network (TOXNET) provides links to databases on toxicology, hazardous chemicals, environmental health, and toxic releases	https://toxnet.nlm.nih.gov
U.S. EPA	Watershed Academy Web	This document provides background information and examples on watershed ecological risk assessments and links to websites about several watershed risk case studies	https://cfpub.epa.gov/watertrain/pdf/modules/wshedecorisk.pdf

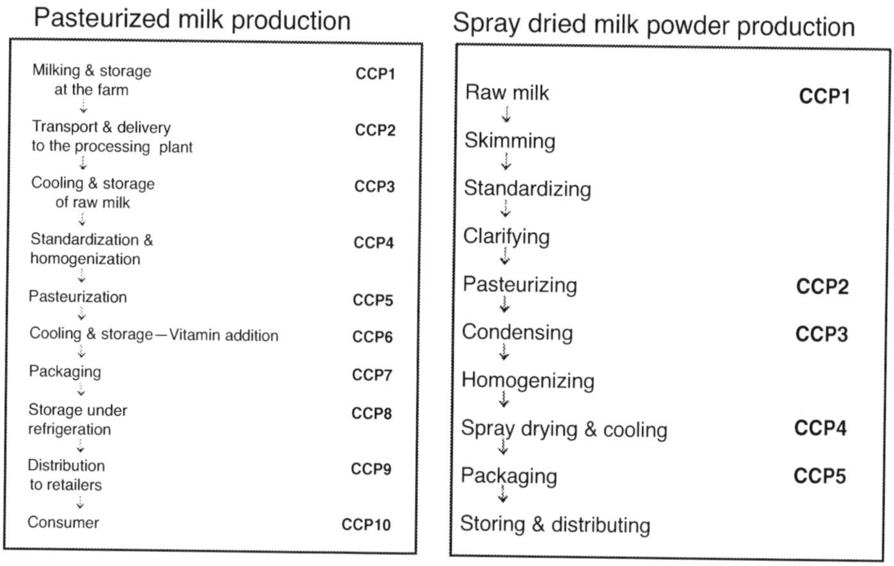

Environmental risk assessment

Pasteurized milk production

Milking & storage at the farm	CCP1
↓ Transport & delivery to the processing plant	CCP2
↓ Cooling & storage of raw milk	CCP3
↓ Standardization & homogenization	CCP4
↓ Pasteurization	CCP5
↓ Cooling & storage—Vitamin addition	CCP6
↓ Packaging	CCP7
↓ Storage under refrigeration	CCP8
↓ Distribution to retailers	CCP9
↓ Consumer	CCP10

Spray dried milk powder production

Raw milk	CCP1
↓ Skimming	
↓ Standardizing	
↓ Clarifying	
↓ Pasteurizing	CCP2
↓ Condensing	CCP3
↓ Homogenizing	
↓ Spray drying & cooling	CCP4
↓ Packaging	CCP5
↓ Storing & distributing	

Fig. 4.8 HACCP for pasteurized milk and spray dried milk powder production [13]. *HACCP*, hazard analysis and critical control points.

shown in Fig. 4.8. The results of the HACCP analysis identified 10 CCPs in the milk pasteurization process while there were five CCPs in the spray dried milk powder production. A CCP can be associated with a process and itself have many CCPs associated with it. As can be observed from the examples, the pasteurization process which has 10 CCPs is identified with only one CCP (CCP2) in the spray-dried milk production process.

Example 2: Cancers are caused due to multiple causes and the primary preventive interventions are in the lifestyle choices and environmental factors. The study by Danaei et al. [14] assessed the comparative risks for 12 types of cancer arising from nine types of risk factors. The study considered the exposure to the minimum levels in Fig. 4.9 as the baseline scenario and the actual levels of exposure around the world in seven regions.

The population attributable fractions (PAF) were defined as follows:

$$\text{PAF} = \frac{\int_{x=0}^{m} \text{RR}(x)P(x)dx - \int_{x=0}^{m} \text{RR}(x)P(x)dx}{\int_{x=0}^{m} \text{RR}(x)P(x)dx}$$

where

PAF: Population attributable fraction.
x: Risk factor exposure level.
m: Maximum exposure level.
$P(x)$: Population distribution of exposure.
$P(x)$: Population distribution of baseline (counterfactual) exposure.
$RR(x)$: Relative risk of mortality from site-specific cancer at exposure level x.

Exposure variable	Theoretical-minimum-risk exposure distribution	Cancer sites affected (age groups assessed)†
Diet and physical inactivity		
Overweight and obesity (high BMI)[a] — BMI (kg/m²)	21 SD 1 kg/m²	Corpus uteri cancer, colorectal cancers (≥30 years), post-menopausal breast cancer (≥45 years), gallbladder cancer, kidney cancer
Low fruit and vegetable intake[a] — Fruit and vegetable intake per day	600 SD 50 g intake per day for adults	Colorectal cancer, stomach cancer, lung cancer, oesophageal cancer (≥15 years)
Physical inactivity[a] — Three categories: inactive, insufficiently active (<2.5 h per week of moderate-intensity activity, or <4000 kJ per week), and sufficiently active. Activity in spare time, work, and transport considered	≥2.5 h per week of moderate-intensity activity or equivalent (400 kJ per week)	Breast cancer, colorectal cancer (≥15 years), prostate cancer
Addictive substances		
Smoking[a] — Current levels of smoking impact ratio (indirect indicator of accumulated smoking risk based on excess lung-cancer mortality)	No smoking	Lung cancer, mouth and oropharynx cancer, oesophageal cancer, stomach cancer, liver cancer, pancreatic cancer, cervix uteri cancer, bladder cancer, leukaemia (≥30 years)
Alcohol use[a] — Current alcohol consumption volumes and patterns	No alcohol use‡	Liver cancer, mouth and oropharynx cancer, breast cancer, oesophageal cancer, selected other cancers (≥15 years)
Sexual and reproductive health		
Unsafe sex[a] — Sex with an infected partner without any measures to prevent infection	No unsafe sex	Cervix uteri cancer (all ages)§
Environmental risks		
Urban air pollution[a] — Estimated yearly average particulate matter concentration for particles with aerodynamic diameters <2.5 microns or 10 microns (PM₂.₅ or PM₁₀)	7.5 μg/m³ for PM₂.₅, 15 μg/m³ for PM₁₀	Lung cancer (≥30 years)
Indoor smoke from household use of solid fuels[a] — Household use of solid fuels	No household solid fuel use with limited ventilation	Lung cancer (coal) (≥30 years)
Other selected risks		
Contaminated injections in health-care settings[a] — Exposure to ≥1 contaminated injection (contamination refers to potential transmission of hepatitis B virus and hepatitis C virus)	No contaminated injections	Liver cancer (all ages)

*New exposure data and epidemiological evidence on disease outcomes and relative risks used when new epidemiological analyses allowed improvements compared with original analyses of Comparative Risk Assessment project—eg, relative risks for site-specific cancers as a result of smoking with better adjustment for potential confounders and new exposure data sources for overweight and obesity. †Italics—outcomes likely to be causal but not quantified because of insufficient evidence on prevalence or hazard size. Corresponding ICD 9 3-digit codes: bladder cancer 188; breast cancer 174; cervix uteri cancer 180; colorectal cancers 153–154; corpus uteri cancer 179, 182; leukaemia 204–208; liver cancer 155; mouth and oropharynx cancer 140–149; oesophageal cancer 150; pancreatic cancer 157; stomach cancer 151; trachea, bronchus, and lung cancers 162; selected other cancers 210–239. Total deaths from cancer are from WHO.¹⁹,²⁰ Methods used by WHO¹⁹,²⁰ are based on a combination of vital statistics, with sample registration or surveillance systems, or without vital registration. ‡Alcohol has benefits as well as harms for different diseases, also depending on patterns of alcohol consumption. A theoretical minimum of zero was chosen for alcohol use because, despite benefits for specific diseases (cardiovascular) in some populations and regional burden of disease due to alcohol use was dominated by its effects on neuropsychological diseases and injuries, which are considerably larger than benefits for vascular diseases. Furthermore, no benefits for neoplastic diseases have been noted from alcohol. §A proportion of HPV infections that lead to cervix uteri cancer are transmitted through routes other than sexual contact. PAF for unsafe sex, as defined in the comparative risk assessment project,⁶ measures current population-level cervix cancer mortality that would be reduced, had there never been any sexual transmission of infection—ie, the consequences of past and current exposure, as we do for accumulated hazards of smoking. By considering health consequences of past and current exposure, nearly all of sexually transmitted diseases are attributable to unsafe sex because, in the absence of sexual transmission in the past, current infections transmitted through other forms of contact would not occur (if infected hosts acquired their infection sexually (and so on in the sequence of past infected hosts).

Fig. 4.9 Cancer risk factors, exposure levels, and theoretical minimum baseline exposure distributions [14]. *(Reproduced with permission from Elsevier Danaei G, Hoorn SV, Lopez AD, Murray CJL, Ezzati M, and the Comparative Risk Assessment group (cancers). 2005. Causes of cancer in the world: comparative risk assessment of nine behavioural and environmental risk factors. Lancet. 366:1784-1793.)*

As it can be seen from the above formula, the researchers have evaluated the risk of cancer as a product of the exposure level (probability) and the increase in mortality (consequence) as discussed earlier in the chapter. The PAF was used to arrive at the number of deaths attributable to an individual risk factor in a region of the world. The authors found that 35% of the 7 million deaths due to cancer could be attributed to the nine risk factors. The results of the analysis are shown in Fig. 4.10 and they indicate that alcohol consumption and smoking are two predominant risk factors for cancers around the world regardless of the income levels.

Example 3: Biofuels are suggested as an alternative to the fossil fuels. While some of the environmental impacts such as greenhouse gas emissions associated with biofuels are lower than fossil fuels, it is not clear if the biofuels pose any environmental risks. Thornley and Gilbert discuss this issue in their paper "Biofuels: balancing risks and rewards" [15]. The authors identified 36 potential biofuel system impacts that cover the ecological, social, and economic domains, as shown in Fig. 4.11. The authors performed a life cycle assessment to estimate the impacts of the biodiesel production from Argentinean Soybeans compared to the reference fossil diesel fuel. The authors then used a risk assessment methodology where the probability and the consequences of the risk factors identified earlier (Fig. 4.11) were combined into risk index scores, as shown in Table 4.5.

The combined risk assessment of the 36 factors was determined by a group of stakeholders. The modal/mean responses from the stakeholders were recorded for the benefits (Maximum benefit = +9) and risks (Maximum risk = −9) for the Argentinean soy biodiesel compared to the reference scenarios of fossil-based diesel. The results presented in Fig. 4.12 indicate the tradeoffs in the risks and rewards of producing biofuels. Additional insights of the authors are that the tradeoffs are different for different biofuels and bioproducts systems. Therefore, the use of "one-size-fits-all" frameworks to assess the sustainability assessment and certification of global bioenergy supply may be counterproductive as it could miss the most important drivers of the environmental impacts in various systems. The authors recommend that the most significant risk factors determined by a comprehensive risk analysis be used for identifying and evaluating the most significant environmental impacts through the LCA. This study is an example of the combined use of LCA and risk analysis in bioenergy systems.

4.6 Perspectives

Traditional risk assessment tools discussed in this chapter are very useful for studying and managing risks that occur relatively frequently in nature. There are other kinds of events called black swan and perfect storm type events which are extremely rare (low probability of occurrence) but catastrophic consequences (consequences are very large), thus resulting in a nonsignificant risk. The challenge with risk assessments for the black

	Total deaths	PAF (%) and number of attributable cancer deaths (thousands) for individual risk factors	PAF due to joint hazards of risk factors
Worldwide			
Mouth and oropharynx cancers	311 633	Alcohol use (16%, 51), smoking (42%, 131)	52%
Oesophageal cancer	437 511	Alcohol use (26%, 116), smoking (42%, 184), low fruit and vegetable intake (18%, 80)	62%
Stomach cancer	841 693	Smoking (13%, 111), low fruit and vegetable intake (18%, 147)	28%
Colon and rectum cancers	613 740	Overweight and obesity (11%, 69), physical inactivity (15%, 90), low fruit and vegetable intake (2%, 12)	13%
Liver cancer	606 441	Smoking (14%, 85), alcohol use (25%, 150), contaminated injections in health-care settings (18%, 111)	47%
Pancreatic cancer	226 981	Smoking (22%, 50)	22%
Trachea, bronchus, and lung cancers	1 226 574	Smoking (70%, 856), low fruit and vegetable intake (11%, 135), indoor smoke from household use of solid fuels (1%, 16), urban air pollution (5%, 64)	74%
Breast cancer	472 424	Alcohol use (5%, 26), overweight and obesity (9%, 43), physical inactivity (10%, 45)	21%
Cervix uteri cancer	234 728	Smoking (2%, 6), unsafe sex (100%, 235)	100%
Corpus uteri cancer	70 881	Overweight and obesity (40%, 28)	40%
Bladder cancer	175 318	Smoking (28%, 48)	28%
Leukaemia	263 169	Smoking (9%, 23)	9%
Selected other cancers	145 802	Alcohol use (6%, 8)	6%
All other cancers	1 391 507	None of selected risk factors	0%
All cancers	7 018 402	Alcohol use (5%, 351), smoking (21%, 1493), low fruit and vegetable intake (5%, 374), indoor smoke from household use of solid fuels (<0.5%, 16), urban air pollution (1%, 64), overweight and obesity (2%, 139), physical inactivity (2%, 135), contaminated injections in health-care settings (2%, 111), unsafe sex (3%, 235)	35%

Fig. 4.10 Individual and joint contributions of risk factors [14]. (Reproduced with permission from Elsevier Danaei G, Hoorn SV, Lopez AD, Murray CJL, Ezzati M, and the Comparative Risk Assessment group (cancers). 2005. Causes of cancer in the world: comparative risk assessment of nine behavioural and environmental risk factors. Lancet. 366:1784-1793.)

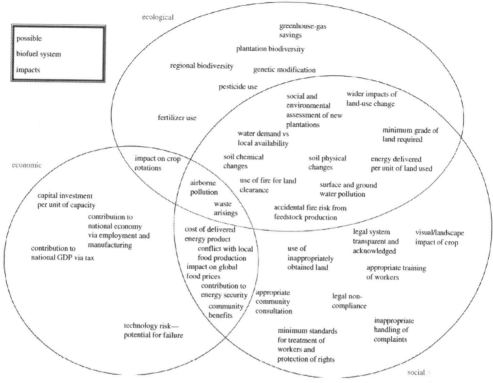

Fig. 4.11 Summary of potential impacts of bioenergy systems (figure 1 from [15]).

Table 4.5 Risk index scores (Table 4.1 in [15]).

Probability of occurrence	Severity of impact	Combined risk assessment
H(3)	H(3)	HH(9)
H(3)	M(2)	HM(6)
M(2)	H(3)	MH(6)
M(2)	M(2)	MM(4)
H(3)	L(1)	HL(3)
L(1)	H(3)	LH(3)
L(1)	M(2)	LM(2)
M(2)	L(1)	ML(2)
L(1)	L(1)	LL(1)

swan and perfect storm type events is that since they are outside the regular expectations of the event, they are not anticipated in advance but since they have huge consequences, they are highly risky. Black swan events are truly unknowable (epistemic uncertainty) and cannot be anticipated in advance and therefore the risk analysis is not possible and most of the consequences need to be managed through risk mitigation-based strategies.

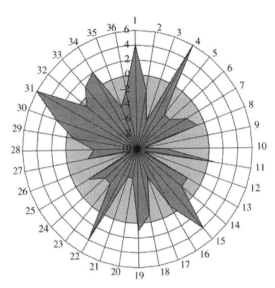

Fig. 4.12 Radar diagram showing overall sustainability assessment of biodiesel from Argentinean soy compared with mineral diesel. Light grey, reference level; dark grey, scores for Argentinean soy system (figure 4 from [15]).

Perfect storm, on the other hand, is rare combination of events that occur regularly and are well known. However, it is the combination of the relatively well-known events that occur with catastrophic consequences that were no anticipated (aleatory uncertainty). In practice, these two types of uncertainties are not distinct and occur together [16]. Global financial crisis in 2008 was such a black swan event that in the hindsight seems obvious but was not anticipated by almost all financial experts. On the other hand, the Earthquake in Japan followed by Tsunami leading to the accident at the Fukushima nuclear plant in Japan is an example of a Perfect storm. The earthquake, tsunami, backup generator failure, flooding of the control room, and cooling system failure were all events that were well known to risk managers and considered in the design of the power plant; however, their rare conjunction was not anticipated leading to the Fukushima nuclear accident. It is for such situations that Taleb [17] identifies six mistakes that are to be avoided by risk managers and are quoted below:

1. Manage risk by predicting extreme events.
2. Studying the past will help us manage risk.
3. Do not listen to advice about what we should not do.
4. Assume risk can be measured by standard deviation.
5. Do not appreciate that what is mathematically equivalent is not psychologically equivalent.
6. Do not realize that optimization makes the systems fragile.

Therefore, the advice of what should not be done is equally important as the advice to pursue a strategy, and the qualitative methods that involve experienced stakeholders who preserve the societal "memory" through expert wisdom must not be discounted in favor of purely quantitative methods in risk management. Additionally, it is the extremely low probability events that have catastrophic consequences ("risk of ruin") lead to the advocacy of precautionary principle. Precautionary principle must be employed in situations where there is a chance (however unlikely) that the risks become catastrophic. For all other situations, regular risk assessment methods discussed in this chapter are adequate.

4.7 Conclusions and perspectives

This chapter focused on environmental risk assessment as a part of the systems analysis. The chapter defined risks, described various types of risks, and methods to classify risks. The conceptual formula for risk was introduced and the risk analysis method was discussed in detail. Various approaches to risk analysis used by different organizations and the detailed methodology for the risk assessment were discussed. Databases, tools, and technical guidance documents for conducting risk assessment, examples of ERA, and perspectives on the ERA were presented.

References

[1] Yoe C. Principles of risk analysis for water resources. Institute for Water Resources. U.S. Army Corps of Engineers, 2017. https://planning.erdc.dren.mil/toolbox/library/iwrserver/2017_R_01_PrinciplesofRiskAnalysisforWaterResources.pdf. [Accessed 16th May, 2021].
[2] Institute for Strategic Studies, Risks and Risk Analysis in the Strategic Context, University of Pretoria, 2008. ISBN: 1-86854-3. https://www.thefreelibrary.com/_/print/PrintArticle.aspx?id=196210233. [Accessed May 5, 2020].
[3] N.N. Taleb, Antifragile: Things that Gain From Disorder, Random House, 2012. ISBN: 1-400-06782-0.
[4] P. Sandman. National Center for Food Protection and Defense/International Food Information Council Risk Communication (NCFPD/IFICRC). "Risk Communicator Training, Module 1: Introduction to National Weather Service. 2008. "Hurricane Ike Warning for Galveston. 2009".
[5] US EPA. Framework for ecological risk assessment. Risk Assessment Forum, Washington, DC. 1992. EPA/630/R-92/001.
[6] US EPA. Ecological risk assessment, 2020. https://www.epa.gov/risk/ecological-risk-assessment. [Accessed May 5, 2020].
[7] US EPA, Science Policy Council Handbook: Risk Characterization, 2000. EPA 100-B-00-02. https://nepis.epa.gov/Exe/ZyPDF.cgi?Dockey=40000006.txt. [Accessed May 5, 2020].
[8] US FDA. Hazard analysis critical control point. 2018. https://www.fda.gov/food/hazard-analysis-critical-control-point-haccp/haccp-principles-application-guidelines#princ. [Accessed May 5, 2020].
[9] I. Lin, D. Petersen, Risk Communication in Action: The Tools of Message Mapping, U.S. Environmental Protection Agency, 2007. (EPA/625/R-06/012). http://nepis.epa.gov/Adobe/PDF/60000IOS.pdf. [Accessed May 5, 2020].
[10] US EPA. https://www.epa.gov/ecobox. [Accessed May 5, 2020].
[11] The National Contaminated Sites Remediation Program. 1996. A framework for ecological risk assessment General guidance PN 1195. https://www.ccme.ca/files/Resources/csm/pn_1195_e.pdf. [Accessed May 5, 2020].

[12] European Chemical Bureau. Technical guidance document on risk assessment. https://echa.europa.eu/documents/10162/16960216/tgdpart2_2ed_en.pdf. [Accessed May 5, 2020].
[13] D.K. Sandrou, I.S. Aryanitoyannis, Implementation of hazard analysis and critical control point (HACCP) to the dairy industry: current status and perspectives, Food Rev. Intl. 16 (2007) 77–111.
[14] G. Danaei, S.V. Hoorn, A.D. Lopez, C.J.L. Murray, M. EzzatiComparative Risk Assessment Group (cancers), Causes of cancer in the world: comparative risk assessment of nine behavioural and environmental risk factors, Lancet 366 (2005) 1784–1793.
[15] P. Thornley, P. Gilbert, Biofuels: balancing risks and rewards, Interface Focus 3 (2013) 20120040.
[16] E. Pate-Cornell, On "Black swans" and "Perfect storms": risk analysis and management when statistics are not enough, Risk Anal 32 (2012) 1823–1833.
[17] N.N. Taleb, D.G. Goldstein, M.W. Spitznagel, The six mistakes executives make in risk management, Harvard Business Rev 87 (2009) 123.

CHAPTER FIVE

Resource assessment

Ganti S. Murthy[a,b]
[a]Biological and Ecological Engineering, Oregon State University, Corvallis, OR, United States, [b]Biosciences and Biomedical Engineering, Indian Institute of Technology, Indore, India

5.1 Introduction

Resources in the form of technology, human resources, financial resources, and natural resources, such as land, water, minerals, and energy are the basis of any process/product systems used in human society. Many of the conflicts in human history can be traced to the control of accessibility to resources and therefore understanding the availability and accessibility of resources is a very critical part of the systems analysis. Resources are generally classified into natural resources and human capital. Available natural resources were much larger than the human societal demand for most of history. With the rapid industrialization in the last few centuries concomitant with the unprecedented rise in human population, the limits of the natural systems have been breached. For example, the human activity today is the major driving force of the materials fluxes in the nitrogen and the mineral cycles in nature. Human activity has also caused large-scale impacts on the water and land throughout the world. With the world population expected to reach 9.0 billion by 2050, it is imperative that the natural resources be managed more sustainably, therefore conducting resource sustainability assessment is critical for the success of any technology or process.

In a resource-constrained world that we live in, it is imperative to understand the limitations posed by the available resources. Therefore, a detailed study of the required resources must be undertaken to verify if the resources are available and if available accessible at acceptable economic, environmental, and societal costs. The economic and environmental aspects of the resources are typically covered in the TEA and the environmental impact phases while the availability and accessibility issues are covered in the resource sustainability analysis. Thus, the resource sustainability analysis is designed to answer the questions, "are the required resources for the project available and if available accessible and sustainable over the lifetime of the project?" This chapter will focus on providing an overview of the resource sustainability of natural resources such as land, water, fertilizers, minerals and metals. A detailed resource sustainability analysis must be conducted for a specific project as part of the systems analysis.

However, it must be noted that systems analysis is inherently multidimensional. This chapter only focuses on the resource assessment aspects while other chapters on various different dimensions of systems analysis.

Biomass, Biofuels, Biochemicals
DOI: https://doi.org/10.1016/B978-0-12-819242-9.00003-8

© 2022 Elsevier BV.
All rights reserved.

5.2 Land resources

The land is one of the most basic natural resources that is needed for any technological enterprise. The exploitation of this resource for food production through rain-fed agriculture is one of the defining moments of human evolution. Use of land for agriculture resulted in moving from nomadic lifestyles to settled societies giving rise to the development the human society as we know it today. The land resources around the world are distributed unequally by geographical locations and income levels (Fig. 5.1). For example, only 15 % of the cultivated land is available in low-income countries which have 38% of the world population (Table 5.1).

Of the total land area of 149 million km^2 of the Earth, 71% (104 million km^2) is habitable and the remaining land is covered by glaciers (10%) and barren land (19%). Fully 50% of the habitable area (51 million km^2) is used for agriculture while 37% of the habitable land is under forests, 11% (12 million km^2) under the shrub. Only 1% (1.5 million km^2) area is the urban and built-up infrastructure while the freshwater sources such as lakes and rivers cover another 1% of the habitable land. Food crops excluding the feed crops account for only 23% (11 million km^2) of the total agricultural land but provide 83% of the global calories and 63% of the global protein supply. On the other hand, 77% (40 million km^2) of the agricultural land is used in the meat and dairy industries to provide just 18% of the global calories and 37% of the global protein supply [2]. Three of the most significant uses of land resources by human society are for housing, food production, and energy production. Overall area for human habitation including the urban areas and built infrastructure is ~1% and hence is not further discussed in this chapter.

Overall, the landmass of earth covers 13.2 billion ha (excluding Antarctica) and can be divided into 1.6 billion ha of arable land (12%), 3.7 billion ha (28%) forests, and 4.6 billion ha of grasslands and woodland ecosystems (35%) [1]. While most of the countries have between 5% and 20% of their land area under cultivation, only a few countries such as India, Bangladesh, Ukraine, and Denmark have more than 50% of their land area suitable for agriculture. There has been a 12% increase in the cultivated land and a 117% increase in the irrigated lands while the per capita cropland availability has decreased from 0.45 to 0.22 ha/person from 1961 to 2009. During this time, the scientific advances in agriculture including the introduction of better seeds increased irrigation and management of water, increased fertilizer use, and better pest and weed control resulted in significant increases in agricultural productivity which increased the food production that is able to support a population that has grown significantly resulting in the reduction of the land needed to support a person to about 0.22 ha/person. Even higher gains in productivities, closing yield gaps and a shift to plant-based diets are needed to meet the food needs of the expected 9 billion human population 2050 without further degradation to our land resources. Since animal-based foods have a large footprint compared to plant-based food products (Fig. 5.2) [2], a shift to plan-based diets is being increasingly recommended by the global scientific community for a sustainable future [3].

Resource assessment

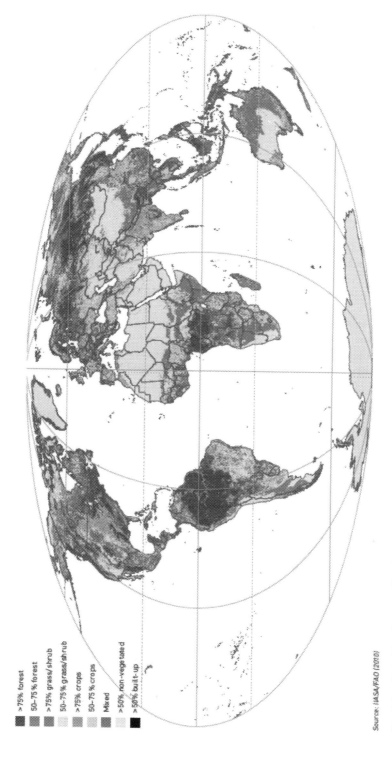

Source: IIASA/FAO (2010)

Fig. 5.1 Global land use and land cover distribution [1].

Table 5.1 Regional distribution of land categories (data from [1]).

Country category	Global share of land, %	Share of global population	Cultivated land Mha	%	Forested land Mha	%	Grassland and woodland systems Mha	%	Sparsely vegetated and barren land Mha	%	Settlement and infrastructure Mha	%	Inland water bodies Mha	%
Low-income	22	38	441	15	564	20	1020	36	744	26	52	1.8	41	1.4
Middle-income	53	47	735	11	2285	33	2266	33	1422	21	69	1	79	1
High-income	25	15	380	12	880	27	1299	39	592	18	31	1	123	4

Resource assessment

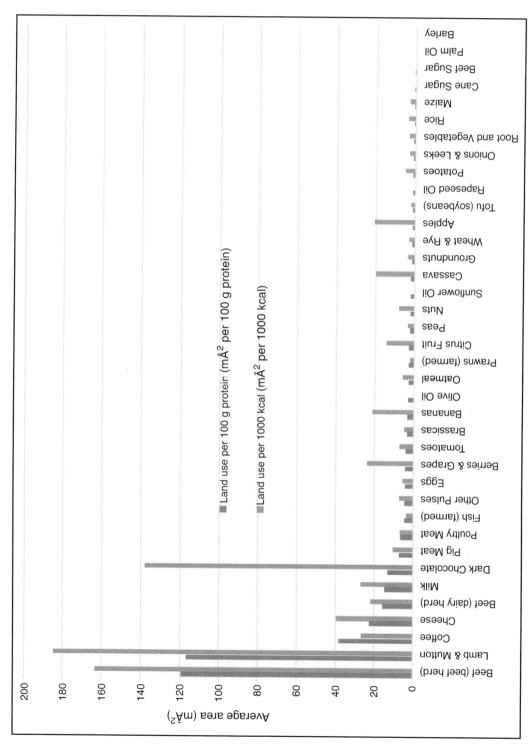

Fig. 5.2 Land footprint of food products (based on data from [2]).

Table 5.2 Key indicators for electricity generation options (data from [5]).

Indicator (per TWh)	Coal	Natural gas	Nuclear	Biomass	Hydro	Wind (onshore)	Solar (PV)
Land Use (km^2)	2.1	1.1	0.1	95	50	46	5.7
GHG Emissions (×1000 t CO$_2$)	1001	469	16	18	4	12	46
Solid Waste (t)	58,600	NA	NA	9170	NA	NA	NA
Dispatchability (A good; B fair; C poor)	A	A	A	B	B	C	C

Land is a critical resource for the production of energy. Coal needs to be mined from open/underground mines, oil and natural gas extraction and transportation pipelines, and the hydropower generation requires building of dams and reservoirs. Associated roads and support infrastructure construction for access and maintain these facilities; all of which require land resources.

The land use for coal, natural gas, nuclear, solar, wind, and hydro energy resources require 4.89, 4.97, 5.08, 17.4, 28.26, and 126.09 ha/MW of installed power capacity [4]. Compared to fossil fuels, the land requirements for renewable energy sources, especially the biomass energy sources are much higher (Table 5.2).

While the land use per unit of power is a useful metric, it does not really reflect the fact that biomass, wind, and solar are renewable and can be produced on the same land for many years while other fossil resources such as coal mines and oil wells need to be relocated after depletion of the sources. Therefore, an approach to time to land use equivalency has been proposed by Trainor et al [6]. The time to land use equivalency was defined as the cumulative land required to produce 1 TWh/year. For renewable energy sources such as wind, solar, and biomass energy technologies, the cumulative land to produce 1 TWh/year is constant whereas for the extractive energy sources such as coal, petroleum and natural gas, the land required expands to account for the depletion of mines, oil and gas wells. Trainor et al. [6] calculated land use equivalency for electricity (Table 5.3) and the liquid fuels sectors (Table 5.4) and concluded that while

Table 5.3 Time to land use equivalency (years) for electricity production in terms of total landscape impacts (data from [6]).

Renewables	Nuclear	Natural Gas Conventional	Shale gas	Tight gas	Coalbed methane	Coal Underground	Surface
Geothermal	39.0	1.8	1.0	1.3	0.6	8.1	0.6
Solar Photovoltaic	114.0	5.2	3.0	3.7	1.9	23.6	1.8
Hydropower	128.0	5.9	3.3	4.2	2.1	26.5	2.1
Solar Thermal	146.1	6.7	3.8	4.8	2.4	30.2	2.4
Wind (Total area)	963.8	44.3	25.0	31.7	15.7	199.5	15.5
Biomass	6149.0	282.9	159.5	202.1	99.9	1272.5	98.9

Table 5.4 Time to land use equivalency(years) for liquid fuels in terms of total landscape impacts (Data from [6]).

Renewables	Conventional oil	Tight oil
Corn	82.7	28.9
Sugar cane	95.9	33.5
Soybean	103.4	36.1
Cellulose	197.5	69.1

all forms of energy have significant land-use impacts, renewable energy sources with the notable exception of biofuels have comparable land use impacts over their lifetimes. Larger land use for biofuels can be primarily attributed to the photosynthesis efficiency of about 2% for conversion of solar energy which is about 10 times lower than the solar PV panels. For example, in UK, the green biomass can harvest about 2 W/m^2 (realistic range is 0.1–1.8 W/m^2) compared to 22W/m^2 for a solar PV panel with 20% conversion efficiency [7]. Therefore, the land-use impacts of biofuels are a significant factor, perhaps the most limiting factor, in any feasibility analysis of biofuels.

5.3 Water resources

Water is the very essence of life. Although more than 70% of the earth surface is covered with water, the usable freshwater sources are about 42,000 km^3/year of which the human use from rivers and aquifers accounts for 3900 km^3/year. About 70% of the total water is used for irrigation (2710 km^3/year) while the industrial and municipal sector uses account for 19% and 11% of the total water use, respectively [1]. Consumptive water use defined as the use of water where it does not return to the local hydrological system by return flows is estimated to be 40% of the total water use.

Agriculture is the largest consumptive water use sector followed by power generation. As can be seen from the data in Table 5.5, there are significant regional differences in the amounts of precipitation and availability of renewable water resources. Total water withdrawal already is already exceeding 50% in many regions of the world especially in the south, western, and central Asia regions while it exceeds the available renewable water resources in Northern Africa indicating an unsustainable use of groundwater resources in that region. The importance of the irrigated agriculture to the overall cereal production can be seen from the percentage of irrigated cereal production to the total production which is almost two times the percentage of land under irrigated agriculture in the tropical regions. Temperate regions of Western Europe and Northern America are some of the most productive rainfed agricultural regions of the world and it is reflected in the relatively lower percentage of irrigated agriculture and high amounts of available renewable water resources in these regions. Globally, industrial water use accounts for the second largest fraction of water use. Variations in industrial water use reflect the current status of industrialization around the world (Table 5.5).

Table 5.5 Water use, availability, and contribution of irrigated agriculture to cereal production of the world (data from [1]).

Continent regions	Municipal use km³/year	%	Industrial use km³/year	%	Agricultural use km³/year	%	Total Withdrawal km³/year	Precipitation mm	Renewable water resources (km³) km³	Irrigation water withdrawal km³	Irrigated land as a % of cultivated land %	Irrigated cereal production as a % of total cereal production %
Africa	21	10	9	4	184	86	215	678	3931	184	5	24
Northern Africa	9	9	5	6	80	85	94	96	47	80	21	75
Sub-Saharan Africa	13	10	4	3	105	87	121	815	3884	105	2	9
Americas	126	16	280	35	385	49	791	1091	19,238	385	10	22
Northern America	88	15	256	43	258	43	603	636	6077	258	11	22
Central America and Caribbean	6	26	2	11	15	64	24	2011	781	15	7	32
South America	32	19	21	13	112	68	165	1604	12,380	112	8	22
Asia	217	9	227	9	2012	82	2456	827	12,413	2012	34	67
Western Asia	25	9	20	7	227	83	271	217	483	227	28	48
Central Asia	5	3	8	5	150	92	163	1602	263	150	30	45
South Asia	70	7	20	2	914	92	1004	634	1766	914	38	70
East Asia	93	14	150	22	434	64	677	2400	3410	434	44	78
Southeast Asia	23	7	30	9	287	84	340	540	6490	287	19	49
Europe	61	16	204	55	109	29	374	81	6548	109	5	8
Western and Central Europe	42	16	149	56	75	28	265		2098	75	9	10
Eastern Europe and Russian Federation	19	18	56	51	35	32	110	467	4449	35	1	4
Oceania	5	17	3	10	19	73	26	586	892	19	7	7
Australia and New Zealand	5	17	3	10	19	73	26	574	819	19	7	7
Pacific Islands	0.01	14	0.01	14	0.05	71	0.1	2062	73	0.05	1	–
World	429	11	723	19	2710	70	3862	809	43,022	2710	17	42
High-income	145	16	392	43	383	42	920	622	9009	383	11	20
Middle-income	195	12	287	18	1136	70	1618	872	26,680	1136	26	49
Low-income	90	7	44	3	1191	90	1324	876	7332	1191	14	55
Low-income food deficit	182	8	184	8	1813	83	2180	881	13,985	1813	26	64
Least-developed	10	5	3	1	190	94	203	856	4493	190	8	38

Similar to the land resources, the water resources are distributed unequally around the globe and the water use has increased by 600% in the last 100 years to meet the food needs of a global population that has grown 4.3 times from 1.8 billion to 7.7 billion during this period. Further population growth to 9.4–10.2 billion by 2050 from the current levels of 7.7 billion is expected to place tremendous pressures on the global water resources and increase by 20%–30% to 5500–6000 km^3/year from the current 46,000 km^3/year [8]. The growth in nonagricultural demand will exceed the increases in the agricultural demand which itself will increase by 60% from the current levels. Additionally, most of the increases will come from already water-stressed countries in Africa and Asia contributing creating a very challenging global situation. For example, industrial water use in Africa is expected to increase 800% from the current negligible levels while in Asia the demand will increase by 250% contributing to the 400% increase in the global water use for the manufacturing sector.

Global water use for energy is expected to increase by 85% by 2050. The domestic water use is expected to increase significantly in Africa and Asia (300%) and Central and South America (200%) by 2050 [8]. Water scarcity can be due to physical or economic water scarcity. Physical water scarcity is defined as the limitation on the amount of physically available water while economic water scarcity is the limitation in the water availability due to the inability to use the available water resources due to economic reasons. Economic water scarcity could be due to lack of infrastructural or institutional support. Major river basins of the world that support high population densities and irrigated agriculture are already under physical water scarcity (Fig. 5.3).

Among the industrial uses of water, electricity production accounts for a major fraction of water use. Globally, there is a move toward biofuels production to mitigate the impacts of climate change by reducing the emissions of greenhouse gases. However, as pointed by Mulder et al. [9], the water intensity of biofuels is much higher compared to conventional energy sources (Table 5.6) and the water resources use with the proposed increases in the biofuels will place additional significant stress on water resources.

As with the land resources, the local water availability and use issues play a major role in assessing the sustainability of water resources. While many water projects around the world are designed to move water over hundreds of kilometers, they are mostly gravity-based and designed to be within the boundaries of the watersheds. On the other hand, the water pumping projects that cross the watershed boundaries are very energy-intensive and the cost-benefit analysis limits their largescale adoption. Therefore, many of the physical water scarcity issues can be studied through study of water resources within a watershed. With the advent of global trade networks that trade in food products, the concept of virtual water first introduced in 1993 by Prof. John Allan (Tony

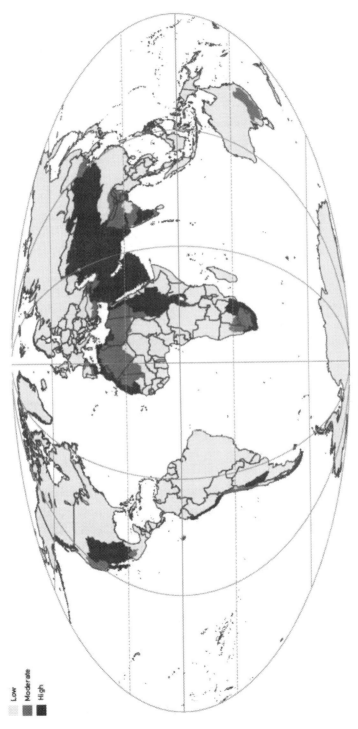

Fig. 5.3 Physical water scarcity of major river basins (figure from [1]).

Table 5.6 Water use for electricity production (data from [9]).

Technology	Key specifications	Water usage (L/MJ)
Nuclear Electric	Once-through cooling	33.25
Nuclear Electric	Recirculating	1.162
Coal Electric	Once-through cooling	28.62
Coal Electric	Recirculating	0.56
Coal Electric	Cooling pond	18.92
Tar Sands	Steam assisted gravity drainage	0.061–0.122
Biomass electric	IGCC Non-irrigated hybrid poplar	0.259
Biomass electric	IGCC Irrigated Hybrid poplar	4.088
Biomass electric	IGCC with various feedstocks irrigated at 400L/kg dry biomass	40
Petroleum electric	25 years expected plant life	0.02
Petroleum diesel	Average US data	0.004
Soy biodiesel	1990 Average Soy production 18.4% oil content.	21.71
Rapeseed Biodiesel		100–175
Corn ethanol	Dry milling Technology, 8.7 MT/ha corn yield, 0.37 L/kg ethanol yield	73–346
Sugarcane ethanol	Non-irrigated production in Brazil with Bagasse burned to process ethanol	38–156
Sugarbeet ethanol		71–188
Wheat ethanol		40–351
Lignocellulosic Ethanol		11–171
Lignocellulosic Methanol		11–138
Lignocellulosic Hydrogen	Prototype only	15–129
Lignocellulosic Electricity	Prototype only	13–195

Allan) [10], [11] becomes important. As per this concept, trade in goods represents not only the physical products themselves s but also the associated water resources that were sued for their production. For example, a kg of wheat requires 1300 L of water during its production (Table 5.1, Chapter 3). Therefore, a country that imports one ton of wheat is importing 1300 tons of water used to produce the one ton of wheat grain. These 1300 tons of water used to produce 1 ton of wheat grain represents the virtual water and the virtual water trade, therefore, mitigates the impacts of water scarcity in a region. A physically water-scarce nation can import water-intensive food products such as cereals, diary, and meat products and still support large populations in a region where the water resources are insufficient to meet all the needs of the population. For example, the energy-rich middle east nations import large quantities of food products that represent large volumes of virtual water flows into the region which enables them to support much larger regional populations in water-scarce region. The concept has been developed further by many researchers discuss the ecological [12] and political [13] implications of the virtual water concept.

5.4 Nutrient resources

Crop yields are determined by the most limiting factor of the growth and therefore adequate nutrients must be supplied to the plants at the right time, in the right amount and in the right form. There are 17 essential nutrients for plants of which carbon, oxygen, and hydrogen are three macronutrients that are supplied by the air and water. Nitrogen, phosphorous and potassium are three macronutrients while Sulphur, calcium, and magnesium are classified as three secondary macronutrients. Iron, zinc, copper, manganese, molybdenum, chlorine, boron, and nickel are eight essential micronutrients. Some of the plants require cobalt and silicon as two additional micronutrients.

The nitrogen, phosphorous, and potassium are three key plant macronutrients that are most commonly supplied through fertilizers. The use of synthetic fertilizers in agriculture and fossil energy sources are critical enabling factors for the ability of modern agriculture to meet the food needs of exponentially growing human populations since the last century [14]. The total nitrogen and phosphorus entering the terrestrial biosphere has doubled since the 1970s. The global consumption of these fertilizers was 245.77 million MT in 2015 and is expected to grow by 11.2% to reach 273.38 million MT in 2020 [15]. Nitrogen fertilizers are the largest source of fertilizers and accounted for over 63% of the total fertilizer supply, followed by almost equal amounts of phosphorus and potassium fertilizers (Table 5.7). Similar to the variability in other natural resources, there are regional imbalances in the fertilizer production and demands (Fig. 5.4).

Nitrogen: Nitrogen is an essential element of life and is part of proteins, DNA and RNA. Nitrogen is the most abundant gas in the atmosphere, yet the usable forms of nitrogen in the soil are relatively scarce due to fierce competition for this nutrient from all forms of life (Fig. 5.5). Naturally, the nitrogen is "fixed" into the soil through natural processes such as lightening or microbial-driven processes such as those by *Rhizobium* sp. in the root nodules of the legumes and cyanobacteria in waterbodies. Historically, nitrogen was supplied through organic matter source most of which contain <5% of nitrogen. The discovery of the Haber-Bosch process for synthesis of ammonia was transformational event that enabled the production of vast quantities of nitrogenous fertilizers using fossil fuels. Among the fertilizers, nitrogen fertilizers are the largest amount of nutrients by mass. The nitrogen is supplied to the plants as urea, anhydrous ammonia,

Table 5.7 World supply of fertilizers (thousands of MT; data from [15]).

Year	2015	2016	2017	2018	2019	2020
Ammonia (NH_3) as N	154,773	158,850	166,402	168,987	169,693	170,761
Phosphoric Acid (H_3PO_4) as P_2O_5	47,424	48,394	49,558	51,190	52,361	53,078
Potash as K_2O	43,571	42,772	44,868	47,249	48,898	49,545
Total ($N+P_2O_5+K_2O$)	245,768	250,016	260,828	267,426	270,952	273,384

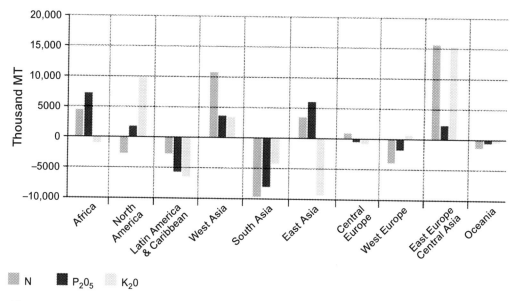

Fig. 5.4 Anticipated nutrient balances in 2020 (figure from [15]).

ammonium nitrate, ammonium sulfate, and mono/di/tri ammonium phosphate. The urea which contains 46% nitrogen is the most widely used fertilizer and is synthetically prepared by reacting ammonia (produced from Haber -Bosch process) with CO_2 at high temperatures and pressures [16].

Phosphorous: Phosphorus is an essential element that is part of the DNA, RNA, and ATP/ADP molecules in all living things. Unlike C, H, N, and O, phosphorous does not exist in gaseous form and is not found in its elemental form in nature due to its high reactivity (Fig. 5.6). Organic matter such as manure, animal bones, and sewage are rich sources of phosphorous and have been used historically as fertilizers. In the modern era, phosphate fertilizers are made from phosphate rocks mined from old marine deposits. Phosphates are sparsely soluble in water and the plants can readily take up only the soluble forms of phosphates. Therefore, the raw phosphate rocks are rarely used without further processing to increase their plant availability. The phosphate rocks are processed by reacting them with sulfuric/phosphoric acids to solubilize the phosphate and increase its availability to plants. Phosphate rocks are processed into various fertilizers such as singe/triple superphosphates, mono/diammonium phosphate, and polyphosphates. Use of phosphate fertilizers increased dramatically after the second world war together with the increases in nitrogen fertilizer use to support the food needs of a rapidly growing population. However, as a consequence of these advances, the global phosphorous nutrient cycle was broken, and humanity is dependent on mined phosphate rocks for phosphorus fertilizers [17]. Unlike nitrogen fertilizers which can be produced if there is access to energy sources, phosphorus fertilizers are nonrenewable and is mined from few

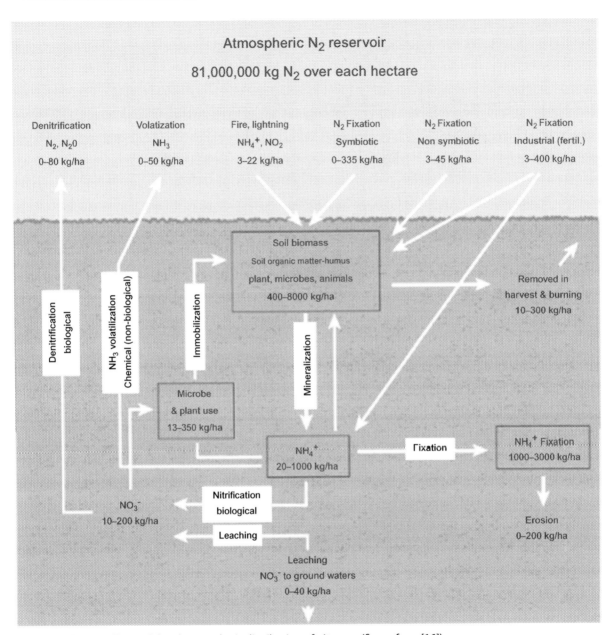

Fig. 5.5 Terrestrial and atmospheric distribution of nitrogen (figure from [16]).

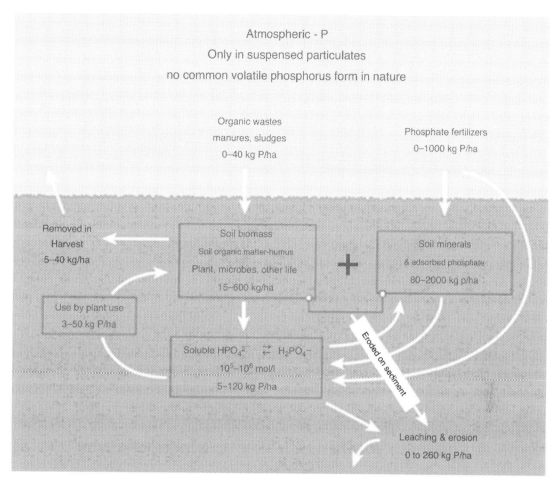

Fig. 5.6 Terrestrial and atmospheric distribution of phosphorus (figure from [16]).

geographic locations. More than 85% of the global phosphorus reserves are contained in five countries: Morocco, United States, China, and South Africa [17]. These reserves are estimated to be sufficient to meet 50–100 years of the current global utilization. This makes the entire modern food production system very fragile to the supply disruptions in the phosphate fertilizer supplies. In a seminal paper, Cordell et al. [18] described the unsustainable dependence of modern society on phosphorus fertilizers and warned of the possibility of peak phosphorus by 2035 when demand would be more than the supply. Most of the phosphorous loss from the agricultural soils occurs through soil erosion, crop residue removal [16]. Thus, preventing soil erosion, recycle of animal manure into farms, and proper treatment of wastewaters are the most effective strategies to minimize the loss of phosphorous.

Potassium: Potassium is a vital element that regulates a number of cellular transport systems, enzymes, and other chemical reactions inside all living cells. Potassium plays an important role in the regulation of the water in the cells, electrolyte balance, and osmotic regulation. Similar to phosphorous, potassium is a mined resource that is found in geological saline deposits and saltwater brines. Potassium fertilizers are applied as potassium chloride (muriate of potash), potassium sulfate, potassium magnesium sulphate, and potassium nitrate. Potassium is abundantly availably in the seawater. Potassium is present in the primary minerals form in the soil and the plant available fractions are typically 0.1%–0.2% as soil solutions (12–15 kg/ha) (Fig. 5.7).

5.5 Metals and minerals

Many of the metals such as copper, zinc, tin, and iron have been used historically, their large-scale production of these metals is a relatively modern phenomena and is at the root of the modern industrial societies. While many of these metals may exist in the Earth's crust at low concentrations, the economically utilizable resources are again like other natural resources restricted to few geographical locations. Therefore, the

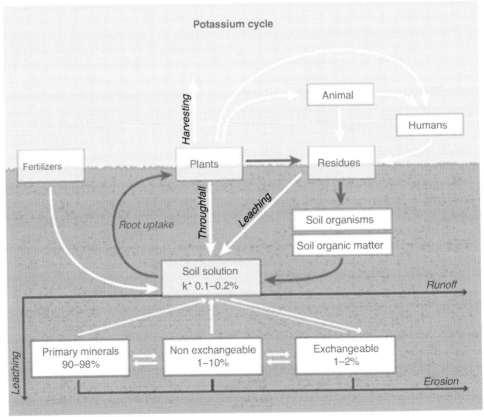

Fig. 5.7 Terrestrial and atmospheric distribution of potassium (figure from [16]).

long-term sustainability aspects of these metals and minerals are an important question that determines the feasibility of any technology. The global production levels of various nonfuels minerals are shown in Table 5.8.

There are about 50 metals that are used in the renewable technologies such as wind, solar, and energy storage technologies [20]. With the increased adoption of the renewable energy technologies around the globe, there will be an increase in the consumption of these materials. Base metals including copper, silver, aluminum, nickel, zinc and platinum are expected to see increases due to shift to renewable energy technologies. Among the rare earths Indium and neodymium are critical for making alloys used in main the super high strength industrial magnets that are used very widely in consumer electronics, industrial motors, and wind turbines. Key metals for the wind generation technologies are aluminum, chromium, copper, iron, lead (for direct-drive technologies), manganese, neodymium, nickel, steel, and zinc. While for solar photovoltaic technologies, important metals are aluminum, copper, indium, iron, lead, nickel, silver, and zin in addition to silicon. Energy storage using lead acid batteries requires lead and steel while for the lithium-ion batteries, aluminum, cobalt, lithium, manganese, nickel, and steel are important metals. There have been concerns with respect to the supply chain security and stability for these critical materials [21].

Fourteen metals (cadmium, dysprosium, gallium, hafnium, indium, molybdenum, neodymium, nickel, niobium, selenium, silver, tellurium, tin, and vanadium) were identified as critical metals for low-carbon energy technologies whose demand is expected to be more than 1% of the world supply per annum in the decade 2020–2030. Of these 14 metals, five metals (neodymium, dysprosium, indium, tellurium, and gallium) were could pose a high risk to the large-scale deployment of solar and wind energy technologies [20]. The global production and reserves for some of the metals used in low carbon energy technologies is presented in Table 5.9.

5.6 Examples

Background: Corn dry grind ethanol process is currently used in the United States to produce 60 billion L/year (~15.8 billion gal/year) of ethanol per year accounting for 58% of the global production. This process uses almost 35% of the US corn production. Each bushel of corn (25.424 kg or 56 lbs) can produce 10.62 L or 2.8 gals of ethanol. The dry-grind corn process consists of the following unit operations: milling/size reduction, liquefaction, simultaneous saccharification and fermentation, distillation and ethanol dehydration using molecular sieves, centrifugation, and drying of Distiller's dried grains with solubles (DDGS).

Problem Statement:
- What is the feasibility of producing corn ethanol to meet the US gasoline needs?
- What is the feasibility of producing 50% of US gasoline needs?
- What is the feasibility of producing 15 billion gallons of corn ethanol?

Table 5.8 World production of nonfuels mineral commodities (data from [19]).

Metals			Industrial minerals	
Alumina, calcined equivalent	119,000	Platinum-group metals	Asbestos, marketable fiber	1310
Aluminum, primary	58,000	Palladium (MT)	Barite	8390
Antimony, Sb content (MT)	143,000	Platinum (MT)	218	Bromine (MT)	301,000
Arsenic trioxide (MT)	37,100	Other (MT)	195	Celestite (MT)	264,000
Bauxite	299,000	Rare earths, rare-earth-oxide	478	Cement, hydraulic	4,100,000
Beryl, gross weight (MT)	6730	(REO) equivalent	130,000	Clay	
Bismuth, refinery (MT)	19,500	Rhenium (MT)	49.7	Bentonite	19,500
Cadmium, refinery (MT)	22,800	Selenium, refinery, Se content (MT)	2190	Fuller's earth	3460
Chromite, gross quantity	31,200	Silver, mine (MT)	27,000	Kaolin	37,800
Cobalt, Co content		Tantalum mineral concentrates (MT)	1200	Diamond (thousand carats)	127,000
Mine (MT)	121,000	Ta content		Diatomite	3140
Refinery (MT)	99,800	Tellurium, Te content (MT)	119	Feldspar	23,100
Copper		Tin, Sn content:		Fluorspar	6090
Mine, recoverable, Cu content	19,200	Mine (MT)	341,000	Graphite, natural	1180
Smelter, gross weight	18,500	Smelter (MT)	354,000	Gypsum	260,000
Refinery	23,200	Titanium mineral concentrates		Iodine, crude (MT)	32,600
Gold, mine (MT)	3090	Ilmenite and leucoxene	6980	Kyanite and related minerals (MT)	408,000
Indium, refinery (MT)	731	Rutile (MT)	760,000	Lime	340,000
Iron ore	2,320,000	Tungsten, W content (MT)	89,700	Magnesite	27,100
Iron and steel		Vanadium, V content (MT)	84,600	Mica (MT)	275,000
Direct-reduced iron	69,700	Zinc		Monazite mineral concentrates (MT) gross weight	6860
Pig iron	1,160,000	Mine, Zn content of mineral concentrates	13,500		
Raw steel	1,620,000	and direct shipping ore		Nitrogen, N content of ammonia	147,000

Resource assessment

Lead			
Mine, concentrates, Pb content	4970	Smelter	13,900
		Zirconium mineral concentrates, gross weight	1520
Refinery	10,800		
Magnesium, primary (MT)	979,000	Peat	26,600
		Perlite, processed ore	4300
Manganese ore, Mn content	17,000	Phosphate rock, gross weight	263,000
Mercury, mine (MT)	2330	Potash, marketable, K_2O equivalent	40,700
Molybdenum, mine, Mo content	288,000	Pumice and related materials	17,100
		Salt, all forms	277,000
Nickel, Ni content		Sand and gravel, industrial, silica	203,000
Mine, recoverable (MT)	2,180,000	Soda ash, natural and manufactured	53,500
Plant (MT)	2,000,000	Sulfur, all forms[46]	73,800
Niobium (columbium) mineral concentrates, (MT)	63,300	Talc and pyrophyllite	7830
		Vermiculite	379

All quantities in thousands of MT unless specified.

Table 5.9 Reserves, production of major metals used in renewable energy technologies [22].

Metal	Reserves Tons (2015)	Production Tons (2015)	Major producer countries	Additional remarks
Iron	230 billion	3.32 billion	China (1380), Australia (824) and Brazil (428)	
Aluminum	55–75 billion	58,300	China (32,000)	Undiscovered sources are estimated to be 3.5 billion tons
Copper	1.8 billion	18.7 million	Chile (5700), China (1750) and Peru (1600)	
Cobalt	25 million	120,400	Congo (63,000)	
Chromium	12 billion	30,500	South Africa (15,000)	More than 120 Million tons on the ocean floors. About 95% of worlds chromium reserves are in Kazakhstan and Southern Africa
Cadmium	–	24,200	China (8090), Republic of Korea (4250)	Cadmium is recovered from Zinc ores and concentrates at 0.03% levels
Indium	–	755	China (370) and Republic of Korea (150)	Indium is recovered from the Zinc-sulfide ore mineral deposits and range from 1–100 ppm
Lead	2 billion	4.271 million	China (2.3 million)	
Lithium	40.7 million	32,500	Chile (11,700), Australia (13,400)	South Africa accounts for about 75% and
Manganese	620 million	18 million	South Africa (6.2 million), Australia (2.9 Million) and China (3 million)	Ukraine about 10% of the global resources
Molybdenum	11 million	267,000	China (101,000), USA (56,300) and Chile (49,000)	Global resources are 19.4 million tons
Nickel	130 million	2.53 million	Australia (0.234 million), Canada (0.24 million)	Other unexplored sources of nickel are the manganese crusts and nodules in the ocean floor
Platinum group of metals	66,000 tons	386 tons	South Africa (198), Russia (103)	Worldwide reserves are estimated to be 100,000 tons; Platinum group of metals consists of Platinum and Palladium
Rare Earth Elements	130 million	124,000	China (105,000)	Bastnasite and monazite are two minerals that are mined for rare earths. Bastnasite deposits are found in China and the United States. Rare earths elements include neodymium and indium
Silicon	Abundant	8.1 million	China (5.5 million)	
Silver	570,000	27,300	Mexico (5400), China (4100), Peru (3800)	
Titanium	740 million	5.61 million	China (0.9 million), Australia (0.72 million), Vietnam (0.54 million)	Ilmenite accounts for 92% of world consumption of titanium minerals. World resources of anatase, ilmenite and rutile are >2 billion tons.
Zinc	200 million	13.4 million	China (4.9 million), Australia (1.5 million) and Peru (1.37 million)	World resources are 1.9 billion metric tons

Data:

Data	References
Average US corn yield: 11.46 tons/ha (181.3 bu/acre)	[23]
US corn, grain- harvested: 32.708 million ha (81.77 million acres)	[23]
US crop land: 152.26 million ha	[24]
Corn to ethanol conversion: 0.4273 L/kg (2.87 gal/bu)	[25]
DDGS production: 0.2928 kg/kg corn or 0.685 kg/L ethanol	[25]
Gasoline:Ethanol energy density equivalency ratio = 1.424	[26]
US gasoline needs: 3.404 billion barrels = 142.98 billion gal = 541.735 billion L/year	[27]
Number of cattle in US: 92 million heads.	[28]
1 gal = 3.785 L	Unit Conversions

Assumptions:
- Based on the Univ. of Georgia Extension publication, each cattle head can be fed up to 1.5 kg (3.3 lbs) of DDGS per day. i.e. each cattle requires 0.548 tons DDGS/year [29].

Calculations:
1. Ethanol equal to 100% current gasoline needs = 541.735 billion L/year × 4.424 = 771.5 billion L/year
2. Ethanol equal to 50% current gasoline needs = 385.75 billion L/year
3. Total corn production, TCP = Average US corn yield × US corn crop area = 11.46 tons/ha × 32.708e6 ha = 374.8 million tons/year
4. Ethanol potential with current corn production = Average US corn yield × US corn crop area × Corn to ethanol conversion
5. = 11.46e3 kg/ha × 32.708e6 ha × 0.4273 L/kg = 160.16 billion L/year (42.31 billion gal/year)
6. Max. Ethanol potential with corn production on all arable lands = Average US corn yield × US arable land area = 11.46e3 kg/ha × 152.26e6 ha × 0.4273 L/kg = 745.6 billion L /year (197 billion gal./year)
7. Area needed to produce ethanol to meet 100% current gasoline needs = Ethanol equal to 100% current gasoline needs /(Average US corn yield × Corn to ethanol conversion) =771.5 billion L/year /(11.46e3 kg/ha × 0.4273 L/kg)=157.6 million ha.
8. Area needed to produce ethanol to meet 50% current gasoline needs = Ethanol equal to 100% current gasoline needs /(Average US corn yield × Corn to ethanol conversion) = 385.75 billion L/year /(11.46e3 kg/ha × 0.4273 L/kg) = million ha.
9. Area needed to produce 15 billion gal/year ethanol = (15 billion gal/year × 3.785 L/gal) /(11.46e3 kg/ha × 0.4273 L/kg) = 11.6 million ha
10. Amount of DDGS produced for 15 billion gal ethanol = 15 billion gal/year × 3.785 L/gal × 0.685 kg/L ethanol = 38.89 million tons

11. Number of cattle that can be fed the DDGS: 38.89 million tons/0.548 tons/head
   = 71 million cattle heads

Results:
- The total area needed to meet gasoline needs of the US (157.6 million ha) is greater than the arable land available (152.26 million ha). The land area needed to meet 50% of the gasoline needs (78.78 million ha) is greater than the crop land dedicated to US corn production (32.708 million ha). Therefore, these two scenarios are unrealistic and infeasible.
- Area needed to produce 56.78 billion L/year (15 billion gal/year) ethanol is 11.6 million ha, which is about 35% of the US corn cropping area. Hence this is a feasible process. As an additional corroboration for this calculation, US produced 59.80 billion L/year (15.8 billion gal/year) ethanol in 2017 as per the renewable Fuels Association [5].
- This process will produce 38.89 million ton of DDGS which can be fed to 71 million cattle heads in various stages of growth. As this represents 77% of the total cattle in the US, there is a potential for market saturation and hence a need for diversification of the

Summary statement:

It is infeasible to meet 100% or even 50% of the US gasoline needs using ethanol from corn. It is feasible to produce 56.78 billion L/year (15 billion gal/year) with existing land resources but the coproduct utilization needs to be looked at carefully to avoid potential market saturation scenarios.

5.7 Conclusions and perspectives

This chapter focused on resource sustainability assessment as a part of the systems analysis. The chapter discussed the importance of assessing resource sustainability through availability in the short term and long term, distribution of resources, supply chains, and economics. The global availability and use of various important resources such as land, water, fertilizer, metals, and minerals resources were discussed. With the increasing adoption of renewable energy technologies, the pressures on already stressed resources will increase and it would be critical to examine the resource sustainability aspects of any proposed technology before large-scale adoption. The resource sustainability analysis along with other dimensions of systems analysis discussed in the earlier chapter must be conducted for comprehensive assessment of any process/technology.

References

[1] FAO. 2011. The state of the worlds land and water resources for food and agriculture (SOLAW)—managing systems at risk. Food and Agriculture Organization of the United Nations, Rome and Earthscan, London.
[2] Our world in data. https://ourworldindata.org/land-use#all-charts-preview. [Accessed May 30, 2020].
[3] W.J. Ripple, C. Wolf, M. Galetti, T.M. Newsome, M. Alamgir, E. Crist, M.I. Mahmoud, W.F. Laurance, World scientists' warning of climate emergency, BioScience 70 (2020) 8–12. https://doi.org/10.1093/biosci/biz088.

[4] STRATA Report. 2017. The footprint of energy: land use of U.S. electricity production. https://www.strata.org/pdf/2017/footprints-full.pdf. [Accessed: 16 May, 2021].
[5] B.W. Brook, C.J.A. Bradshaw, Key role for nuclear energy in global biodiversity conservation, Conserv. Biol. 29 (2014) 702–712.
[6] A.M. Trainor, R.L. McDonald, J. Farigone, Energy Sprawl is the largest driver of land use change in United States, PLoS One 11 (2016) e0162269.
[7] MacKay D. Sustainability without the hot air. https://www.withouthotair.com/c6/page_38.shtml. [Accessed May 30, 2020].
[8] K. Mulder, N. Hagens, B. Fisher, Burning water: a comparative analysis of the energy return on water invested, AMBIO 39 (2010) 30–39.
[9] A. Boretti, L. Rosanpj, Reassessing the projections of the World Water Development Report, Clean Water 2 (2019) 15.
[10] J.A. Allan, Fortunately there are substitutes for water: otherwise our hydropolitical futures would be impossible. In: priorities for Water Resources Allocation and Management, Overseas Development Administration, London, 1993, pp. 13–26.
[11] J.A. Allan, Virtual water: a strategic resource. Global solutions to regional deficits, Ground water 36 (1998) 545–546.
[12] Z. Yang, X. Mao, X. Zhao, B Chen, Ecological network analysis on global virtual water trade, Environ. Sc. Technol. 46 (2013) 1796–1803.
[13] D. Roth, J. Warner, Virtual water: virtuous impact? the unsteady state of virtual water, Agric. Human Values 25 (2008) 257–270.
[14] V. Smil, Global population and the nitrogen cycle, Sci. Am 277 (1997) 76–81.
[15] FAO. 2017. World fertilizer trends and outlook to 2020. Summary Report. http://www.fao.org/publications/card/en/c/cfa19fbc-0008-466b-8cc6-0db6c6686f78/. [Accessed 16 May, 2021].
[16] Reetz Jr H.F. 2016. Fertilizers and their efficient use. International Fertilizer Industry Association. www.fertilizer.org. [Accessed May 31, 2020].
[17] k. Ashley, D. Cordell, D. Mavinic, A brief history of phosphorus: from the philosopher's stone to nutrient recovery and reuse, Chemopshere 84 (2011) 737–746.
[18] D. Cordell, J-O. Drangert, S. White, The story of phosphorous: global food security and food for thought, Global Environ. Change 19 (2009) 292–305.
[19] USGA. 2020. National minerals information center. Statistical Summary. https://www.usgs.gov/centers/nmic/statistical-summary. [Accessed May 31, 2020].
[20] R.L. Moss, E. Tzimas, H. Kara, P. Willis, J. Kooroshy, The potential risks from metal bottlenecks to the deployment of strategic energy technologies, Energy Policy 55 (2013) 556–564.
[21] K. Nansai, K. Nakajima, S. Kagawa, Y. Kondo, S. Suh, Y. Shigetomi, Y. Oshita, Global flows of critical metals necessary for low-carbon technologies: the case of neodymium, cobalt and platinum, Env. Sc. Technol. 48 (2014) 1391–1400.
[22] World Bank Group. 2017. The growing role of minerals and metals for a low carbon future. http://documents.worldbank.org/curated/en/207371500386458722/pdf/117581-WP-P159838-PUBLIC-ClimateSmartMiningJuly.pdf. [Accessed May 31, 2020].
[23] US Corn yields. 2020. https://www.nass.usda.gov/Statistics_by_Subject/result.php?B10EB84A-5970-335E-A882-5B42E549DF66&sector=CROPS&group=FIELD%20CROPS&comm=CORN. [Accessed May 31, 2020].
[24] FAOSTAT. 2020. http://www.fao.org/faostat/en/#data/RL. [Accessed May 31, 2020].
[25] Renewable Fuels Association. GREET. 2020. https://ethanolrfa.org/how-ethanol-is-made/. [Accessed May 31, 2020].
[26] EIA petroleum and other liquids consumption. 2020. https://www.eia.gov/dnav/pet/pet_cons_psup_dc_nus_mbbl_a.htm. [Accessed May 31, 2020].
[27] Alternative Fuels Data Center. 2020: https://www.afdc.energy.gov/fuels/fuel_comparison_chart.pdf. [Accessed May 31, 2020].
[28] USDA NASS. 2018.: https://usda.mannlib.cornell.edu/usda/current/USCatSup/USCatSup-06-24-2016.pdf. [Accessed 5 October, 2018].
[29] University of Georgia Extension Publication. 2018. http://extension.uga.edu/publications/detail.html?number=B1482&title=Using%20Distillers%20Grains%20in%20Beef%20Cattle%20Diets. [Accessed May 31, 2020].

CHAPTER SIX

Policy, governance, and social aspects

Ganti S. Murthy[a,b]
[a]Biological and Ecological Engineering, Oregon State University, Corvallis, OR, United States, [b]Biosciences and Biomedical Engineering, Indian Institute of Technology, Indore, India

6.1 Introduction

While technical feasibility, economic viability, and environmental sustainability are necessary conditions for the success of a technology or a strategy to address some of the systems challenges, it is not sufficient. A conducive social environment, supportive policy, and stable governance is the fundamental factor for the success of any technology or process. The complexities of modern life mean that many of the challenges that require systems analysis techniques for their solutions are complex and fall under the so-called "wicked" problems. The policy prescriptions for "tame" and "wicked" problems are markedly different [2]. The case of "tame" problems are those problems which have a fundamental convergence in the technological and scientific realm and mitigation paths are clear and direct. The task of the policymakers is to find an optimum policy mechanism to follow the path indicated by the scientific consensus while mitigating the effects on other stakeholders through a combination of incentives and disincentives for noncompliance. On the other hand, the "wicked problems" defy such easy and defined paths for solutions and there is often difficulty in even formulating the scientific consensus and the mitigation paths. The challenges for policy-making for these problems become more challenging as there are multiple solutions pathways with different impacts on various stakeholders. Naturally, different stakeholders will champion different solutions that may even be at odds with each other while aiming to solve the same problem.

6.2 Complexities of policy making

Policy making and governance involve any complexities. Ideally, the policy-making process will have one clear goal or desired outcome which is achieved through a focused and dedicated single policy instrument. While policy making requires a clear purpose, well-defined objectives, and targeted policies to achieve the objectives, realities of policy making, and the governance processes require strategic ambiguities, acceptance of the "sufficing" solutions. As well articulated by Lawrence [13], "The principles of good policy making require precision and hiding costs, accepting second-best justifications, and packaging policies together to further broaden support."

In general, it has been seen that mandates are not a good policy compared to direct incentives and taxes as the cost of the policy is not clear with mandates. While this ambiguity in who pays may be beneficial to achieve broad-based support for the policy, the results are unlikely to be achieved as the true costs and benefits of the policy are realized by various stakeholders. This can lead to the building of distrust of institutions which is not good for society. The requirements for good policy making are often violated in practice and there are several examples of such policies that tried to accomplish too much through a single policy instrument that ended up accomplishing very little in the end. For example, the advanced biofuels mandates in the US formulated through the revised Renewable Fuel Standards (RFS) in 2007 aimed to replace the petroleum with 36 billion gals of biofuels by 2022 of which 21 billion gals were advanced biofuels from second and third-generation feedstocks. The stated goals of the RFS were to achieve energy security through reducing imports of fossil fuel, reduce GHG emissions from the transportations sector, and encouraging rural economies. Pursuing three policy goals with one policy instrument has resulted in nonaccomplishment of all three goals in a substantial way as envisioned originally. Lawrence [13] argues that these policy goals could have been met with three policy instruments such as tariffs on imported oil and increased fuel efficiency for reducing dependence on foreign oil and increasing the energy security, a Pigouvian tax or Cap and trade for reducing carbon dioxide emissions, and targeted subsidies for supporting farmers and rural communities. Using three targeted policy instruments for three goals could have had a much better chance of success than trying to achieve three stated policy goals using one policy instrument of revised RFS. Bundling all the three stated policy goals into mandates while is politically expedient by making the costs of the policy nontransparent while highlighting the benefits to select groups of stakeholders allowed the passage of mandates. However, after 14 years of the passage of the Revised RFS the policy failed to incentivize the advanced biofuels industry adequately, reduce the dependence on foreign oil or reduce GHG emissions in a significant way. Similar concerns were expressed regarding the biofuels policies of the Asian countries which tried to reduce fossil fuel consumption, offset high oil prices and win the political support of the farmers (multiple targets) by encouraging biofuels (single policy instrument) [16].

Oliverial et al. [14] are even more scathing in their criticism of the biofuel policies around the world for using inefficient policy mechanisms to achieve misguided priorities without regard to the ecological blind spots. They point out that the biofuels industry in the world is primarily driven by the policies for bioethanol in the US and Brazil and the biodiesel policies enacted in the EU. They point out how the stated goals for encouraging the biofuels policies were changing in the United States, Brazil, and EU zone countries have been changing throughout the last century and are reflective of the incoherent biofuel politics in these countries. Other authors have also criticized the biofuels policies around the world stating that most of the biofuel policies are based on politics and with very little focus on achieving tangible environmental goals which is an often-stated goal of these policies [4, 15, and 16]. Policies often have unintended consequences which are difficult to foresee [15]. For example, the food price crisis in 2009 around the world due to policies encouraging first-generation biofuels in the US and EU, the large-scale destruction of Amazon rainforests due to bioethanol policies of Brazil and Borneo rainforests in Indonesia and Malaysia due to the biodiesel policies of EU. Best policy making occurs when there is convergence between stakeholders, concurrence on the objectives, and the effective policy instrument.

Policy making challenges vary depending on the type of problem being addressed. As alluded to earlier, "tame problems" are often easy to characterize, have scientific consensus on the core challenges, and a clear path for mitigation with clearly identified stakeholders. Addressing such tame problems with clear identification of the core issues and the consensus among the stakeholders, it is relatively easy to formulate technology or process-centric policies. The Montreal protocol banning the production of CFC around the world in 1987 to address the issue of Ozone layer depletion is an example of such a problem that was successfully addressed [2]. The scientific consensus about the major contribution of CFCs to around ozone depletion, the concentration of the worldwide CFC production among a few major multinational companies located in few developed countries, already emerging societal acceptance of non-CFC based solutions all contributed to the development of scientific, political, and societal consensus which resulted in Montreal Protocol of 1987. Climate change is similar to ozone depletion in that both the problems involve a locally produced pollutant that is dispersed globally, there is strong data-driven scientific consensus that human activity is the major contributor to the accumulation in the environment [17]. However, unlike the ozone depletion causing CFCs where the role of natural processes is minimal, the role of natural cycles of carbon is crucial and very critical for addressing the GHG emissions. There are still large uncertainties regarding the drivers, the sensitivity of various components of the333 complex earth atmospheric systems to various mitigation measures is still not completely understood. Additionally, unlike the CFC production, the GHG production is distributed in both developed and developing countries that have different policy priorities. Therefore, the challenges for policy making for "wicked problems" like climate change are much more than for "tame problems."

6.3 Commonly used policy making models

The difference between the policies required to address the tame and wicked problems necessitates an understanding of the policy making process. Policy making process can be studied from multiple perspectives. Various policy making models have been proposed in the literature. One approach advocated by Collingridge and Douglas [18] involves looking at the policy making from the importance assigned to the expert opinion. This is an important aspect of policy making that must be especially understood by technical experts as there is a widespread belief among the scientific community that the scientific information and knowledge flow linearly to the policy making process and result in the action on the ground despite multiple refutations of the same [2].

Collingrdge and Douglas [18] proposed three models of policy making, namely, synoptic rationality, disjointed incrementalism, and mixed scanning. In the synoptic rationality model of policy making, the policy makers acquire a comprehensive view of the problem before the formulation of policies and hence the expert opinion plays a very important and huge role at the beginning of the policy making process. Such a model of policy making is advocated for scenarios where the technological basis for the problem is very well known, and there is a large body of scientific literature to arrive at an expert consensus about the underlying causes, possible remediation methods, limits, and alternative approaches. The policy making for the Montreal protocol discussed above is an example of the Synoptic rationality model of policy making. In the disjointed incrementalism model, the policies are made under information deficient scenarios where the cost of inaction is much higher than the cost of conservative action to mitigate a potential problem. Advocates of precautionary principle can be thought of as an example of the application of the disjointed incrementalism model. The mixed scanning model is a process that combines the characteristics of both the synoptic rationality and the disjointed incrementalism models.

In addition to the above models which focus on the needed expert inputs, the classic way of looking into the public policy making model approaches are described by Hahn [1]. They are Institutionalism, systems theory, pluralism, elitism, process models, rationalism, and incrementalism. The brief characteristics of each of these approaches are summarized below.

Institutionalism:
- Classical approach
- Focus on structures, organization, duties, and functions of government institutions.
- "What unit of Govt. is responsible for what?"
- "What are the lines of authority and accountability?"

Systems theory
- Emphasis on the environment of political systems, inputs, outputs, and feedback.
- Often widely used implicitly.
- Helps understand external linkages.

Pluralism
- Policy making is seen as the result of influencing groups.
- Identifying groups in conflict and competition is an important aspect of this method.

Elitism
- Recognizes that most people are uninterested, uninvolved, and uninfluential in policy making.

- Elites have a disproportionate impact on policy making.
- Elites act on behalf of themselves or other groups.
- Elites are not homogenous.

Process models
- Generalized sequence of steps or actions that occurs as policy issues are raised, debated, and resolved.
- "Focus on what happens, when, and how than on who the participants are and why particular outcomes occur."

Rationalism
- Looks at policy making as a rational exercise involving clarifying and ranking goals, identifying alternatives, and predicting consequences.

Incrementalism
- Developed as a reaction to the rationalism model.
- Better describes the reality and is a prescriptive model
- Decision makers more likely to move away from the problems than toward the goals.

It is pertinent to point out here that none of these models are sufficient to describe the systems of policy making. They must be considered as snapshots of a dynamic system viewed from viewpoints. In that context, all the above models are correct and inadequate at the same time and are complementary viewpoints of a complex dynamic system.

6.4 Policy making frameworks

While various perspectives and models of policy making as discussed in the previous section are very important, it is essential to analyze the policy making under different frameworks to identify the points of interventions and the possible responses. Various policy making frameworks have been developed and studies to understand the policy making processes [12]. One of the first policy frameworks was the PSR framework developed by OECD in the early 1970s for assessing the environmental issues (Fig. 6.1A). The framework divides the policy making into the interacting components of pressures, state and the responses. Many environmental issues have been addressed using the PSR framework.

Three components of the PSR framework are pressures that factor such as population, ground water withdrawal, exploration, and exploitation of natural resources which result in the changes to the state of the environment, such as the available forest land, degradation levels of the arable land, status of the ground water aquifers, state of the rivers, etc. The decisions are those interventions that can be policy-driven and impact both pressures and the states. For example, the laws protecting the forests, incentivizing soil erosion prevention and adoption or more efficient irrigation practices are various decisions that impact both the pressures and the state variables. Through this framework, the

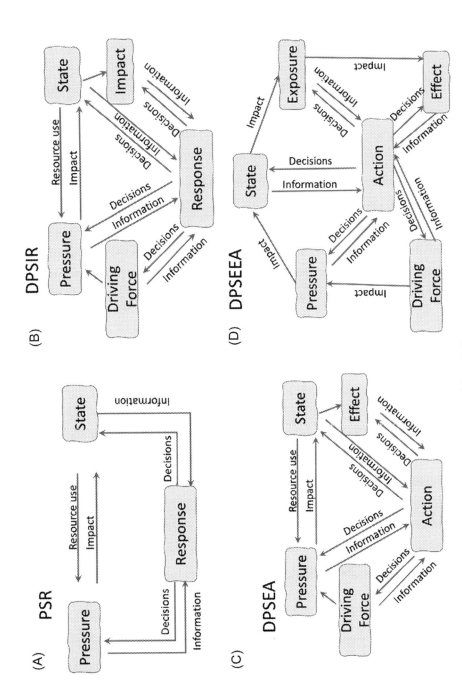

Fig. 6.1 Important policy making frameworks (adapted from [12]).

impact of various pressures, state, and the policy interventions through the decisions can be modeled. Such models have been widely applied in conjunction with optimization techniques to identify the best possible environmental outcomes. Many variants of the policy frameworks were developed by incorporating additional factors such as driving forces, impacts over time culminating in the DPSIR framework.

While the PSR and DPSIR frameworks were used for addressing environmental issues, the World Health Organization developed DPSEA models to understand the linkages between health and environmental impacts. The five components of the DPSEA framework refer to the driving forces impact the pressures on the environment which in turn impact the state of the environment which has an impact on the human health. The information flows from the driving forces, pressures, state of environment, and its effects on human health lead to the formulation of action which lead to decision that change all the states of the system. For example, increased urbanization (driving force) can cause increased vehicular traffic in the cities (pressure on environment) leading to increased air pollution and smog (state of environment). The increased smog and air pollution can lead to adverse health impacts such as increased incidences of Asthma and other respiratory diseases (effect). Reduction of vehicular traffic by increasing availability of timely and efficient public transport system, regulation of vehicular traffic by increasing road taxes, removing older vehicles all are series of actions that can impact all the components of the system. The information about the various components of the system is essential for effective action. In the DPSEEA framework, the exposure to the state is explicitly modeled to characterize the dependence of health effects to the exposure to the state.

It is to be noted that while these (and other) frameworks are an example of studying the policy making in the environmental and health sectors, there are many factors that are ignored completed or considered only at a partial level. For example, many of these frameworks are static, ignore carrying capacity constraints, assume linearity when the real-life models are highly nonlinear, neglect historical trends, and fail to consider multiple trajectories for the evolution of the policies. Therefore, it is to be understood that these frameworks only aid to understand the policy making and provide a better picture for formulating clear policies by identifying various components of a highly dynamic, nonlinear stochastic socio-political-economic system.

6.5 Social and governance aspects

The effective implementation of policies requires good governance and buy-in from society. While the societal aspects should be taken into consideration at the policy making phase through participatory processes, the good governance is critical for the effective implementation of the policies. The UN defines governance [19] as "Governance refers to the exercise of political and administrative authority at all levels to manage a country's affairs. It comprises the mechanisms, processes and institutions, through which citizens and groups articulate their interests, exercise their legal rights, meet their obligations and mediate their differences." Good governance is "participatory, consensus-oriented, accountable, transparent, responsive, equitable and inclusive, efficient and effective and follows the rule of law" [20]. Good governance is important to have functional society where multiple stakeholders can interact in confidence that the rule of law will be upheld. Therefore, good governance will strengthen the confidence of the stakeholders in institutions and the rule of law through various measures. There is a strong linkage between the performance, adaptability, and the stability of the institutions and the good governance principles of participation/inclusion nondiscrimination/equality, and rule of law/accountability [21]. Therefore, it is not surprising that many comprehensive indicators of sustainability do include measures of governance. Governance metrics of sustainability often are designed to measure transparency, equality and fairness, efficiency, corruption, and institutional access.

The social aspects of sustainability are critical for the success of any policy or technology. In democratic societies, societal participation is critical for the success of any policy. The cultural variations in the world, the perceptions, social interactions, and the human factors influence the social aspects of sustainability. Reflecting this irreducible complexity, the social aspects are often defined in ambiguous terms and are general social policy goals rather than concrete indicators of progress toward sustainability. There are more than 200 types of social metrics of sustainability used by public and private sectors [22]. While the public sector agencies emphasize safety and health, population, infrastructure availability, budgets and expenditure and the education, the private sector often emphasizes the human rights and resources, performance of products, and the metrics associated with the production and supply chains.

6.6 Case studies

The above discussion introduces the different aspects of the complex issues of policy making, social, and governance. To concretize the discussion presented above, three case studies are presented below which represent a case of successful policy implementation at a national level, the inability of policies to mask the underlying economic realities, and the how social factors can lead to inability to achieve policy goals even with good technologies and effective policy implementation.

6.6.1 Case Study 1. Biofuels in Brazil

Brazil is the second-largest producer of bioethanol behind the United States. Together Brazil and the United States account for over 90% of the world bioethanol production and the primary feedstocks for Brazilian bioethanol is sugarcane which is cultivated extensively in Brazil due to favorable agroclimatic conditions and is used to produce sugar, bioethanol, and electricity from the bagasse. The electricity production from the over 360 sugarcane processing plants that produce over 34 billion L/year ethanol and contribute over 25% to the national electricity production [10].

History of bioethanol in Brazil goes back over a century to early 1900s when ethanol was used a fuel for the Ford T model cars [8]. Beginning in 1931, a program for mandatory blending of ethanol at 5% in imported gasoline was introduced which was extended to all gasoline by 1938. During the World War II period, the percentage of ethanol blends increased to 42% due to petroleum supply constraints but post-war prosperity around the world saw a wane in the interest in the ethanol blending programs in Brazil. The oil crisis of 1973 again revived the interest in ethanol-blended transportation fuels to reduce the import dependence led to the formulation of the Proálcool program by the government. In 1975 [7], this policy tool was used to create both the push and pull in the ethanol markets. The supply-side incentives included guaranteed producer prices and financing of newer plants for increasing ethanol production capacity. The demand side incentives were mandatory lower ethanol pricing compared to gasoline, obligating the fuel stations to sell ethanol, and reduction in the taxes for vehicles that run completely on hydrous ethanol. Additionally, the shocks to the supply and demands were managed through the creation of the strategic ethanol reserves. As a result of this program, ethanol production jumped from 600 Million L/year in 1976 to 3.4 billion L/year in 1980 and 12.3 Billion L/year by 1987. The first 100% ethanol cars were introduced in 1978 and the minimum ethanol blend content increased from <5% in 1975 to 15% in 1979 which further increased to 22% by 1987. The program received increasing support due to the increasing oil prices in the 1980s when the oil imports accounted for over 42% of the Brazilian imports. The later decade on 1987–1997 with decreasing oil prices and increasing sugar prices provided counter shocks to the ethanol sector. Increasing demands and lack of new capacity addition resulted in importing of over 1 billion L/year ethanol from the United States in 1997. These developments led to a policy revision and elimination of subsidies and complete deregulation of the bioethanol sector in 2000 [8]. As a result of these policy changes and increasing demand for renewable fuels globally and the United States in particular, the Brazil ethanol production capacity increased to 28 Billion L/year in 2008, supporting over 1 million direct jobs in over 60,000 sugarcane producers and 400 processing plants.

Stable government policies, planning for flexibility from the beginning, planning for an appropriate scale-up of programs, planning and developing measures to minimize supply fluctuations in the market, minimizing environmental impacts, and involving private sector were identified as the five key lessons from the successful implementation of the biofuels programs in Brazil [11]. Brazilian biofuels sector represents a successful application of policies, technologies, and economics for producing biofuels.

6.6.2 Case Study 2. Biogas in the European Union

The European Union countries started several programs on renewable energy program as a policy response to climate change. Among the many programs, the German programs in solar photovoltaic, wind energy, and biogas programs were considered the flagship programs for reducing the dependence on foreign natural gas imports while meeting climate change obligations [5]. Twin goals of energy security and reducing the GHG emissions to combat climate change were proposed to be achieved through the policy of encouraging biogas production from the agricultural biomass, primarily corn. Forage maize was used as a renewable feedstock for the production of biogas which was either upgraded to compressed renewable natural gas and/or combusted for production of renewable electricity and heat in efficient CHP plants. These plants were not economically competitive at the existing market rates and hence were supported by policy-driven agreements for a guaranteed purchase of the renewable electricity at much higher rates with feed in tariffs of up to 22 ct/kWh for 20 years. First policies such as Germany's Renewable Energy Act 2000 to support the renewable electricity were formulated in 2000 [5]. These policies resulted in significant rise in the contribution of biogas to the electricity and the heating sectors of Germany from ~1000 plants in 2000 to over 9000 biogas plants in 2020 contributing about one-fifth of the renewable electricity.

A 2014 change in policy to cut the incentives from new biogas plants resulted in almost an immediate drop in the construction of new plants. Many of the new plants were built to use biogenic waste rather than bioenergy crops as the feedstocks due to this policy change. The number of new biogas plants has continued to decline since the changes and with the expiry of the 20-year agreements for guaranteed power purchases, it is expected that Germany may see a loss of a significant number of plants in coming years [3]. The new Renewable Energy Act 2021 removed the incentives for biogas plants and has technology-agnostic incentives for renewable energy sources. It is expected that the growth in the renewable energy sector will be primarily focused in the solar and wind energy sector even according to the government projections [4]. The dramatic growth, stabilization, and the expected decline of the biogas-based renewable energy sector in Germany is a case study of how policies can only go so far in supporting economically unviable technologies. Therefore, ensuring underlying techno–economic viability is important for the successful implementation of technology. Employing systems analysis approaches can help in presenting a more complete picture of the sustainability of a solution from multiple perspectives.

6.6.3 Case Study 3. Biogas in India

Biogas production through anaerobic digestion has huge potential in India which has the largest cattle population in the world and where most of the population resides in the rural areas. Until the last decade, most of the rural population depended on inefficient wooden and charcoal cookstoves which caused indoor pollution having an adverse impact on the health of rural population with women and children facing disproportionately larger adverse impacts. The concept of converting animal and human wastes into a clean cooking fuel and fertilizer byproduct was very appealing and a looked like a perfect match between the problems of waste management and energy supplies on paper. Therefore, Government of India supported by several large loans from the World Bank embarked on several ambitious programs for converting the animal and human wastes into biogas through anaerobic digestion process starting in 1981 [6]. Several innovative designs of small-scale anaerobic digesters were developed, tested, and operated very successfully under Indian conditions by many researchers in India. The government implemented policies to encourage the building of decentralized small-scale anaerobic digester systems in rural India, incentivizing their adoption through heavy subsidies. These policies addressed many of the technical, economic, and institutional challenges to that are faced by developing countries in implementing the biogas technologies [9].

The policy tools were focused on addressing the high capital cost of biogas technologies but were often insufficient to overcome the financial barriers in capital deficit majority of rural households. In addition, the biogas technologies are maintenance intensive and require constant attention for optimal performance [6]. Therefore, the plants although operated efficiently initially faced neglect and fell into disrepair due to lack of maintenance. Additionally, some of the unique social issues contributed to the failure of many digesters. Marriages and other life event celebrations often result in excess food waste generation which resulted in over-feeding events causing failure of anaerobic digestion process due to acidification of the reactor. Additionally, during these events, the families often visit their families in other villages, and the biogas digesters are not properly fed during this time resulting in the failure of biogas digesters. Lack of adequate training with respect to the operation and maintenance of the digesters contributed to overfeeding, underfeeding or suboptimal use of feed: water ratios resulting in suboptimal performance and even failure of the small-scale anaerobic digesters.

There is a social stigma attached to the use of human wastes for cooking gas. This resulted in the changes to the feedstock from animal and human waste-based digesters to animal manure and agricultural residue-based digesters. Additionally, the top-down implementation of many schemes resulted in pilferage and inefficient implementation of the subsidies. Parallel efforts to encourage the use of bottle LPG for cooking through subsidies led to increased adoption of the LPG for cooking as this option did not have the disadvantages such as requirement of regular maintenance, social stigma attached with the use of waste generate gas for cooking, less hassles in procurement and operation, less dependence on other external factors. Therefore, despite nearly forty years of policy support through various subsidies and incentives the contribution of the biogas to the rural energy mix remains insignificant [6].

6.7 Conclusions and perspectives

This chapter focused on complexities of the policy making, governance, and the societal involvement. Understanding these complexities and addressing them head-on is very critical aspects for the successful implementation of any technology in the real world. Multiple case studies from across the world showing policies with various degrees of success were presented.

This critical analysis of the policy making with a focus on biofuels points out that the policy making is a complex endeavor in which the inputs from the technology developments provide options to the policy makers to achieve larger societal goals. The role of economically viable technologies is to provide options to policy makers to incentivize practically feasible and economically viable strategies for achieving larger societal goals. The role of various stakeholders such as the industry, society, and the governance structure is critical in ensuring that the technological options identified by the technologists are successful. To be successful, a policy must have a clear goal and use one policy instrument for achieving that goal. Although politically painful in the short-term, successful policies can only be implemented if there is a clear statement of costs and benefits to various stakeholders. The acceptance of solutions and polices that may be painful in the short-term but are beneficial in the long term to the society and the nature is only possible through outreach to various stakeholders. For this involvement of all parts of society in the making and implementing of policies through participatory mechanisms is critical.

References

[1] A.J. Hahn, Policy making models and their role in policy education, in Increasing Our Understanding of Public Problems and Policies, 1987, 222-235. https://ageconsearch.umn.edu/record/17854?ln=en. [Accessed: 16th May, 2021].

[2] R. Grundmann. "Ozone and climate governance: an implausible path dependence." Comptes Rendus Geoscience, 30th Anniversary of the Montreal Protocol: From the safeguard of the ozone layer to the protection of the Earth Climate, 350 (7) (2018), 435–41. doi:10.1016/j.crte.2018.07.008.

[3] Clean Energy Wire. 2020a. Germany to lose biogas capacity amid struggle to achieve 2030 renewables target. https://www.cleanenergywire.org/news/germany-lose-biogas-capacity-amid-struggle-achieve-2030-renewables-target. [Accessed January 30, 2021].

[4] Clean Energy Wire. 2020b. What's New in Germany's Renewable Energy Act 2021. https://www.cleanenergywire.org/factsheets/whats-new-germanys-renewable-energy-act-2021. [Accessed January 30, 2021].

[5] S. Theuerl, C. Herrmann, M. Heiermann, R. Grundmann, N. Landwehr, U. Kreidenweis, A. Prochnow, The future agricultural biogas plant in Germany: a vision, Energies 12 (2019) 396. https://doi.org/10.3390/en12030396.

[6] S. Mittal, E.O. Ahlgren, P.R. Shukla, Barriers to biogas dissemination in India: a review, Energy Policy 112 (2018) 361–370. https://doi.org/10.1016/j.enpol.2017.10.027.

[7] J.A.M Costa, From sugarcane to ethanol: the historical process that transformed Brazil into a biofuel superpower, Mapping Politics 10 (2019).

[8] S.L.M. Salles-Filho, P.F. Drummond de Castro, A. Bin, C. Edquist, A.F.P. Ferro, S. Corder, Perspectives for the Brazilian bioethanol sector: the innovation driver, Energy Policy 108 (2017) 70–77. https://doi.org/10.1016/j.enpol.2017.05.037.

[9] T. Nevzorova, V. Kutcherov, Barriers to the wider implementation of biogas as a source of energy: a state-of-the-art review, Energy Strateg. Rev. 26 (2019) 100414. https://doi.org/10.1016/j.esr.2019.100414.

[10] USDA Foreign Agricultural Service. 2020. Biofuels Annual. Report number: BR2020-0032. Global Agricultural Information Network, US Department of Agriculture: https://apps.fas.usda.gov/newgainapi/api/Report/DownloadReportByFileName?fileName=Biofuels%20Annual_Sao%20Paulo%20ATO_Brazil_08-03-2020 [Accessed January 2021].

[11] M. Moraes, Perspective: lessons from Brazil, Nature 474 (2011) S25. https://doi.org/10.1038/474S025a.
[12] H. Meyar-Naimi, S. Vaez-Zadeh, Sustainable development based energy policy making frameworks, a critical review, Energy Policy 43 (2012) 351–361. https://doi.org/10.1016/j.enpol.2012.01.012.
[13] R.Z. Lawrence. How good politics results in bad policy: the case of biofuel mandates. Discussion Paper 2010-10, Belfer Center for Science and International Affairs; CID Working Paper No. 200. Center for International Development, Harvard University Cambridge, MA, 2010.
[14] G.L.T. Oliveira, B. McKay, C. Plank, How biofuel policies backfire: misguided goals, inefficient mechanisms, and political-ecological blind spots, Energy Policy 108 (2017) 765–775. https://doi.org/10.1016/j.enpol.2017.03.036.
[15] E-International Relations. 2008. The politics of renewable energy: unintended consequences of biofuel policies. https://www.e-ir.info/2008/02/10/the-politics-of-renewable-energy-unintended-consequences-of-biofuel-policies/. [Accessed January 30, 2021].
[16] Nikkei Asia. 2021. Asian biofuel policies produce a volatile political cocktail. https://asia.nikkei.com/Spotlight/Asia-Insight/Asian-biofuel-policies-produce-a-volatile-political-cocktail. [Accessed January 27, 2021].
[17] DA. Smith, K. Vodden, Global environmental policy: the case of ozone depletion, Canadian Public Policy/Analyse de Politiques 15 (4) (1989) 413. https://doi.org/10.2307/3550357.
[18] D. Collingridge, J. Douglas, Three models of policymaking: expert advice in the control of environmental lead, Soc. Stud. Sci. 14 (1984) 343–370. https://doi.org/10.1177/030631284014003002.
[19] UN System Task Team on the Post-2015 UN Development Agenda. https://www.un.org/millenniumgoals/pdf/Think%20Pieces/7_governance.pdf [Accessed February 6, 2021].
[20] Drishti IAS. 2020. Good governance. https://www.drishtiias.com/to-the-points/paper4/good-governance-2. [Accessed February 6, 2021].
[21] T. McKulka, 8 Governance principles, institutional capcaity and quality. UNDP. New York, USA. ISBN 978-92-1-126333-6. https://www.undp.org/content/dam/undp/library/Poverty%20Reduction/Inclusive%20development/Towards%20Human%20Resilience/Towards_SustainingMDGProgress_Ch8.pdf. [Accessed 16th May, 2021].
[22] S.A. Cohen, S. Bose, D. Guo, K. DeFrancia, O. Berger, B. Filiatraut, A.C. Miller, M. Loman, W. Qiu, and C.H. Zhang. The growth of sustainability metrics (Sustainability Metrics White Paper Series: 1 of 3). 2014. https://doi.org/10.7916/D8RN36RW. [Accessed February 6, 2021].

CHAPTER SEVEN

Resilience thinking

Ganti S. Murthy[a,b]
[a]Biological and Ecological Engineering, Oregon State University, Corvallis, OR, United States, [b]Biosciences and Biomedical Engineering, Indian Institute of Technology, Indore, India

7.1 Introduction

Sustainability as a concept for addressing many issues facing humanity today has seen vigorous adoption in the scientific literature particularly after the definition of the term by the UN Brundlandt Commission in 1987 [1]. The sustainability concept has been incorporated into the Sustainable Development Goals of the UN adopted in 2015 builds on the Millennium Development Goals adopted in 2000 [2]. Greater emphasis on sustainability is a recognition that sustainable development is at the core of a successful human society. In recent years, sustainability as a concept has been used in the formulation of developmental goals, national policies, corporate vision documents among many declarations.

Given the irreversible changes occurring in the Anthropocene, authors have argued that the concept of sustainability is not sufficient [3]. Simply sustaining patterns of consumption that degrade natural resources is simply not possible and it may be better to realize that sustainable development may not be achievable without drastically changing the patterns of human activity. This recognition is not new. In fact, one of the first reports on the unsustainable trends of resource use was by Meadows et al. [4] in their pioneering book "Limits to growth." Unfortunately, many of the trends first identified in this pioneering book were still holding in a 30-year update of this book [5]. Many authors have raised the question of what is the sustainable use of resources? What is to be sustained? Is it the human use of resources? Consumption patterns, state of natural resources? Importantly the argument against sustainability focuses on the fact that adherence to the sustainability paradigm over the last few decades has not meaningfully changed the behaviors of humanity that have resulted in the Anthropocene [3,6].

It is in such a background that resilience is proposed as a concept that can help in meaningfully adapting to the drastic changes in all human and natural systems expected in Anthropocene. Sustainability is framed in terms of maintaining the equilibrium of system parameters whose knowledge is known, i.e., We have a priori knowledge of what to aspect(s) of the system sustain and know the capacity needed to sustain those aspects. In contrast to sustainability, resilience emphasizes the ability to retain the same structure and function in the face of stressors to a dynamic system. Practices that enhance sustainability and resilience are inherently present in all cultures around the world in the form of traditional wisdom. The concept of resilience was first reported in scientific literature by Holling in a classic paper in the context of ecological systems in 1973 [7]. The discovery of the multiple basins of attractions, and adaptive management in the face of inherent uncertainties in complex ecological systems formed the underpinnings of the resilience concept [8]. The concept was investigated in other domains and after 2000, there has been a large increase in the application of the resilience concept in the diverse fields [8].

Despite the wide application of these concepts, the inherent multidisciplinary approaches require clear definitions of the terms [9]. As the concepts of sustainability and resilience are proposed in scientific literature, clarity in the definition and scope of the terms can help frame these discussions in their proper context and avoid any confusion. This becomes very important as the debates and discussions are framed in a multidisciplinary context.

Sustainability: the word is derived from the Latin root word *sub* (from below) *tenere* (to hold) *sustinere* (hold/support) implying the holding up the current state or trajectory of systems.

Stability: the word comes from the Latin root word *stabilis* which means to stand firm or steady. The stability of a system indicates the ability of the system to remain in the current state regardless of the external/internal disturbances affecting the system.

Robustness: latin word *robustus* meaning strong is the root word for robustness. Robustness implies the ability of the system to withstand perturbations to the system without changing the system output.

Vulnerability: the word derived from the Latin root word *vulnus* implies the susceptibility of the system to perturbations. A vulnerable system is sensitive to even small perturbations and has a low adaptive capacity to perturbations.

Resilience: resilience refers to the ability of the system to rebound or come back to its steady-state after a perturbation is applied to the system. Latin word *resilio* meaning to rebound is the root of this word.

Fragility: fragility comes from the Latin root word *fragilis* meaning "to break." The fragility is a characteristic of the system in which the ability of the system to respond to perturbations decreases over time due to exposure to small perturbations. It is important to note that the magnitude and the frequency of the perturbation is important determinants of fragility. Mostly nonliving systems and many overoptimized human systems such as the financial systems before the 2008 crash are fragile. For example, subjecting an aircraft wing to small vibrations over time makes the aircraft wing more vulnerable to catastrophic failure due to the accumulation of microscopic cracks which increase the fragility of the system.

Antifragility: Antifragility was first defined by Nasim Taleb to describe systems that become stronger in the face of 'small' perturbations [10]. Most living systems are antifragile as a byproduct of the evolutionary process which removes any antifragile systems.

For example, the wings of a bird will be strengthened over time due to repeated exposure to small exertions. This is similar to how weightlifters increase their ability to lift weights through lifting weights during their practice sessions. As can be observed, the ability of the birds wing or the weightlifter is only strengthened if the magnitude of the perturbations is well within the ability of the system to handle and there is a sufficient period of rest between the perturbations which allows the system to recover.

These definitions can be further explained with the following illustration (Fig. 7.1). Consider a system in steady-state 1, as shown in Fig. 7.1. System in steady-state 1 is stable to perturbations of magnitude up to perturbation 1 as the region of attraction around the system is able to maintain the system in the same steady-state. However, as the magnitude of the perturbations increases to perturbation 2, the system is no longer stable and moves to a new steady-state 2 which still has the same structure and function as the system in steady-state 2. Thus, the system is robust to perturbations up to perturbations 2. When the magnitude of the perturbations increases further to perturbations 3, the system moves to an entirely new steady-state 3 with a completely different structure and function. The system is said to be vulnerable to perturbation 3 as there is a loss of structure and function. Over time, if the system is subjected to "small" perturbations which do not change the steady-state of the system but weaken the structure and function of the system ultimately making it vulnerable to even perturbation 1, such a system is called "fragile." In contrast, if the system when subjected to the "small" perturbations becomes stronger and maintains the steady-state not only for perturbation 1 but also perturbation of magnitude 2, the system is said to be "antifragile." The system in steady-state 5 can absorb perturbations of higher magnitudes (up to Perturbation 2) and still maintain the structure and function compared to system steady-state 1. Therefore, the system in steady-state 5 is more resilient compared to system in steady-state 1. Similarly, the system in steady-state 4 is less resilient than system in steady-state 1 as indicated by the shrinking size of the basins of attractions indicated in a darker color. Most natural systems are anti-fragile while many of the nonliving systems are fragile.

It is very well known in control systems theory that the total robustness/fragility of the system is constant [6,11]. Thus, an increase in robustness to some parameters makes the system more fragile to other parameters. The only way to increase robustness to variations while decreasing the fragility for all parameters of the system is through increasing the usage of resource inputs to the system.

7.2 Understanding and quantifying resilience

Resilience as an aspect of complex natural systems was understood in multitude of ways since its introduction to the scientific literature by Holling in 1973 [7]. When discussing resilience, the concepts of adaptation and transformation assume critical importance [6]. The adaptation refers to the adjustments to the system that allow it

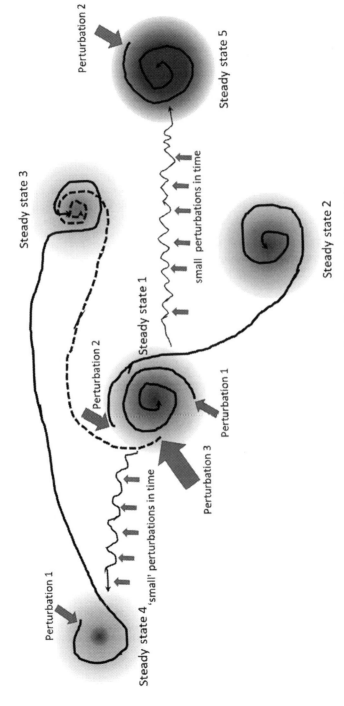

Fig. 7.1 Concepts of stability, robustness, vulnerability, fragility, and antifragility. See text for details.

to function while remaining in the same basin of attraction while the transformations involve moving to new basins of attractions. In any complex system, there are regulatory feedbacks that are responsible for the selforganization, nonlinearity, multiple stable attractors, ability to adapt and transform the system in response to the perturbations. The spatio-temporal dynamics of these regulatory systems especially can be studied using control theory approaches. These regulatory feedback systems often operate at multiple spatio-temporal scales. For example, cells respond to various external stimuli often occur at multiple time scales. The allosteric regulation of enzymes in the cells occurs at milliseconds time scale while the processes to increase/decrease the amount of enzyme is regulated at minutes scales while the cell replication occurs at hours–days scale. Similar variations in the spatio-temporal scales can also be observed in the socio-economic systems. The daily variations in the crop prices in commodity markets, crop cycles spanning few months, agricultural practices policy changes happening over few years, and the ecosystem changes happening over few years to decades.

The dynamics of complex systems can be considered from multiple time scales, with the caveat that the exact timescale definitions are dependent on the system under consideration. Anderies et al. [6] describe the three-time scales where the concepts of resilience can be used to address challenges in complex systems. A short time scales, the system adaptations to change resilience can be considered equivalent to changing the robustness of the system. The focus is on understanding the robustness-fragility trade-offs, managing the uncertainties while maintaining the outputs of the system, basins of attractions, and managing the regime shifts. Referring to the agricultural systems example above, the short-term resilience is to maintain the crop yields for given weather during the cropping season or responding to the pest pressure through the use of pesticides. At intermediate scales, the adaptability reflecting the ability of the systems to adapt without fundamentally changing the structure and function of the system is emphasized. This is similar to a farmer adapting to the drought through a change in the selection of crops for the following season or installing a drip irrigation system to better manage the water resources. The long-term transformations imply a complete change in the systems including the changes to the basins of attractions, nature of uncertainties, inputs and the outputs. It is important to note that the transformations can be very slow to notice in the beginning and may transition to the new basin of attraction quite rapidly in the final stages of the transformations [6,8]. Similar considerations can also be applied to the application of the resilience concept at different spatial scales such as the field, farm, and regional/global spatial scales [12].

It is clear from the above description that resilience is not about preserving all aspects of the system and it is about conserving critical aspect of the system related to the structure and function. This brings up a question as to what are the aspects of the system that are conserved in the resilient systems? Lundberg and Johansson [13] define "core-goals" of the system as its identity which will be preserved in resilient systems even at the cost of other non-core goals of the system. After going through an adaptive transformation, a resilient system must retain its core goals while the flexible noncore goals may be discarded, or new ones added.

Understanding the factors that enhance the resilience of a system is critical to develop appropriate scientific, technological, implementation, management, and policy strategies. Many studies of socio-economic systems, ecological studies have shown that redundancies, systems diversity, and decentralized architectures enhance the resilience of a system [9,14–16]. Enhancing the redundancies will increase the capacity of the system to absorb unpredictable shocks by increasing the basin of attraction. Additionally, redundancies also decrease the vulnerability of the system. Diversity in the systems also provides many pathways for absorbing the disturbances and thus increases the resilience of the system. Decentralized structures reduce the risk of failure due to the failure of the critical nodes and thus increases the resilience of the system and its ability to reorganize after a perturbation event.

Measuring the resilience of a system has been investigated extensively and can be classified into the quantitative and qualitative methods [17]. In control systems literature, the performance of the feedback controller in a closed-loop is measured using parameters such as rise time, overshoot, settling time, and steady-state error. Many resilience studies have applied similar concepts (albeit with different terminologies) to characterize the resilience of a system. Recovery length (similar to the settling time) has been suggested as a measure of the system resilience [14]. The recovery time increases and the variability in the system increase in response to disturbances as the system resilience decreases. Increasing the size of the basins of attractions of the system, called the safe operating space of the system results in an increasing the resilience of the system [15]. A system with larges safe operating space can tolerate larger variances to the system and still return to the original state. This is a concept similar to the overshoot parameter of feedback control systems. It is to be noted that while a system's resilience can be enhanced by decreasing the response time and increasing the scale of response (i.e., increasing the "gain of the feedback controller"), the system's resilience for signals with different frequencies than the input disturbances for which the performance was optimized can deteriorate significantly [6]. This is the classic tradeoff between the robustness and fragility that is discussed in the control system literature. This is not a simply theoretical consideration but has significant practical implications. For example, what should be the frequency of changes in the crop selection in response to weather and market events? When should they be ignored as noise and when should the market event be considered serious enough to warrant a policy response? For a farmer, with a standing crop, the optimum frequency may be to implement weekly changes in irrigation amount, while the optimum frequency for making decisions regarding the irrigation method (flood irrigation, vs sprinkler vs drip irrigation), and the crop choices maybe every 2–5 years while for policy makers the medium to long term (5–10 years) may be an appropriate time scale to respond to changes in the weather patterns, market conditions.

Table 7.1 Some of the qualitative assessment frameworks reported in literature [17].

Domain	Procedures/dimensions/steps/guiding principles of the assessment framework
Socio-ecological systems [18]	1. Define and understand the system. 2. Identify the appropriate scale of the system. 3. Identify the internal and external drivers of the system. 4. Identify key players. 5. Develop conceptual models needed for the recovery activities.
Socio-ecological systems [19]	Three dimensions of resilience: buffer capacity, self-organization, and ability to adapt
Generic [20]	Eight guiding principles: threat and hazard assessment, robustness, consequence mitigation, adaptability, risk-informed planning, risk-informed investment, harmonization, and comprehensive scope
Communication networks [21]	Six factors: defend, detect, diagnose, remediate, refine, and recover
Communication networks [22]	Resilience properties: reliability, safety, availability, confidentiality, integrity, maintainability, performance, and interactions
Generic [23]	Resilience defined as a function of absorptive capacity, adaptive capacity, and restorative capacity

Many qualitative frameworks (Table 7.1) have been proposed for measuring resilience using qualitative methods [17]. The Resilience Alliance [18] describes a generic five-step assessment framework for socio-ecological systems (Fig. 7.2). Most of these frameworks while conforming to the generic principles also include domain-specific assessment steps as can be noted from the descriptions in Table 7.1.

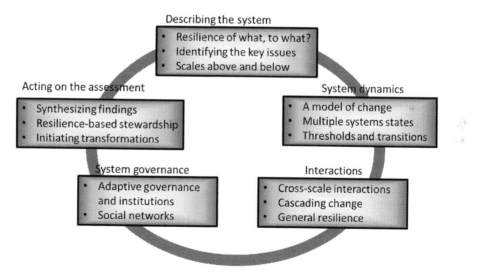

Fig. 7.2 A generic resilience assessment framework (figure adapted from [18]).

Lundberg and Johansson [13] developed a SysRes model for envisioning various resilience methods and metrics in the context of event-based constraints, functional dependencies, adaptive capacity, and strategy. SysREs Model captures six essential functions of adaptive systems including anticipation, monitoring, response, recovery, learning, and self-monitoring which are critical for assessing the system's resilience. The progress toward the center of the spiral increases the system resilience while the disturbances and nonlinear responses of the system move the trajectory away from the center of the spiral (Fig. 7.3).

There have been several developments toward quantitative metrics for assessing resilience [17]. Unsurprisingly, most of the quantitative methods are developed for assessing resilience in engineering domain applications. A fee of the quantitative metrics proposed in the literature is summarized in Table 7.2.

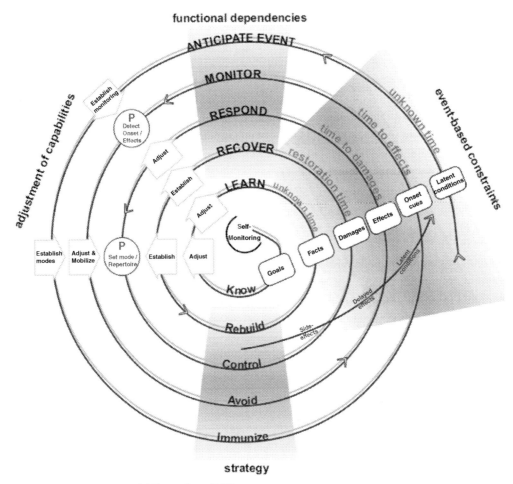

Fig. 7.3 The SysRes Model (figure from [13]).

Resilience thinking

Table 7.2 Qualitative assessment frameworks reported in literature (figures from [17]).

Resilience concept	Resilience formula
	$$RL = \int_{t_0}^{t_1}[100 - Q(t)]dt$$ Where: Time at which disruption occurs t_0 and the time at which the system returns to the normal state is t_1 as indicated by the performance measure Q(t). The resilience measure is RL
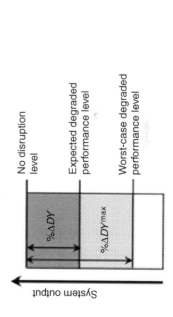	$$R(X,T) = \frac{T\star - XT/2}{T\star} = 1 - \frac{XT}{2T\star}s$$ Where: R is resilience X ε is the percentage functionality lost after a disruption T ε [0,T★] is time T★ is a suitably long interval
	$$R = \frac{\%\Delta Y^{max} - \%\Delta Y}{\%\Delta DY^{max}}$$ Where: %ΔDY is the difference in non-disrupted and expected disrupted system performance %ΔDY^{max} is the difference between the nondisrupted and worst case disrupted system. The R is the resilience measure

(continued)

Table 7.2 (Cont'd)
Resilience concept   **Resilience formula**

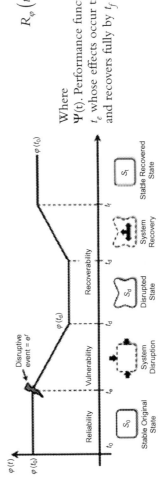

$$DR = \sum_{i=1}^{N} SO_{HR}(t_i) - SO_{WR}(t_i)$$

Where
DR is the resilience measure
SO_{HR} is the output of the system under hastened condition
SO_{WR} us the recovery without hastened condition

$$R_\varphi(t|e^j) = \frac{\varphi(t|e^j) - \varphi(t_d|e^j)}{\varphi(t_0) - \varphi(t_d|e^j)}$$

Where
$\Psi(t)$. Performance function at the time t. Disruptive event e^j at time t_e whose effects occur till t_d and the system starts recovering at time t_s and recovers fully by t_f

7.3 Resilience thinking in systems analysis

Importance of the concept of resilience in systems analysis cannot be overemphasized. The resilience thinking refers to the application of resilience principles to systems [9]. Folke [8] defines resilience thinking as, "Resilience thinking is about how periods of gradual changes interact with abrupt changes, and the capacity of people, communities, societies, cultures to adapt or even transform into new development pathways in the face of dynamic change."

With the increasing complexity of socio-ecological systems, tighter integration of various socio-economic systems such as financial sectors, energy, food and global manufacturing sectors, the complexity of human systems has increased tremendously in the last millennia. The socio-ecological systems in multiple regions can be connected through telecouplings thus further complicating the already complex connections and impacting resilience of the systems [24]. Large-scale human activity is leading to homogenization of the global hydrologic cycle resulting is loss on resilience and increased vulnerability [25]. Such telecouplings and impacts on global scale have become more common with globalization of trade and free movement of people and goods. With a focus on increasing the efficiency of these systems, these systems are increasingly operating under highly optimized conditions that tend to maximize the returns for the stakeholders. An unintended consequence of such over-optimization is the increased fragility of these systems through their susceptibility to "black swan events." A black swan event is a highly improbable and unpredictable disturbance that has the potential to disrupt the systems completely which appears to be quite predictable in the hindsight [26]. Taleb [10] specifically warns against over-optimization in complex systems and describes the redundancy in the systems is an important factor in reducing the system fragility.

In the past when regional socio-ecological systems could adapt and transform when subjected to local disturbances, such as local droughts, financial collapses, epidemics, and wars. These systems could use support from other regions and external resources to aid them in these transitions. An unintended consequence of tighter integration of the regional socio-ecological systems as a result of globalization is that the crises when they happen can impact all the regions of the globe simultaneously causing significant harm. Liu et al. [27] proposed three deep causes for the synchronous failure. First, dramatic increase in the human population causing a dramatic increase in the scale of human activity related to the scale of natural processes and the natural systems. For example, the scale of human interventions in the global nitrogen cycle is already comparable to the scale of the natural processes. Second, increasing density, capacity, and the transmission speed of the material, energy and information as a result of the technology developments has increased the connectivity and interdependencies between the systems. Third, there is a decrease in diversity among human cultures, institutions, practices, and technologies. The democratization of technologies, and rapid dissemination of information globally is both the cause and consequence of the increase connectedness among the various systems. These three trends together contribute to the decreasing resilience of the global systems and increase the chances of synchronous failure. Firstly, these three trends generate multiple stresses throughout the various regional socio-ecological systems. Secondly, they also create increase the risk of synchronous failure by propagating the disturbances throughout the global network of systems. Continuous optimization of the global networks to increase efficiency and economic profitability has reduced the overall capacity of the global system for adaptation and adjustments. Liu et al. [27] describe how the three causes described above can cause synchronous failure in three types of processes. The Long Fuse Big Bang processes occur when the stresses slowly build up in the system over time and precipitate a rapid failure when the system undergoes a nonlinear shift in the system state due to its inability to cope up with stresses due to the loss of adaptive and coping mechanisms. The simultaneous stresses (SS) scenarios occur when there are SS in a one part of the connected network and cause a synchronous failure of the whole network. The causal dependence/independence of these stresses and additive/synergistic combination of their effects play a major role in determining the thresholds of these multiple stresses that can precipitate asynchronous failure. Ramifying cascade type processes occur in highly connected systems where the severe perturbation in one of the nodes is rapidly propagated through the network. The global financial-food-energy crisis of 2007–2008 can be analyzed using this paradigm, as shown in Fig. 7.4.

7.4 Conclusions and perspectives

This chapter focused on introducing the concepts of resilience, resilience thinking, fragility and antifragility, and their application to systems analysis. Managing systems in the Anthropocene with all the inherent complexities requires a move from simplistic "sustainable systems" to more nuanced resilient and antifragile systems, as discussed in this chapter. There has been a tremendous increase in the publications in the last decade reflecting the intensive research into these aspects. Going forward, the importance of resilience thinking in the systems analysis will increase.

Resilience thinking 125

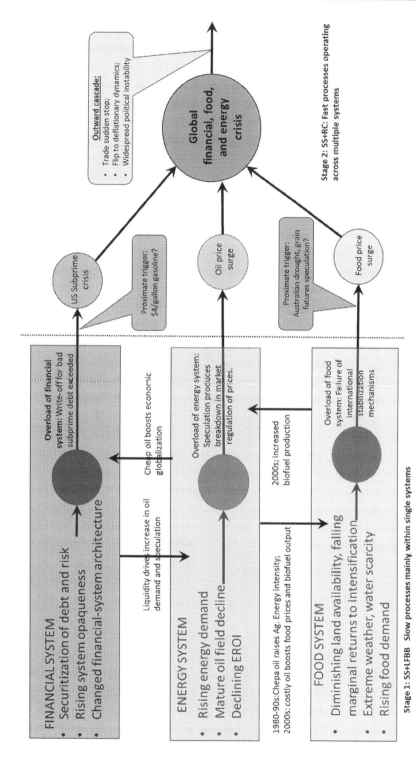

Fig. 7.4 Synchronous failure in complex systems. Global financial, food, and energy crisis case (figure adapted from [27]).

References

[1] Brundtland G.H. 1987. Report of the world commissions on environment and development. Our common future. https://sustainabledevelopment.un.org/content/documents/5987our-common-future.pdf. [Accessed February 9, 2021].

[2] UN. Transforming our world. The 2030 agenda for sustainable development. UN Report number A/RES/70/1. 2015. https://sustainabledevelopment.un.org/content/documents/2125203cAgenda%20for%20Sustainable%20Development%20web.pdf. [Accessed February 9, 2021].

[3] M.H. Benson, R.K. Craig, The end of sustainability, Soc. Natural Res. 27 (2014) 777–782. https://doi.org/10.1080/08941920.2014.901467.

[4] D.H Meadows, D.L. Meadows, J. Randers, W.W. Behrens III, Limits to Growth. A Report for the Club of Romes Project on the Predicament of the Mankind, A Potomac Associates Book, Falls Church, VA, USA, 1972, ISBN: 978-0876632222.

[5] G. Turner, A comparison of the limits to growth with 30 years of reality, Global Environ. Change 18 (3) (2008) 397–411.

[6] J.M. Anderies, C. Folke, B. Walker, E. Ostrom, Aligning key concepts for global change policy: robustness, resilience, and sustainability, Ecol. Soc. 18(2) (2013) 8. http://dx.doi.org/10.5751/ES-05178-180208.

[7] C.S. Holling, Resilience and stability of ecological systems, Annu. Rev. Ecol. Syst. 4 (1973) 1–23.

[8] C. Folke, Resilience (Republished), Ecol. Soc. 21 (2016) 44.

[9] N. Urruty, D. Tailliez-Lefebvre, C. Huyghe, Stability, robustness, vulnerability and resilience of agricultural systems. A review, Agron. Sustain. Dev. 36 (2016) 15.

[10] N.N. Taleb, Antifragile. Things that Gain from Disorder, Random House and Penguin, New York, USA, 2012. ISBN 9780812979688.

[11] K. Hiroaki, Towards a theory of biological robustness, Mol. Syst. Biol. 3 (2007) 137.

[12] J.M. Bullock, K.L. Dhanjal-Adams, A. Milne, T.H. Oliver, L.C. Todman, A.P. Whitmore, R.F. Pywell, Resilience and food security. Rethinking an ecological concept, J. Ecol. 105 (2017) 880–884.

[13] J. Lundberg, B.J.E Johansson, Systemic resilience model, Reliab. Eng. Syst. Safety 141 (2015) 22–32.

[14] S.R. Carpenter, Spatial signatures of resilience, Nature 496 (2013) 308–309.

[15] S.R. Carpenter, W.A. Brock, C. Folke, E.H. Nes, M. Scheffer, Allowing variance may enlarge the safe operating space for exploited ecosystems, Proc. Nat. Acad. Sci. 112 (2015) 14384–14389.

[16] G.S. Cumming, G.D. Peterson, Unifying research on social–ecological resilience and collapse, Trends Ecol. Evol. 32 (9) (2017) 695–713.

[17] H. Seyedmohsen, K. Barker, J.E. Ramirez-Marquez, A review of definitions and measures of system resilience, Reliab. Eng. Syst. Safety 145 (2016) 47–61.

[18] Resilience Alliance. 2010. Assessing resilience in social-ecological systems. Workbook for practitioners. Version 2.0. https://www.resalliance.org/resilience-assessment. [Accessed 16th May, 2021].

[19] J.H. Kahan, A.C. Allen, J.K. George, An operational framework for resilience, J Homel. Secur. Emerg. Manag. 6 (2009) 1–48.

[20] C.I. Speranza, U. Wiesmann, S. Rist, An indicator framework for assessing resilience in the context of social-ecological dynamics, Glob. Environ. Change 28 (2014) 109–119.

[21] J.P.G. Sterbenz, E.K. Cetinkaya, M.A. Hameed, A. Jabbarand, J.P Rohrer, Modeling and analysis of network resilience, Proceedings of the IEEE COMSNETS, Bangalore, 2011.

[22] P. Vlacheas, V. Stavroulaki, P. Demestichas, S. Cadzow, D. Ikonomou, S. Gorniak, Towards end-to-end network resilience, Int. J. Crit. Infrastruct. Prot. 6 (2013) 159–178.

[23] E.D. Vugrin, D.E. Warren, M.A. Ehlen, Framework for infrastructure and economic systems: quantitative and qualitative resilience analysis of petrochemical supply chains to a hurricane, Process Saf. Prog. 30 (2011) 280–290.

[24] J. Liu, V. Hull, J. Luo, W. Yang, W. Liu, a. Viña, C. Vogt, et al., Multiple telecouplings and their complex interrelationships, Ecol. Soc. 20 (2015) 44.

[25] D.F. Levia, I.F. Creed, D.M. Hannah, K. Nanko, E.W. Boyer, D.E. Carlyle-Moses, N. Giesen, et al., Homogenization of the terrestrial water cycle, Nature Geosci. 13 (2020) 656–658.

[26] N.N. Taleb, The Black Swan. The Impact of the Highly Improbable, Random House, New York, USA, 2007. ISBN 9780812973815.

[27] T. Homer-Dixon, B. Walker, R. Biggs, A. Crépin, C. Folke, E.F. Lambin, G.D. Peterson, et al., Synchronous failure. The emerging causal architecture of global crisis, Ecol. Soc. 20 (2015) 6.

CHAPTER EIGHT

General logic-based method for assessing the greenness of products and systems

Edgard Gnansounou

Ecole Polytechnique Fédérale de Lausanne (EPFL), School of Environment, Civil Engineering, and Architecture, Institute of Civil Engineering, Bioenergy and Energy Planning Research Group, Lausanne, Switzerland

8.1 Introduction

Humankind heavily depends on nonrenewable materials for the production processes and end-uses. The impacts in terms of depletion of resources, decrease of biodiversity, and climate change are hotspots leading to policy measures for promoting sustainable development at global and local levels. Sustainability becomes the cornerstone of every strategic and operational decision. However, its assessment remains challenging since it is a multi-level, multidimensional, multiactors, spatial, and temporal evaluation process. As it would be intractable to cope front-on this complex issue, all existing assessment methods are simplified approaches focusing in a certain extent on aggregate indexes. The aim is to give instruments to decision-makers for improving the sustainability at design as well as implementation and operational levels. Nevertheless, the way to estimate sustainability is not neutral and may influence the efficacy of improvement measures. Beyond the conventional definition of sustainability with its three pillars, the tradeoff between them or the dependency from one to the other are questions that cannot result in a unique response. Strong sustainability precludes any substitution between the reduction of natural environment capital and the increase in human wellbeing. Weak sustainability tolerates a certain tradeoff largely in favor of economic and societal pillars. Due to the emergency of environmental concerns, policy makers begin stating adoption of strong sustainability in policy acts but at operational level, they are still obliged to rely on a weak sustainability concept because of the necessity to continue maintaining the economic growth to serve unsatisfied social needs. In the perspective of strong sustainability, the "Economy" would be considered as part of the "Social" that at its turn would be included in the "Natural Environment." However, irrespective of the conception of Sustainability, the choice of the relevant indicators and the implementation of their relationships in the assessment methods are challenging.

Depending on the problematic, sustainability assessment may consist in comparison of the product or the system under study with a reference, or in characterizing that product or system from a sustainability viewpoint. In the case of comparison, the reference is often less sustainable and the problem consists in assessing how far the product or the system under study can improve the reference from a sustainability point of view. In the case of characterization, the objective is to assess how far the product or the system is sustainable with respect to a given referential case defined, for instance, as the best practice. The issue in this chapter is a characterization of the sustainability of products or systems and grant them grades according to their greenness. One of the questions that are discussed is how the indicators at different levels should interrelate to allow assessing the global sustainability of a product or system in a relevant and consistent way. But first, what are Green-Economy and its constituent elements–the Green Products or Systems?

According to the United Nations Environmental Programme Green-Economy is the *"one that results in improved human well-being and social equity, while significantly reducing environmental risks and ecological scarcities"* [1]. This definition derived from an "enlightened anthropocentrism" paradigm since it is people-centric. The German Cooperation (GIZ), the Global Green Growth Institute, and the Green Economy Coalition proposed five principles of inclusive Green-Economy as (1) wellbeing, (2) justice, (3) planetary boundaries, (4) efficiency and sufficiency, and (5) good governance [2].

In this chapter, a "Green Product or System" is defined considering its ability to contribute to building a Green-Economy. The European Commission proposed an operational definition of green products that highlights the environmental performances including more efficient use of resources compared to counterfactual products. Georgeson and Maslin estimated the US green economy sales revenue to amount $1.3 trillion annually employing about 9.5 million workers [3]. In fact, the need to incorporate sustainability into the products and systems gave rise to the concept of "greenness" and several initiatives exist that aim at identifying "green" organizations or products. For example "The Dow Jones Sustainability World Index" launched in 1999 benchmarks for greenness the top 10% of the largest 2500 companies in the Standard & Poor's based on long-term economic, environmental, and social criteria [4]; the selected big companies then claim for greenness through their brands. Considering that the number of hundreds ecological labels contributes to scramble the evidence making of greenness, the European Commission adopted in 2013 an initiative aiming at *"Building the Single Market for Green Products Facilitating better information on the environmental performance of products and organisations"* [5]. The initiative was based on two methods. The first aims at assessing quantitative indicators of products' environmental footprint on their life cycle for removing ambiguity on greenness evidencing. The second method applies to companies. Both methods use a life cycle thinking approach and seek to harmonize and simplify the application of life cycle assessment (LCA) to products and organizations. The European Commission intended to operationalize both methods by two aspects: (1) the way to assess the environmental impacts and (2) the specific rules that will allow benchmarking products/organizations. The test phase of the European initiative results in two operational methods, i.e., the "Product Environmental Footprint (PEF)," and the "Organization Environmental Footprint (OEF)."

The PEF and OEF methods require that for making comparisons, product/organization must be confronted to best cases that will be developed based on selected impact categories.

Since there are a lot of impact categories, the pilot phase of the European initiative aimed at developing "Product Environmental Footprint Category Rules (PEFCR)" and "Organization Environmental Footprint Sector Rules (OEFSR)" focusing, for each group of products and sectors, on the most relevant processes and the three or four most significant category impacts to consider when assessing the environmental Footprint. After the end of the pilot phase in 2018, a transitional phase has started consisting in pursuing the elaboration of models for other groups of products and sectors and evaluating the efficacy of these models for further inclusion in existing and future environmental policy instruments. Then the second phase concerning the future policy that will consider these instruments is planned to start in 2021. In the two examples of "The Dow Jones Sustainability World Index" and European "PEF/OEF)," the term "green" is used with different meanings, i.e., sustainability in the former and environmentally friendly in the latter. Regardless of this difference, all methods are confronted with the following difficulty: how to aggregate the selected indicators into a final score of greenness? Furthermore, contrary to products that are traded by big companies in global competitive markets, the ones that focus on the short-circuit economy have to struggle for competitiveness. Several barriers that need to alleviate for the entry of these companies into the market include the following factors: possible expensiveness of local raw materials and other inputs, lack of scale economy, competition with products from global markets, insufficient awareness of the consumers, and decision-makers about the short-circuit-circular-economy specificities. Greenness is often expected to generate some economic advantages due to efficient resource use and technological innovation. However, the picture is more blurred since producing with less and renewable resources may cost more even though it more and more involves new opportunities and technological deals.

8.2 The sustainability value added

As far as entry into the market is concerned, several kinds of the product must be distinguished. For instance, the products of category I are those traded by big companies on the global competitive market. They must be competitive and do not need any support from the stakeholders. The category II products are traded by small and medium-sized enterprises and target local as well as global markets and may need support in their infancy to remove barriers to enter into the markets. Finally, the category III products claim for a local short circuit and often need for a certain time supports from public authorities and consumers in order to alleviate the barriers to entry into the market. Their challenge is to establish sustained progress of their markets after a reasonable time of start-up. The counterpart of the needed supports would be their "Sustainability Value Added (SVA)," which would be the answer to the questions: what and how much additional values these products provide owing to their sustainability?

Economists first tackled the issue of how to quantify "Sustainable Value Added." One method to assess the relation between sustainability and the efficiency of a product is the eco-efficiency. The World Business Council mainstreamed this concept as a relative reduction of environmental impact in the course of value creation [6]. Two types of indicators are frequently used, the ratio of economic value to the environmental impact [7] or its inverse [8]. The former expresses the increasing function of eco-efficiency with the decrease of environmental impact and with the increase in the value of the product or the system. The economic value used in the eco-efficiency indicators can be estimated from demand or supply-side, e.g., value-added. The assessment of environmental impact can be based on partial environmental effects, e.g., emissions of pollutants, mid-point, end-point, or full aggregated environmental impact categories. The concept of eco-efficiency was operationalized by the WBCSD through the three following steps [6]:

(1) Reduction of the resource consumption by lowering the materials, energy, water inputs in the production process, and land use; the measures include recycling, reuse, durability of products, end-of-life management within an integrated efficient conceptual and operational management along the life cycle of the product.

(2) Reduction of the environmental impacts, which includes measures such as the use of renewable resources, lowering discharges of pollutants in the air, soil, water, and avoidance of toxic matters in the products.

(3) Increasing the value of the product and service by enhancing the consumers' surplus through focusing their needs, addressing the flexibility, modularity, and functionality of the products that would lead the consumers to increase their willingness to pay.

The main criticism of eco-efficiency is that due to the potential rebound effect, the increase in eco-efficiency could result in a decrease of effectiveness. This effect can be explained by the fact that the success of an eco-efficient product and the subsequent increase in volume of its demand compared to the counterfactual product may more than compensate for the reduction of relative environmental impact. As an alternative, the United Nations Environmental Programme promoted the decoupling between the consumption of resources, the environmental impact, and value generation, which means soberness in natural resources and absolute decrease of environmental impact [9] that is termed in the present study as "eco-effectiveness" [1].

Another approach consists in internalizing the environmental impacts into the economic value. Figge and Hann, (2004) expressed the "*Sustainable Value Added (SVA)*" as:

Absolute Sustainable Value Added = Value Added − external environmental and social cost + relative Sustainable Value Added [10].

Where the *Value added* is the turnover minus the input costs, the *External environmental and social cost* is the monetary internalized cost of corresponding negative externalities, and the *Relative Sustainable Value Added* is the positive environmental and social externalities.

The merit of the SVA concept is the intention to gather quantitative and qualitative variables of value creation and losses through a monetization approach. However, this merit is limited by the intractability of monetization process. Furthermore, from system thinking point of view, this approach sets the boundary of the corporate in a way that could be discussable in today's emergency of environmental concerns. Internal costs and benefits versus externalities will be more and more difficult to discern since the Sustainability of Corporate is evolving toward including upstream supply chain and downstream delivering subsystem of the products to consumers, considering social responsibility into an overall Corporate Sustainability disclosure. Yang et al. (2017) shared this conception of Corporate Sustainability and proposed a sustainable value analysis tool that aims at helping companies to tap "values captured and uncaptured" and turn them into opportunities for value creation across the whole life cycle of the products [11]. Beyond an instrumental tool, the sustainable value analysis tool is a systematic methodology where the participants of a series of workshops analyze the values captured and uncaptured across the life cycle of the studied products and identify related opportunities and measures aiming at converting them into profit.

8.3 The logic-based model

Sustainability is assessed by selecting relevant indicators and aggregating them mostly by weighting. Sala et al. (2018) reviewed in the context of the LCA various weightings using quantitative models [12]. The logic-based model (LBM) is a rule-based method that combines qualitative and quantitative values to assess the overall performance of a product or a system with regard to imprecise concepts such as sustainability, vulnerability, or future scenarios.

8.3.1 Classification of the sustainability indicators

The method used in the LBM to score the sustainability of a product or a system follows a hierarchical structure [13,14]. At the bottom of the structure, specific indicators evaluate the higher-level indicators that are termed "general indicators." In their turn, the general indicators' values combine and explain the values of the "determinants." Finally, the combinations of values of the determinants are allotted to the values of "sustainability." Due to the existence of correlations between variables, few combinations may not be feasible; they can be avoided by adding appropriate constraints.

Let us assume at first that all combinations are possible, each grade is quantified using a score in a chosen set of consecutive integers, e.g., 1–3. In this case, 1–3 would represent the linguistic values "low," "moderate," and "high" performances of the variable.

Let S the Specific Indicators and IS their values. The values $IS_{i,j,k}$ are the successors of values $IG_{i,j}$ of General Indicator $G_{i,j}$ that at their turn are the successors of the Determinants D_i. The combinations of the values $IS_{i,j,k}$ will be allotted between the possible grades of their predecessor's value $IG_{i,j}$ using as metric the distance to the maximum of the values $IS_{i,j,k}$. The values of upper variables are allocated into the grades of their predecessors by the same method.

Exclusion and consistency rules may be included in the approach that preclude or enforce fewer combinations of specific indicators values, combinations values of general indicators and determinants. In general, the use of the quantitative classification method is just for initiating the design of the knowledge base.

If the exclusion or consistency rules constrain the values of specific indicators, the users are assisted by the system during the input phase. Warnings appear when the user violates these constraints. Once all the relevant values of the specific indicators are entered, the software using conflict resolution procedures treats additional qualitative rules that may contradict the quantitative classification.

The quantitative classification method is given by Eqs. (8.1–8.3) for the general indicators, determinants and sustainability, respectively.

$$IG_{i,j} = \text{ROUND}\left(\left[\text{MaxIG}^{\star}_{i,j} - \text{MinIG}^{\star}_{i,j}\right] \star \left[1 - \sqrt{\frac{\sum_{k=1}^{K_{i,j}}\left(\text{MaxIS}^{\star}_{i,j,k} - IS_{i,j,k}\right)^2}{\sum_{k=1}^{K_{i,j}}\left(\text{MaxIS}^{\star}_{i,j,k} - \text{MinIS}^{\star}_{i,j,k}\right)^2}}\right]\right) + \text{MinIG}^{\star}_{i,j} \quad (8.1)$$

$$ID_i = \text{ROUND}\left(\left[\text{MaxID}^{\star}_i - \text{MinID}^{\star}_i\right] \star \left[1 - \sqrt{\frac{\sum_{j=1}^{J_i}\left(\text{MaxIG}^{\star}_{i,j} - IG_{i,j}\right)^2}{\sum_{j=1}^{J_i}\left(\text{MaxIG}^{\star}_{i,j} - \text{MinIG}^{\star}_{i,j}\right)^2}}\right]\right) + \text{MinID}^{\star}_i \quad (8.2)$$

$$SC = \text{ROUND}\left(\left[\text{MAxSC}^{\star} - \text{MinSC}^{\star}\right] \star \left[1 - \sqrt{\frac{\sum_{i=1}^{N}\left(\text{MaxID}^{\star}_i - ID_i\right)^2}{\sum_{i=1}^{N}\left(\text{MaxID}^{\star}_i - \text{MinID}^{\star}_i\right)^2}}\right]\right) + \text{MinSC}^{\star} \quad (8.3)$$

where the variables are defined as follows:

$IS_{i,j,k}$ is the score of the specific Indicator $S_{i,j,k}$ of the general Indicator $G_{i,j}$ of the determinant D_i, $k = 1, ..., K_{i,j}$. Max $IS\star_{i,j,k}$ is the maximum score for $IS_{i,j,k}$, $k = 1, ..., K_{i,j}$, Min $IS\star_{i,j,k}$ is the minimum possible score for $IS_{i,j,k}$, $k = 1, ..., K_{i,j}$, $K_{i,j}$ is the maximum number of specific indicators $S_{i,j,k}$ for the general indicator $G_{i,j}$, $IG_{i,j}$, is the score of the General Indicator $G_{i,j}$ of the Determinant D_i, $j = 1, ..., J_i$, Max $IG\star_{i,j}$, is the maximum score of $IG_{i,j}$, $j = 1, ..., J_i$, Min $IG\star_{i,j}$ is the minimum score for $IG_{i,j}$, J_i is the maximum number of general indicators $G_{i,j}$ of the Determinant D_i, ID_i is the score of the determinant D_i, Max $ID\star_i$ is the maximum score of ID_i, Min $ID\star_i$ is the minimum score of ID_i, SC is the Sustainability score,

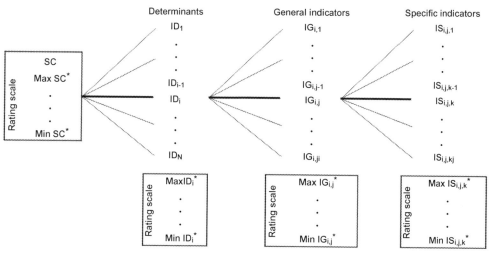

Fig. 8.1 Indicators and determinants of sustainability.

Max SC★ is the maximum score of SC, Min SC★ is the minimum score for SC, and N is the number of sustainability determinants. These variables are illustrated in Fig. 8.1.

8.3.2 Outlining the development and utilization of the LBM

A six-step approach structures the development and utilization of the LBM (Fig. 8.2). From steps 1 to 4, the basic model is built. Steps 5 and 6 are related to the utilization of the LBM. The development starts with the selection of the determinants throughout a brainstorming that results in a clear narration of the context, the goal, and the scope. This stage is important since the other variables derived from the determinants. They are the most general variables. Their identification and selection would depend on the value systems that inspire the concept of sustainability. For instance, in the conventional concept of sustainability, the determinants could be the following: environment, social, and economy known as the three pillars of sustainability. The value system in the background of such a concept is "enlightened anthropocentrism" that could be expressed as follows: humankind looks primarily for its own wellbeing but takes care to a certain extent to the retroactions of its environment. Adversely, with an alternative concept of sustainability, the determinants could be social and ecology. For each determinant, the general indicators are selected, the same for the specific indicators that are listed as drivers of each general indicators. The qualitative rating scales are chosen for each variable and the possible scores are associated with the grades.

The system makes use of forwarding and backward chaining. In the forward chaining, the allocation of the values is undertaken considering all additional rules, solving all the possible conflicts and the final sustainability score is estimated.

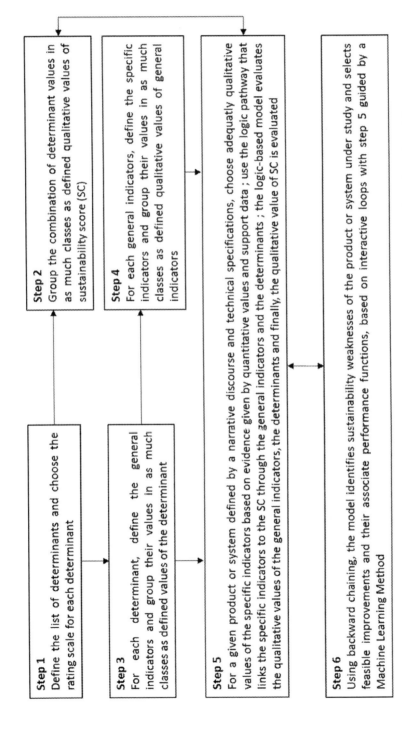

Fig. 8.2 Outline of the development and utilization of the LBM for sustainability assessment. *LBM*, logic-based model.

In backward chaining, the system can assist the user in justifying the final and intermediate performances by identifying the rules that have been activated. Gnansounou and Nouatin (1997) implemented this approach; they developed the MAGES tool aiming at elaborating consistent scenarios for assessing future energy demand [15]. An extension of this tool may consist in generating the appropriate improvement target for a given performance based on the cost/advantage of feasible improvements using a machine learning approach.

8.4 Application for assessing the sustainability of products and systems

Since there is a large variety of products and systems, it would be unpractical to fix the variables (determinants, general indicators, specific indicators) once for all. Nevertheless, from one product to the other the goals are similar: avoiding abuse of natural resource consumption, enhancing environment conservation, social, and economic progress. Hence, it may be relevant to develop a general model for products and systems that can be instanced in particular cases. This instantiation may consist in adapting the variables and the rules. It may also consist in putting more emphasis in producing more specific support data. The starting point of the development of a general LBM for assessing the sustainability of products and systems is the clarification of the sustainability concept that could be appropriate for such an assessment. Two main conceptions are debated: the conventional conception and the environment-driven sustainability.

8.4.1 Extending the conventional conception of sustainability

Fig. 8.3 illustrates a more flexible interpretation of the conventional conception of sustainability where the sustainability concept is defined as tradeoffs between six disjoint subsets. The seventh disjoint subset (ABC) is more abstract. It involves the background relations that link the three pillars. The tradeoffs are implemented through rules.

The methodological design of one version of the general LBM is based on this interpretation of sustainability. It is worth noting that even disjoint, the six subsets are not independent; with the seventh subset (ABC) they constitute a partition of the global sustainability system. In this modeling except (ABC) each subset is modeled as a determinant. (ABC) governs how the values of the six determinants relate to those of global sustainability.

8.4.1.1 Determinants
8.4.1.1.1 Determinant 1—A_0: ecology
In the LBM for products and systems,"Ecology" is one of the environmental determinants. It includes the use of natural resources, local, and global environmental damages. Pollution of air, soil, and water, emissions of greenhouse gases, reduction of biodiversity,

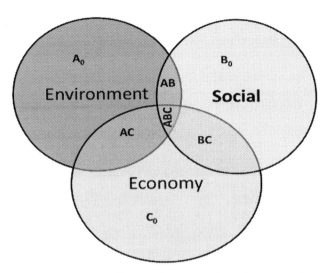

Fig. 8.3 Sustainability concept for products and systems. A_0 ecology: use of natural resources; local ecological damages, global ecological damages. AB environmental-social: social impact of environmental regulations; consumers awareness of greenness. AC environmental-economy: economic impact of environmental regulations; economic opportunities related to environmental regulations. B_0 social wellbeing: social appropriateness; social integration. BC social economy: working conditions; employment, security of supply. C_0: economic profitability: competitiveness; supports from stakeholders. ABC: background relationship between the three pillars.

and destruction of the stratospheric ozone are the concerns that are considered in the model to select the general indicators related to this determinant. Support data for the specific indicators can be provided by the results of LCA.

1. Use of natural resources

Intensification of the use of natural resources in a product or a system value chain contributes to their depletion except in the case where the resources are renewable. Furthermore, this general indicator includes the soberness of natural resource consumption even renewable. The natural resources could be energy, water, land, minerals particularly rare earth. Possible specific indicators are (1) depletion of abiotic resources, (2) depletion of biotic resources. Support data could include the description of the flows and balance of energy, water, and materials uses, and damage factors.

2. Local ecological damages

The product may trigger local impacts in several countries along its value chain and its life cycle. The assessment of the local ecological damages considers also the exposure of organisms to the concentration of pollutants. Specific indicators are the following: impacts on the quality of (1) local air, (2) local water, and (3) local soil. Other possible specific indicators are damages to the local ecosystem and to the landscape.

3. Global ecological damages

The global ecological damages are of planetary scale. The carbon footprint of the product or the system is used as indicator of the climate change. The contribution of the product value chain to the loss of global biodiversity and to the depletion of the stratospheric ozone is considered as well.

8.4.1.1.2 Determinant 2—AB environmental social

This second determinant includes the environmental dimensions of the social pillar or the social dimensions of environmental pillar depending on the perspective. At a product or a system-level, one of the concerns is how the impacts of the environmental regulation are socially distributed through the value chain of the product or the system. Another consideration is about the participation of the peoples to the environmental assessment of the facilities related to products or systems.

1. Social distribution of environmental quality

Depending on their spatial location, people may be impacted by environmental burdens differently. The model emphasizes the cases of three types of harm, i.e., the air pollution that is the main cause of respiratory diseases in urban areas, the noise, and industrial waste. Thus, the specific indicators are the following: (1) social distribution of air pollution, (2) social distribution of noise, (3) proximity to industrial waste disposals.

2. Social distribution of human health due to end-use of products

The end-use of products can affect favorably the health or be the direct or indirect cause of disease. For instance, excess use of sugar and salt can cause chronic diseases and the use of some chemicals can be harmful to the health. Since the consumption or end-use of these products are not necessary socially homogeneous, the issue of how its consequences are distributed within the social categories plays an important role in the environmental justice. The following specific indicators are considered in the LBM: (1) general impact on human health, (2) impact of the health of vulnerable people.

3. Social impact of environmental policy cost

Environmental incentives or disincentives through the price and access to products or services induce social dimensions. Their impact on the purchasing power is emphasized in the model with a differentiation between the overall social that measures the impact on the private consumption and the particular impact on the purchasing power of the vulnerable peoples.

(1) General financial impact of environmental regulation on the purchasing power, (2) impact on the purchasing power of vulnerable peoples.

4. Social participation to the environmental assessment

The participation of the impacted parties to the environmental impact assessment is mandatory. However, beyond the compliance with the law, companies should take the steps that will contribute to its social integration, which includes social responsibility.

That leads to the two specific indicators: (1) participation before and during the industry establishment, (2) social responsibility.

8.4.1.1.3 Determinant 3—AC environmental economy

All environmental policy encourages efficient use of natural resources and enacts incentives and penalties. The appropriateness of the product to the principles of the environmental policy matters as well as its ability to grasp the potential opportunities created by that policy.

1. Appropriateness of the product to the environmental policy

The two specific indicators comprise the relative environmental efficiency (eco-efficiency) as well as the effectiveness that is the absolute efficiency (degree of presence of rebound effect). Compared to the specific indicators of the first general indicator of determinant 1 (use of natural resources) the ones considered for this general indicator relate the environmental performance of the products or system with their market volume and assess the coupling or decoupling (lack of rebound effect).

2. Environmental value captured and uncaptured

Yang et al. (2017) defined the concept of value captured and uncaptured as, respectively, the benefit delivered to the organization and its stakeholders including tangible and nontangible [11]. The latter benefit can be for instance clean production beyond the requirements of the regulation. The value uncaptured is the potential value like the nonutilization of an overcapacity, potential cost reduction, and potential resources savings due to recycling or reuse. The two specific indicators are the following: (1) environmental value captured, (2) environmental value uncaptured.

8.4.1.1.4 Determinant 4—B0 social wellbeing

The social wellbeing is related to how peoples value the product or the system. Wellbeing is a general concept that means the good quality of the life and is often considered as a synonym of happiness. In that sense, the term can encompass most of other indicators that contribute to the health, the equity, the integration, the self-esteem, and human dignity. The term is sometimes separated into economic, social, and environmental wellbeing that points out the source of improvement of the quality of life. With regard to the characteristics of a product or a system and its sustainability, two general indicators are selected in the model with respect to the two following questions: does the product meet the needs of users or does it create new and futile needs that can divert resources for more important needs? Does the product or the system contribute to social integration, particularly to the poor and vulnerable peoples?

1. Appropriateness to general social needs

The issue of appropriateness was much discussed in the case of some technology transfers to the developing countries that proceed more from mimetic mode. However, it could be extended to the promotion of consumerism by aggressive advertisement. The two specific indicators are in different value orders. "Consumerism-prone" is considered as negative while obviously "good health driver" is positive.

2. Social integration

This general indicator highlights the contribution of the product or the system to social equality. The distribution system of a product, its marketing, its price can contribute to social discrimination. The vulnerable peoples are prioritized throughout the two following selected specific indicators: (1) appropriateness to the needs of vulnerable peoples, (2) social accessibility.

8.4.1.1.5 Determinant 5—BC social economy

The social value of the economy is to some extent ambivalent since it can be understood as in the utilitarian concept of social utility that proceeds from the aggregation of individual utilities. That definition is an intraeconomical one. The definition adopted here is from sociology. Corporates are social organizations before being profit generation prone. Their governance and more particularly the working conditions matter. The creation or upholding of qualified jobs and their leadership for promoting gender equality, diversity, and security of supply of local areas are criteria that are taken into account in the two general indicators.

1. Corporate governance and working conditions

Internal corporation governance is of concern in the model. That is a collection of procedures and social relations in the organization of the company. They assure the wellbeing of the workers while the company seeks to achieve its strategic and operational objectives. Accountancy and transparency, integration of the human resources of the company in the decision processes, promotion of a collective identity are few qualities waited from a good corporate governance. The three following specific indicators are considered in the model: (1) safety of the workers, (2) social protection, and (3) participation of the workers.

2. Employment

Besides its role of income redistribution, employment has a social value. The loss of job and a long period of jobless is a source of poverty, reduction of social interactions, and decrease of self-esteem of the unemployed peoples. Employment is a trigger of social inclusion. Hence, the creation of jobs in an area generally contributes to the enrichment of its social life depending to the quality of the jobs. Qualified jobs contribute to the social wellbeing of the workers contrary to unqualified and painful works. Finally, even exceeding the scope of employment, gender equality in the enterprises throughout the salaries, the gender balance in the internal governance is most important for social equity.

3. Security of supply

The pandemic disease of Covid 19 experienced in 2020 and 2021 highlighted the vulnerability of a world organized predominantly through the globalization of the economy. During this crisis, industrialized countries lacked respirators and masks and had to urgently resort to massive imports of medical masks from China and India, whereas they have the industrial capacity to produce them. This shows the need for a better balance between mass products and services intended for the international market and those belonging to short circuits. The specific indicators related to the security of supply are (1) local share of the supply chain, (2) contribution to supply local consumers. Depending on the context of study, "local" may address different levels such as group of countries, e.g., European Union, country, or part of a country. Security of supply involves additional costs that could be challenged with the social costs of extreme dependence to a limited number of long-distance supply chains.

8.4.1.1.6 Determinant 6—C_0 economic profitability

Economic profitability is economic return in relation with investment. It is generally considered as the major economic indicator of a company. Its drivers are the production cost and the market price. It influences the market development. In global market-oriented economy profitability is one of the most important objectives and competiveness the most important driver of the development of companies. Adversely the small and medium-sized enterprises may be less competitive and would not be economically sustainable without the supports of the stakeholders.

1. Competitiveness

The market share of a company is an indicator of its competiveness even though a very high market share would involve a reduction of the competition on the market. Thus, the term "competitiveness" in the model only means that the product can be traded without financial supports from public authorities. The two indicators used in the model are (1) production cost compared to market price, (2) market development.

2. Supports from the stakeholders

This general indicator particularly addresses the local companies working on short-circuit where the production costs are higher compared to others and for which the support of the stakeholders may be indispensable for a more or less long period. Two kinds of stakeholders were considered, namely, the public authorities, and the consumers. (1) Support from public authorities, (2) consumer's willingness to pay.

Improving the information provided to the stakeholders increases their awareness of the sustainability performance of the products once these performances are enhanced by appropriate measures.

Table 8.1 summarizes all the 59 variables.

Table 8.1 Determinants, general Indicators, and specific indicators.

6 Determinants	16 general indicators	37 specific indicators
D1: ecology	G1.1: use of natural resources	S1.1-1: depletion of abiotic resources
		S1.1-2; depletion of biotic resources
	G1.2: Local ecological damages	S1.2-1: local air quality
		S1.2-2: local water quality
		S1.2-3: local soil quality
	G1.3 Global ecological damages	S1.3-1: climate change
		S1.3-2: stratospheric ozone
		S1.3-3: biodiversity
D2: environmental social	G2.1: Social distribution of environmental quality	S2.1-1: social distribution of air pollution
		S2.1-2: social distribution of noise
		S2.1-3: proximity to industrial waste disposals
	G2.2: Social distribution of human health due to consumption	S2.2-1: general impact on human health
		S2.2-2: impact on the health of the vulnerable peoples
	G2.3: Social impact of environmental policy cost	S2.3-1: general financial impact of environmental regulation on the purchasing power
		S2.3-2: impact of environmental regulation on the purchasing power of the vulnerable peoples
	G2.4: Social participation to the environmental impact assessment	S2.4-1: participation before and during the industry establishment
		S2.4-2: social responsibility
D3: environmental economy	G3.1: Appropriateness of the product to the environmental policy	S3.1-1: eco-efficiency of the product
		S3.1-2: eco-effectitiveness
	G3.2: Environmental value captured and uncaptured	S3.2-1: environmental value captured
		S3.2-2: environmental value un-captured
D4: social wellbeing	G4.1: Appropriateness to general social needs	S4.1-1: consumerism—prone
		S4.1-2: good health driver
	G4.2: Social integration	S4.2-1: appropriateness to the needs of vulnerable peoples
		S4.2-2: social accessibility
D5: social economy	G5.1: corporate governance and working conditions	S5.1-1: safety
		S5.1-2: social protection
		S5.1-3: participation of the workers
	G5.2: Employment	S5.2-1: jobs creation
		S5.2-2: impact on employment structure
		S5.2-3: gender equality
	G5.3: security of supply	S5.3-1: local share of the supply chain
		S5.3-2: contribution to supply local consumers
D6: economic profitability	G6.1: competitiveness	S6.1-1: production cost compared to market price
		S6.1-2: market development
	G6.2: supports from the stakeholders	S6.2-1: support from public authorities
		S6.2-2: consumer's willingness to pay

8.4.1.2 Rule base

The rule base of this version of general LBM comprises 4759 rules in its basic state. In total, 231 rules define the general indicators by combinations of the specific indicators' values. In total, 432 rules define the determinants in function of the general indicators and 4096 rules are used for associating the determinants' values to the scores of the sustainability. This high number of rules for the final stage in the definition of sustainability is due to the number of determinants (6).

8.4.2 Alternative paradigm of sustainability

Fig. 8.4 illustrates an alternative paradigm centered on the environment that structures the second general LBM model of sustainability assessment of green products. With this pareadigm "Economy" is a subset of "Social" that is at its turn a subset of "Environment." Hence, the whole sustainability is environmental sustainability. "Environment" is composed of "Social" and "Ecology." "Social" is composed of "Economy" and "Sober wellbeing" which means a wellbeing restrained by the balance between "hedonism" and "Ecology." Two aspects of "Ecology" are considered: the local and the global scales. This paradigm is alternative because "Economy" and "Sober wellbeing" restrain each other; the same for "Ecology" and "Social."

The structure of the second general model is summarized in Table 8.2. There are 71 variables composed of two determinants D1—ecology and D2—social, three levels of general Indicators amounted to 30 variables and 39 specific indicators.

The rule base is composed of 778 rules. In total, 234 rules define the general indicators, 496 rules link the general indicators from one level to the other, 32 rules and 16 rules evaluate, respectively, the determinants and sustainability.

In order to illustrate the difference between the two models, three cases are compared in Table 8.3. In this comparison, it is worth noting that the term "Social" has different meanings. In Model 1, "Social" involves the determinants D2, D4, and D5 whereas "Economy" involves the determinants D3 and D6. In Model 2, "Economy" is

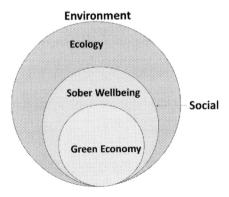

Fig. 8.4 Environment-centered sustainability.

General logic-based method for assessing the greenness of products and systems

Table 8.2 Determinants, general indicators, and specific indicators in the alternative paradigm.

(2) Determinants	(4) General indicators I	(9) General indicators II	(17) General indicators III	(39) Specific indicators
D1: ecology	E1.1: local ecosystem	F1.1: pressure on local natural resources	G1.1.1.1: use of local natural resources	S1.1.1.1-1: depletion of local abiotic resources S1.1.1.1-2: depletion of local biotic resources
		F1.1.2: local ecology quality	G1.1.2.1: local ecological damages	S1.1.2.1-1: local air quality S1.1.2.1-2: local water quality S1.1.2.1-3: local soil quality
	E1.2: global ecological issues	F1.2.1: pressure on global natural resources	G1.2.1.1: use of global natural resources	S1.2.1.1-1: depletion of global abiotic resources S1.2.1.1-2: depletion of global biotic resources
		F1.2.2: global ecology quality	G1.2.2.1 global ecological damages	S1.2.2.1-1: climate change S1.2.2.1-2: stratospheric ozone S1.2.2.1-3: biodiversity
D2: social	E2.1: sober wellbeing	F2.1.1: environmental social	G2.1.1.1: social distribution of environmental quality	S2.1.11-1: social distribution of air pollution S2.1.11-2: social distribution of noise S2.1.11-3: proximity to industrial waste disposals
			G2.1.1.2: social distribution of human health due to consumption	S2.1.1.2-1: general impact on human health S2.1.1.2-2: impact on the health of the vulnerable peoples
			G2.1.1.3: social impact of environmental policy cost	S2.1.1.3-1: general financial impact of environmental regulation on the purchasing power S21.1.3-2: impact of environmental regulation on the purchasing power of the vulnerable peoples

(*continued*)

Table 8.2 (Cont'd)

(2) Determinants indicators I	(4) General indicators I	(9) General indicators II	(17) General indicators III	(39) Specific indicators
		F2.1.2: social wellbeing	G2.1.1.4: social participation to the environmental impact assessment	S2.1.1.4-1: participation before and during the industry establishment S2.1.1.4-2: social responsibility
			G2.1.2.1: appropriateness to general social needs	S2.1.2.1-1: consumerism—prone S2.1.2.1-2: good health driver
			G2.1.2.2: social integration	S2.1.2.2-1: appropriateness to the needs of vulnerable peoples S2.1.2.2-2: social accessibility
		F2.1.3: social economy	G2.1.3.1: corporate Governance and Working conditions	S2.1.3.1-1: safety S2.1.3.1-2: social protection S2.1.3.1-3: participation of the workers
			G2.1.3.2: employment	S2.1.3.2-1: jobs creation S2.1.3.2-2: impact on employment structure S2.1.3.2-3: gender equality
			G2.1.3.3: security of supply	S2.1.3.3-1: local share of the supply chain S2.1.3.3-2: contribution to supply local consumers
E2.2: green economy		F2.2.1: environmental economy	G2.2.1.1: appropriateness of the product or the system to the environmental policy	S2.2.1.1-1: eco-efficiency of the product S2.2.1.1-2: eco-effectiveness
			G2.2.1.2: environmental value captured and uncaptured	S2.2.1.2-1: environmental value captured S2.2.1.2-2: environmental value un-captured
		F2.2.2: economic profitability	G2.2.2.1: competitiveness	S2.2.2.1-1: production cost compared to market price S2.2.2.1-2: market development
			G2.2.2.2: supports from the stakeholders	S2.2.2.2-1: support from public authorities S2.2.2.2-2: consumer's willingness to pay

Table 8.3 Illustration of the difference between the two models.

	Performances			Sustainability	
Cases	Ecology	Social	Economy	Model 1	Model 2
1	Moderate	High	High	High	Moderate
2	High	High	Moderate	Moderate–High	High
3	High	Moderate	High	Moderate–High	Moderate

included in "Social" that involves all determinants but D1. The comparison shows that contrary to Model 1, in Model 2, to get the sustainability high, both the performances of "Ecology" and "Social" must be high.

A new definition of "Green Product or System" derives from the results of this work that is:

A Green Product or System is the one that is granted with a high sustainability performance under the alternative paradigm of Sustainability where "Economy" is included into "Social" and "Social" into "Environment."

Several shades of green can be defined as well. As examples, besides the green itself, light green and blue-green could grant, respectively, the products and systems that grade high in the first paradigm but moderate in the second paradigm, or moderates in both paradigms.

8.5 Conclusions and perspectives

Sustainability is a qualitative concept and it is pointless to seek full quantitative models for assessing the global sustainability of a product or a system even though using quantitative tools to generate support data are of utmost importance. These tools are for example *Process design and Simulation, LCA, Environmental Impact Assessment, Techno-economic Assessment, Vulnerability Assessment,* and other *Decision Making Support Systems*. However, based on the support data, transparent rules must be designed in order to assign the combinations of the values of the indicators to the grades of Sustainability. The application of the LBM contributes to meet the need of consistent methodological frameworks for combining quantitative and qualitative data to assess the sustainability of products and systems.

References

[1] UNEP Green-Economy. https://www.unenvironment.org/explore-topics/green-economy/about-green-economy, 2020. [Accessed March 28, 20].
[2] Partners for inclusive green economy. Principles, priorities and pathways for inclusive green economies: economic transformation to deliver the SDGs. https://www.greengrowthknowledge.org/initiatives/partners-inclusive-green-economy, 2020. [Accessed March 28, 20].
[3] L. Georgeson, M. Maslin, Estimating the scale of the US green economy within the global context. Palgrave Commun. 5 (2019) 1–12.
[4] C. Searcy, D. Elkhawas, Corporate sustainability ratings: an investigation into how corporations use the Dow Jones Sustainability Index. J. Clean. Prod. 35 (2012) 79–92.

[5] EC. Communication from the Commission to the European Parliament and the Council. Building the single market for green products facilitating better information on the environmental performance of products and organisations, European Commission, Brussels, 9.4.2013 COM/2013/0196 final (2013).
[6] M. Lehni, Eco-efficiency. Creating more value with less impact, WBCSD, Conches-Geneva, 2000.
[7] S. Schmidheiny, Changing Course: A Global Business Perspective on Development and the Environment. Business Council for Sustainable Development, MIT Press, Cambridge, MA, 1992.
[8] I. Callens, D. Tyteca, Towards indicators of sustainable development for firms: a productive efficiency perspective, Ecol. Econ. 28 (1999) 41–53.
[9] M. Fischer-Kowalski, M. Swilling, E.U. von Weizsäcker, Y. Ren, Y. Moriguchi, W. Crane, F. Krausmann, N. Eisenmenger, S. Giljum, P. Hennicke, P. Romero Lankao, A.S. Sewerin. Siriban Manalang, Decoupling natural resource use and environmental impacts from economic growth, A Report of the Working Group on Decoupling to the International Resource Panel, United Nation Programme Environment, Nairobi (2011).
[10] F. Figge, T. Hann, Sustainable value added—measuring corporate contributions to sustainability beyond eco-efficiency, Ecol. Econ. 48 (2004) 173–187.
[11] M. Yang, D. Vladimirova, S. Evans, Creating and capturing value through sustainability, Res.-Technol. Manag. 60 (3) (2017) 30–39.
[12] S. Sala, AK. Cerutti, R. Pant, Development of a Weighting Approach for the Environmental Footprint, Publications Office of the European Union, Luxembourg, 2018.
[13] E. Gnansounou A. knowledge-based method for simulating the long-term evolution of electricity consumption Technical Report EPFL-LASEN, Lausanne, (1997) (in French).
[14] E. Gnansounou, Assessing the sustainability of biofuels: A logic-based model, Energy 36 (2011) 2089–2096.
[15] E. Gnansounou, T. Nouatin, LAPSEL: A decision support system for the prospective of the electric power sector. Technical Report EPFL-LASEN, Lausanne, (1997) (in French).

CHAPTER NINE

A systems analysis of first- and second-generation ethanol in the United States

Ganti S. Murthy[a,b]
[a]Biological and Ecological Engineering, Oregon State University, Corvallis, OR, United States, [b]Biosciences and Biomedical Engineering, Indian Institute of Technology, Indore, India

9.1 Introduction

Biofuels have a long history in the United States. First ethanol plants were built as an addition to the already established corn wet milling industry. With the 1973 oil crisis, the ethanol from corn received an impetus and the dry milling technology for corn ethanol production started in the early 1980s. Initially, ethanol was used as a gasoline oxygenate to replace the carcinogenic fuel additive Methyl Terbutyl Ether. After 2000, the ethanol production from corn witnessed a steep increase from 3% of the >11% in 2011. Currently, United States produces about 59.8 billion L of ethanol from corn and most of it produced using the dry-grind ethanol technology. Currently, there are no commercial scale second-generation biofuels in operation in the United States.

There are two processes used to produce ethanol from corn. Most of the ethanol production from corn in the United States uses dry grind ethanol process in newer plants while older plants mostly use corn wet milling process.

9.1.1 Dry milling corn ethanol technology

Corn dry milling technology consists of five key steps: size reduction, liquefaction, simultaneous saccharification and fermentation, distillation, and coproduct processing (Fig. 9.1).

- Corn receiving and size reduction: incoming corn is weighed, its moisture content is determined, and is checked for the presence of mycotoxin-producing fungi. The particle size of the corn is reduced using hammer mills. Typically, a 2 mm sieve is used in the hammer mill.
- Jet cooking and liquefaction: the slurry of cornflour, process water, and fresh makeup water is heated rapidly by injecting steam directly into the process flow. The rapid rise of temperatures and shear forces completely gelatinize the starch. In the liquefaction tank, the starch is partially hydrolyzed by the action of α-amylase. Typical process conditions in the liquefaction tank are 90°C, 6.5 pH, and 90 min of residence time. In

Biomass, Biofuels, Biochemicals
DOI: https://doi.org/10.1016/B978-0-12-819242-9.00012-9

© 2022 Elsevier BV.
All rights reserved. 147

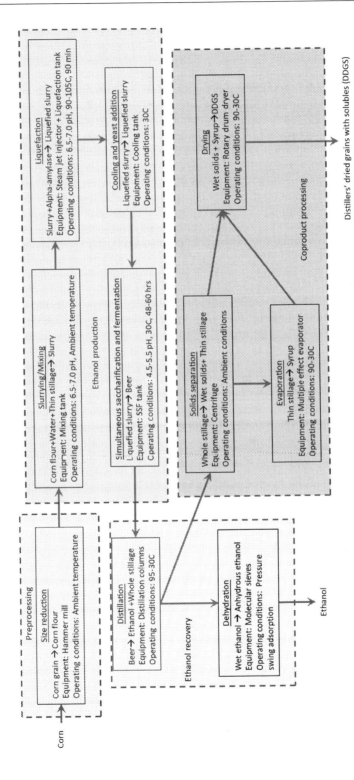

Fig. 9.1 Corn dry grind ethanol process.

most industrial applications, the dosage of alpha-amylase is split into before jet cooking (10 % of α-amylase dose) and after jet cooking (90 % of α-amylase dose). The dextrose equivalent of the liquified mash is between 12 and 22.

- Simultaneous saccharification and fermentation: liquified mash after adjustment of pH to ~4.5–5.0 is further hydrolyzed by the action of glucoamylase. Overall liquefaction and saccharification efficiencies are in the range of 70%–90%. Glucose thus produced is utilized by the added yeast to produce ethanol with a typical fermentation efficiency of 95%. Since the saccharification and fermentation proceed simultaneously, the glucose concentrations do not increase significantly thus reducing opportunities for bacterial contamination. The SSF process typically is conducted at 30–33°C, 4.5–5.0 pH, and 48–60 h of residence time.
- Ethanol recovery: first step of ethanol recovery consists of three distillation columns: beer boiler, rectification, and stripper columns. In these distillation steps, beer (fermented corn mash) is distilled to recover a concentrated ethanol stream (~94%–95% ethanol). Due to the formation of ethanol–water azeotrope a second step consisting of molecular sieves is used to recover anhydrous ethanol.
- Coproduct processing: corn mash after ethanol removal is called whole stillage. The whole stillage is centrifuged to separate into thin stillage and wet grains. The thin stillage is concentrated in multiple effect evaporators to obtain a syrup. The syrup is mixed with the wet graind and dried in a drum drier to obtain distillers dried grains with solubles (DDGS). In some cases, the syrup is directly sold as a cattle feed additive to minimize drier loads and the dried wet grains are knows as distillers dried grains (DGS).
- Water and ethanol form a positive azeotrope that precludes the use of a simple distillation for complete separation of ethanol in corn dry-grind ethanol process. Molecular sieves are used in a pressure swing process to separate the anhydrous ethanol from ethanol–water mixture (95% ethanol) after two distillation steps. The commercial molecular sieve used in ethanol dehydration has pores of 3Å size (water molecule 2.8Å, ethanol 4.4Å). It is made by partial substitution of sodium ions with potassium ions in zeolite.
- On average one bushel (25.4 kg) of corn produces 10.2 L (2.7 gal) of ethanol, 8.2 kg of CO_2, and 8.2 kg of DDGS. DDGS has high fiber content (>10%) and is therefore used mostly in cattle rations and its use in swine and chicken rations is limited. Technologies such as dry degerm defiber, quick germ quick fiber, enzymatic milling are used to modify the conventional dry grind corn process to recover germ, pericarp fiber, and endosperm fiber from the front end of the dry-grind ethanol plants.

Corn arrives in railroad cars and trucks and is stored onsite in large grain silos for up to 3 weeks. The corn is cleaned and ground to flour in a hammer mill in the first step of the process. In the second step, the cornflour is mixed with fresh water and some amount of recycle streams to form a slurry which is heated by direct injection of steam to 90–95°C for 90 min. Alpha-amylase enzyme is added to the slurry to break down the long chains of glucose into shorter fragments in pseudo-random attacks on the α (1→4) bonds to hydrolyze the amylose and amylopectin.

The slurry viscosity rapidly decreases during the process and enables easy pumping in the plant. The liquefied slurry with a DP of ~10–12 is cooled and the pH is adjusted to 4.5–5.0 and sent to the SSF process. In the SSF process, active dry yeast, nitrogen source for yeast (generally urea), and glucoamylase enzyme is added. The pH and temperature are maintained between 4.5 and 5.5 and 27–33°C, respectively, during the SSF process which occurs for between 48–60 h. The glucoamylase breaks the oligosaccharides into individual glucose molecules that are then fermented by the yeast into ethanol. Typically, the ethanol concentration at the end of the SSF process is between 13% and 18% v/v. The fermented stream called beer is distilled in three steps to recover >99.5% ethanol. The remaining fraction called whole stillage is centrifuges and separated into a solids fraction (wet solids) and a liquid fraction (thin stillage). Thin stillage is evaporated to concentrate it into syrup which is optionally mixed back with the wet solids and dried in a rotary drum drier to produce distiller's dried grains with solubles (DDGS).

The DDGS is used as cattle feed. Roughly, the dry grind process produces ethanol, DDGs, and carbon dioxide in equal amount by mass. The actual dry grind ethanol process yields 0.42 L of ethanol and 0.29 kg of DDGS per kilogram of corn processed. The water requirements for the process are about 2.5–4.0 L water/L ethanol during the dry-grind corn processing.

9.1.2 Wet milling technology

The primary product of corn wet milling technology is the production of high-quality starch. Corn gluten meal, corn gluten feed, corn germ (corn oil and corn oil cake) are produced as coproducts of the process. This is a very established and old technology that has been used since the beginning of the 20th Century. The process is very complex and can be divided into five sections: corn steeping, germ recovery, fiber recovery, protein recovery, and starch washing.

9.1.3 Second-generation ethanol

The primary difference in the first and the second generation ethanol production technologies is in the feedstocks. The first generation bioethanol processes use feedstocks that are used as human food such as corn, wheat, tapioca, broken rice and potatoes, and other starch-rich feedstocks. Second-generation ethanol refers to ethanol produced from non-food feedstocks such as agricultural residues, dedicated energy crops, waste biomass, and municipal solid wastes. The cellulosic ethanol process generally consists of feedstock preparation, pretreatment, hydrolysis, fermentation, ethanol recovery, and downstream processing steps (Fig. 9.2). A brief description of the important steps is provided below.

- Cleaning and size reduction: biomass is cleaned of debris such as sticks, rocks, soil, and any metal objects. Knife mills are used to reduce the particle size of the incoming

A systems analysis of first- and second-generation ethanol in the United States 151

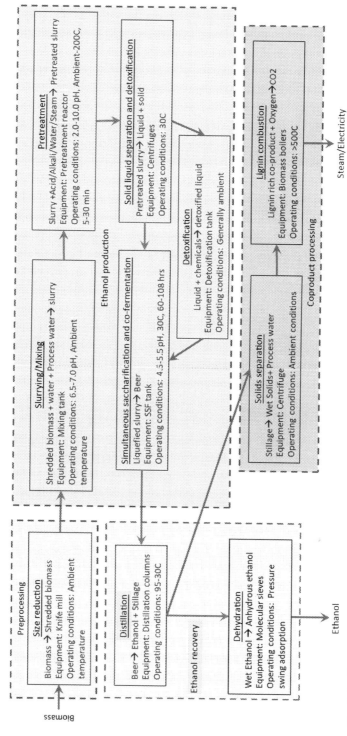

Fig. 9.2 Generic cellulosic ethanol process.

biomass. Knife mills are preferred for size reduction since cellulosic biomass is fibrous in nature.
- Pretreatment: primary goal of any pretreatment process is to facilitate enzyme action by making the substrate available. Pretreatment processes are critically dependent on type of feedstock. Ideal pretreatment removes the barriers to effective hydrolysis, preserves sugars, minimizes inhibitor formation, is independent of the feedstock particle size, minimizes energy and resource use, is cheap and safe. There are many types of pretreatment processes classified into physio-chemical (steam explosion, liquid hot water, ammonia fiber expansion AFEX, ionic liquids, etc.) and chemical pretreatment (dilute acid, dilute alkali, organosolv, etc.) methods. Of these dilute acid (using 0.5%–3% H_2SO_4), liquid hot water, and steam explosion are the most commonly used pretreatment methods. In general, the most pretreatments are conducted in the temperature range of 160–200°C with a residence time between 10 and 30 min. Alkali-based pretreatment methods such as AFEX process have lower temperature (90°C) requirements. Pretreatment methods using lime with aeration is different from above-mentioned methods as it is conducted at near ambient temperatures (25–60°C) and pressures (1 bar) with long residence times (2 weeks–2 months). The severity index is an indicator of the combined effect of temperature (T,°C), reaction time (t, min). This is specific for a particular feedstock and specific pretreatment method.

$$CS = \log_{10}\left(te^{((T-100)/14.75)}\right)$$

A modification to the above definition of the severity factor to incorporate the effects of pH is called combined severity factor and is defined as follows:

$$mCS = \log_{10}(te^{((T-100)/14.75)}) - |pH - 7|$$

Increasing severity factors beyond the value for maximum sugar release results in the formation of inhibitors. More than 70 different cell growth inhibitors have been detected in pretreated hydrolyzate. Furfural and hydroxymethyl furfural (HMF) formed from the degradation of xylose and glucose respectively, lignin degradation products such as cinnamaldehyde, p-hydroxybenzaldehyde, and syringaladehyde and hemicellulose degradation products such acetic, formic, glucuronic, and galacturonic acids are important inhibitors. The inhibitors are classified into aliphatic acids, furan compounds, and aromatic aldehydes. The action of inhibitors is primarily through chemical interference with cell maintenance functions, inhibition of ethanol production pathways, and osmotic effects on cells. Suppression/removal of inhibitor effects can be accomplished by detoxification or bioreduction strategies. Evaporation, extraction with organic solvents, ion exchange resins, adsorption on activated charcoal, alkaline detoxification (overliming), enzyme detoxification are some of the detoxification strategies. Bioreduction strategies include strain selection, long-/short-term adaptation, targeted genetic modification, and improved process design. Of these approaches, the overliming process, in which the pH of the pretreated hydrolyzate is increased to 10–12 pH and subsequently reduced to 5.5 pH using H_2SO_4 or sodium sulfite (more effective) is one of the best-known methods for detoxification although the exact detoxification mechanism is not completely understood. Microbial detoxification is focused mostly on the lignolytic enzymes such as laccases, Mn peroxidases, heme peroxidases, blue copper oxidases, and phenol oxidases produced by white rot, brown rot and soft rot fungi. Although the microbial detoxification methods are effective, they are not commercially practiced due to the unavailability of lignolytic enzymes at industrial scales.

- Enzymatic hydrolysis and fermentation: after pretreatment, the solids stream consisting mainly of cellulose and lignin is enzymatically hydrolyzed using a synergistically acting mixture of cellulase enzymes. There are two general schemes of fermentation followed in cellulosic ethanol processing. In the first scheme, called separate saccharification and fermentation, the glucose released by enzymatic hydrolysis is fermented separately from the pentose-rich stream obtained from the detoxified liquid hydrolyzate. This approach allows for optimization of pentose and hexose fermentations separately. In the second scheme, the pentose-rich stream is combined with simultaneous saccharification and fermentation of hexose sugars and is therefore called simultaneous saccharification and cofermentation process.
- Ethanol recovery technologies: the ethanol recovery technologies can be divided into end-of-pipe or slip stream alcohol recovery technologies. In the end-of-pipe technologies, exemplified by the distillation process, ethanol is recovered after the completion of fermentation while in the slipstream technologies, ethanol is simultaneously recovered from a slipstream during fermentation. Distillation, gas/steam stripping, liquid–liquid extraction, adsorption, and pervaporation are some of the technologies that are/can be used for ethanol recovery. Based on the energetics of distillation, >5 wt% ethanol concentration in the feeds stream is required to minimize the energy use in distillation. The values for other technologies varies from >2 wt% to >5 wt% based on the design and operating parameters.

9.2 Systems analysis of ethanol technologies

Biofuels have been proposed as a replacement for fossil fuels. In particular, the first/second-generation ethanol from biobased feedstocks has been proposed as alternatives to gasoline used in the transportation sector. It is important to determine the technical feasibility, economic viability, resource sustainability of these proposals from a systems perspective. The systems analysis approach is used to assess the feasibility and resource requirement of meeting the 100%, 50%, or 15 billion gallons of corn ethanol in the United States. The second case study uses the similar approach to assess the feasibility of producing 100%, 50% of the US gasoline needs or 35 billion gal. of second-generation ethanol.

9.2.1 Corn ethanol case study

Background: Corn dry grind ethanol process is currently used in the United States to produce 60 billion L/year (~15.8 billion gal/year) of ethanol per year accounting for 58% of the global production (https://ethanolrfa.org/resources/industry/statistics/). This process uses almost 35% of the United States corn production. Each bushel of corn (25.424 kg or 56 lbs) can produce 10.62 L or 2.8 gal of ethanol.

Goal and Scope:

The goal of this analysis is to:
1. Determine the levels of ethanol that can be produced without major disruption in the corn markets or expanding the area of cultivation significantly.
2. Understand the technical feasibility, economics, environmental impacts, use of resources, environmental risks, and policy support for corn ethanol.

The scope of the analysis will be limited to the operations between the farm to the burning of ethanol in an E10 mix in a regular sedan passenger vehicle. Emissions resulting from use in other types of vehicles or at higher blends will not be part of this analysis. The impact of weather, soil properties, management practices on soil emissions will not be considered nor will be modeled using detailed biogeochemistry models. Agricultural emissions will be estimated using the IPCC factors.

Data sources:
1. USDA National Agricultural Statistical Service.
2. Peer-reviewed journal papers
3. Industry outlook reports
4. US DOE reports

9.2.1.1 Technical feasibility analysis of corn ethanol in the United States

Problem Statement:
- What is the feasibility of producing corn ethanol to meet all the US gasoline needs?
- What is the feasibility of producing corn ethanol to meet 50% of US gasoline needs?
- What is the feasibility of producing 15 billion gallons of corn ethanol?

Data: Table 9.1

Assumptions:
- Based on the University of Georgia Extension publication, each cattle head can be fed up to 1.5 kg (3.3 lbs) of DDGS per day, i.e., each cattle require 0.548 tons DDGS/year [7].

Calculations:
1. Ethanol equal to 100% current gasoline needs = 541.735 billion L/year × 4.424 = 771.5 billion L/year.
2. Ethanol equal to 50% current gasoline needs = 385.75 billion L/year.
3. Total corn production, TCP = Average US corn yield × US corn crop area = 11.46 tons/ha × 32.708e6 ha = 374.8 million tons/year.

Table 9.1 Data sources for the technical feasibility analysis of US corn production.

Data	References
Average US corn yield: 11.46 tons/ha (181.3 bu/acre)	[1]
US corn, grain-harvested: 32.708 million ha (81.77 million acres)	[1]
US crop land: 152.26 million ha	[4]
Corn to ethanol conversion: 0.4273 L/Kg (2.87 gal/bu)	[5]
DDGS production: 0.2928 kg/kg corn or 0.685 kg/L ethanol	[5]
Gasoline: Ethanol energy density equivalency ratio = 1.424	[3]
US gasoline needs: 3.404 billion barrels = 142.98 billion gal = 541.735 billion L/year	[2]
Number of cattle in US: 92 million heads.	[6]
1 gal = 3.785 L	Unit conversions

4. Ethanol potential with current corn production = Average US corn yield × US corn crop area × Corn to ethanol conversion.
5. =11.46e3 kg/ha × 32.708e6 ha × 0.4273 L/kg =160.16 billion L/year (42.31 billion gal/year).
6. Max. Ethanol potential with corn production on all arable lands = Average US corn yield × US arable land area = 11.46e3 kg/ha × 152.26e6 ha × 0.4273 L/kg = 745.6 billion L /year (197 billion gal./year).
7. Area needed to produce ethanol to meet 100% current gasoline needs = Ethanol equal to 100% current gasoline needs /(Average US corn yield × Corn to ethanol conversion) =771.5 billion L/year /(11.46e3 kg/ha × 0.4273 L/kg) = 157.6 million ha.
8. Area needed to produce ethanol to meet 50% current gasoline needs = Ethanol equal to 100% current gasoline needs /(Average US corn yield × Corn to ethanol conversion) = 385.75 billion L/year /(11.46e3 kg/ha × 0.4273 L/kg)$ = million ha.
9. Area needed to produce 15 billion gal/year ethanol = (15 billion gal/year × 3.785 L/gal) /(11.46e3 kg/ha × 0.4273 L/kg) = 11.6 million ha.
10. Amount of DDGS produced for 15 billion gal ethanol = 15 billion gal/year × 3.785 L/gal × 0.685 kg/L ethanol = 38.89 million tons.
11. Number of cattle that can be fed the DDGS: 38.89 million tons/0.548 tons/head = 71 million cattle heads.

Results

- The total area needed to meet gasoline needs of the United States (157.6 million ha) is greater than the arable land available (152.26 million ha). The land area needed to meet 50% of the gasoline needs (78.78 million ha) is greater than the cropland dedicated to US corn production (32.708 million ha). Therefore, these two scenarios are unrealistic and infeasible.
- Area needed to produce 56.78 billion L/year (15 billion gal/year) ethanol is 11.6 million ha, which is about 35% of the US corn cropping area. Hence this is a feasible process. As an additional corroboration for this calculation, United States produced 59.80 billion L/year (15.8 billion gal/year) ethanol in 2017 as per the renewable Fuels Association [5].

- This process will produce 38.89 million tons of DDGS which can be fed to 71 million cattle heads in various stages of growth. As this represents 77% of the total cattle in the United States, there is a potential for market saturation and hence a need for diversification of the DDGS and coproducts markets.

It is infeasible to meet 100% or even 50% of the US gasoline needs using ethanol from corn. It is feasible to produce 56.78 billion L/year (15 billion gal/year) with existing land resources but the coproduct utilization needs to be looked carefully to avoid potential market saturation scenarios. The third scenario for producing 56.78 billion L/year will be explored in full detail below.

9.2.1.2 Techno-economic analysis of corn ethanol in the United States

Problem Statement: Calculate the Payback period, NPV, and internal rate of return (IRR) of a 151 million L/year (40 million gal/year) corn ethanol plant.

Data: The USDA model for corn dry-grind ethanol process [8] was used for conducting the techno-economic analysis.

Assumptions: The assumptions for the techno-economic analysis are summarized in Table 9.2. A composite factor 0f 4.0 was used to estimate the direct Fixed Capital costs for the plant. Adjusted Basic rate of 69.0 $/h for operators and a lumped rate of 52 $/h was assumed for all other labor requirements in the plant.

Calculations: The simulations were performed in SuperPro V 8.5 by Intelligen Inc.

Results: The plant uses about 367.1 million MT of corn/year as feedstock and produces 147.7 million L/year ethanol and 121,200 MT/year of DDGS (Table 9.3). The capital and operating costs for the 147.7 million L/year corn ethanol plant are $84.06 million and $89.07 million per year, respectively (Table 9.4). The return of investment (ROI) is 48.06 % while the payback period is 2.08 years. The IRR is 742% before taxes

Table 9.2 Project economic evaluation parameters.

Year of analysis	2020		NPV interest		
Construction start	2020		Low	7	%
Construction period	30	Months	Medium	9	%
Startup period	4	Months	High	11	%
Project life time	20	Years	Depreciation	10	Years
Inflation	4	%	Salvage Value	5	%

Table 9.3 Stream flow analysis.

Material	kg/year	Material	kg/year
Corn	367,097,738.00		
Lime	438,190.00	Caustic	18,423,742.00
Liq. Ammonia	733,337.00	Air	275,268,861.00
Alpha-Amylase	257,139.00	Yeast	96,466.00
Glucoamylase	371,408.00	Water	159,930,861.00
Sulfuric Acid	733,337.00	Octane	2,383,429.00

Table 9.4 Economic analysis summary.

Profitability analysis (2020 prices)

A.	Direct fixed capital	84,060,000.00	$
B.	Working capital	0.00	$
C.	Startup cost	0.00	$
D.	Up-front R&D	0.00	$
E.	Up-front royalties	0.00	$
F.	Total investment (A+B+C+D+E)	84,060,000.00	$
G.	Investment charged to this project	84,060,000.00	$
H.	Revenue rates		
	DDGS (Revenue)	121,203,953.00	kg /yr
	ETHANOL (main revenue)	147,712,701.00	L(STP) /yr
I.	Revenue price		
	DDGS (revenue)	0.10	$/kg
	ETHANOL (main revenue)	0.91	$/L(STP)
J.	Revenues		
	DDGS (revenue)	12,649,000.00	$/yr
	ETHANOL (main revenue)	134,419,000.00	$/yr
	Total revenues	147,068,000.00	$/yr
K.	Annual operating cost (AOC)		
	AOC	89,070,000.00	$/yr
L.	Unit production cost/revenue		
	Unit production cost	0.60	$/L(STP) MP
	Unit production revenue	1.00	$/L(STP) MP
M.	Gross profit (J–K)	57,997,000.00	$/yr
N.	Taxes (40%)	23,199,000.00	$/yr
O.	Net profit (M–N+depreciation)	40,400,000.00	$/yr
	Gross margin	39.44	%
	Return on investment	48.06	%
	Payback time	2.08	years

MP = Total flow of stream "ETHANOL"

and 416% after taxes. The NPV is $40.92 million ($I = 5.0\%$), $34.38 million ($i = 7.0\%$) and $29.23 million ($i = 9.0\%$) (Table 9.5). The plant has a payback period of 2.08 years.

Conclusion: The economic analysis of the dry-grind corn ethanol process as assessed by the positive NPV under various discount rate scenarios, high IRR compared to NPV, positive cash flows, and a short payback period of 2.08 years indicates that the technology is economically viable under various scenarios.

9.2.1.3 *Environmental impact assessment of corn ethanol in the United States*

The Environmental Impact Assessment was performed using GREET [9] which is a tool recommended by the US EPA for conducting life cycle emission calculations for transportation fuels in the United States. The standard GREET pathway for the US dry-grind corn ethanol (Fig. 9.3) consists of corn production, corn transportation, dry grind

Table 9.5 Cash flow analysis (thousand $).

Year	Capital investment	Debt finance	Sales revenues	Operating cost	Gross profit	Loan payments	Depreciation	Taxable income	Taxes	Net profit	Net cash flow
1	−25,218	20,174	0	0	0	0	0	0	0	0	−5044
2	−33,624	26,899	122,556	76,139	46,417	7386	5602	39,030	15,612	29,020	22,295
3	−25,218	0	147,068	89,070	57,997	7386	5602	50,611	20,244	35,968	10,750
4	0	0	147,068	89,070	57,997	7386	5602	50,611	20,244	35,968	35,968
5	0	0	147,068	89,070	57,997	7386	5602	50,611	20,244	35,968	35,968
6	0	0	147,068	89,070	57,997	7386	5602	50,611	20,244	35,968	35,968
7	0	0	147,068	89,070	57,997	7386	5602	50,611	20,244	35,968	35,968
8	0	0	147,068	89,070	57,997	7386	5602	50,611	20,244	35,968	35,968
9	0	0	147,068	89,070	57,997	7386	5602	50,611	20,244	35,968	35,968
10	0	0	147,068	89,070	57,997	7386	5602	50,611	20,244	35,968	35,968
11	0	0	147,068	89,070	57,997	7386	5602	50,611	20,244	35,968	35,968
12	0	0	147,068	89,070	57,997	0	5602	57,997	23,199	40,400	40,400
13	0	0	147,068	89,070	57,997	0	5602	57,997	23,199	40,400	40,400
14	0	0	147,068	89,070	57,997	0	5602	57,997	23,199	40,400	40,400
15	0	0	147,068	89,070	57,997	0	5602	57,997	23,199	40,400	40,400
16	0	0	147,068	89,070	57,997	0	5602	57,997	23,199	40,400	40,400
17	0	0	147,068	83,468	63,599	0	0	63,599	25,440	38,160	38,160
18	0	0	147,068	83,468	63,599	0	0	63,599	25,440	38,160	38,160
19	0	0	147,068	83,468	63,599	0	0	63,599	25,440	38,160	38,160
20	0	0	147,068	83,468	63,599	0	0	63,599	25,440	38,160	38,160
IRR before taxes	742.42%			Interest %				5.00	7.00		9.00
IRR after taxes	416.48%			NPV				409,267	343,833		292,338

IRR, internal rate of return.

A systems analysis of first- and second-generation ethanol in the United States 159

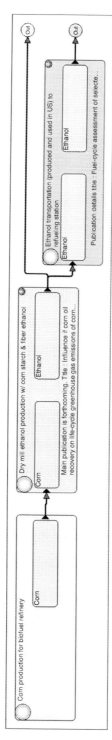

Fig. 9.3 Dry grind corn ethanol production pathway.

ethanol production, and ethanol transportation processes. Of these, the corn production and ethanol production processes contribute most to the environmental impacts.

Goal and scope: The goal of the analysis is to determine the life cycle emissions for a system that consists of corn production, corn transportation, dry grind ethanol production, and ethanol transportation processes. The scope of the analysis is the use of the results for assessing the life cycle impacts of corn to ethanol for determining the limited environmental impacts which are regulated by the US EPA for transportation fuels. The system boundary consists of all processes included in the GREET pathway for the corn production, corn transportation, ethanol production, and ethanol transportation and blending (Fig. 9.3). The functional unit for the analysis is 1000 MJ of corn ethanol at the pump.

Life Cycle Inventory and emissions: The life cycle inventory for two main processes in the dry-grind corn ethanol pathway is shown in Table 9.6 for 1000 MJ ethanol produced and made available at the pump. The category emissions for the inventory in Table 9.6 are shown in Table 9.7. The CO_2 emissions from the production processes

Table 9.6 Inventory for two main processes in the corn ethanol pathway.

	Corn production			Dry grind ethanol process	
Inputs	Quantity	Units	Inputs	Quantity	Units
Nitrogen	32.95	g	Corn	8.38992	Kg
Phosphoric Acid P_2O_5	139.29	g	Alphaamylase	2.33	g
Potassium Oxide	146.41	g	Gluco Amylase	5.05	g
Calcium Carbonate	1290.21	g	Water_process	11.367	L
Herbicides	5.85	g	Sulfuric Acid	17.17	g
Insecticides	12.49	mg	Ammonia	16.64	g
Corn	4.22674	Kg	Calcium Oxide	9.95	g
Irrigation water	553.194	L	Cellulase	3.34	g
Energy	7305.584	KJ	sodium hydroxide	20.87	g
Diesel for Non Road Applications	49.23%		Yeast	0.8	g
Gasoline Blendstock	14.70%				
Liquified Petroleumn Gas	17.93%		Energy Use	28,375.79	KJ
Natural Gas	13.55%		Natural Gas	88.96%	
Electricity	4.59%		Electricity	11.04%	
Outputs			Outputs		
Corn	25.424	Kg	Ethanol	3.789	L
NO_x	2.4	g	DGS	1.66618	Kg
N_2O	2.73	g	Corn Oil	0.1135	Kg
			VOC	2.24	g
			PM10	0.86	g
			PM2.5	0.15	g
			CO_2	-83.48	g
			CO_2_LandUseChange	600.74	g

Table 9.7 Emissions the corn ethanol pathway.

Total emissions		Unit	Urban emissions		Unit
CO_2 Total	41.52	kg	CO_2 Total	2.52	kg
CO_2	34.11	kg	CO_2	2.52	kg
CO_2_Biogenic	−0.06	kg	CO_2_Biogenic	−8.97E-04	kg
CO_2_LandUseChange	7.46	kg	CO_2_LandUseChange	0	kg
VOC	57.22	g	VOC	13.06	g
CO	41.71	g	CO	1.84	g
NO_x	92.23	g	NO_x	3.82	g
PM10	15.6	g	PM10	0.19	g
PM2.5	5.9	g	PM2.5	0.15	g
SO_x	64.33	g	SO_x	3.48	g
CH_4	70.1	g	CH_4	3.36	g
N_2O	33.85	g	N_2O	98.97	mg
NO_2	0	kg	NO_2	0	kg
SO_2	0	kg	SO_2	0	kg
H_2	0	kg	H_2	0	kg
BC	0.69	g	BC	14.81	mg
POC	1.14	g	POC	49.61	mg
Groups			Groups		
GHG-100	52.84	kg	GHG-100	2.69	kg
Biogenic carbon mass ratio	100	%			

account for over 82% of the total CO_2 emissions while the land-use change contributes about 18% of the emissions. The biogenic carbon sequestration is minimal in established agricultural systems and is reflected in the very small negative value of biogenic CO_2 emissions which indicates a small increase in the soil carbon. The localization of emissions is an important factor as the actual impacts of the emissions that are in less populated areas may not be as severe on human populations as in urban areas. Therefore, the urban emissions due to dry-grind corn ethanol are also summarized in Table 9.7. A comparison of the emissions especially the CO_2 for fossil fuels is shown in Table 9.8 that indicates a lower emission value for the fossil fuels overall (0.11 vs 0.09 kg/MJ of CO_2 for E100 and E10 respectively) but when the biogenic emissions are considered (-0.07 vs. 0.0 kg/MJ of CO_2 for E100 and E10 respectively) the overall emissions (0.04 vs. 0.09 kg/MJ of CO_2 for E100 and E10 respectively) are lower for the E100 compared to the E10 fuel pathway.

9.2.1.4 Resource use for corn ethanol in the United States

The land, water, and nutrients are important resources that could restrict the ability for large-scale implementation of biofuel technologies. Land and nutrient resource use discussed earlier in the technical feasibility section of this case study. Table 9.9

Table 9.8 Comparison of life cycle emissions (well to wheel) for the gasoline (E10) and ethanol (E100).

	Total emissions		Urban emissions	
Emission category	E100	E10	E100	E10
VOC (kg/MJ)	6.47E−05	6.20E−05	1.94E−05	3.89E−05
CO (kg/MJ)	2.21E−04	7.00E−04	1.29E−04	4.82E−04
NOx (kg/MJ)	1.01E−04	6.02E−05	9.21E−06	2.56E−05
PM10 (kg/MJ)	1.57E−05	4.27E−06	1.69E−07	1.78E−06
PM2.5 (kg/MJ)	5.79E−06	3.17E−06	1.36E−07	1.56E−06
SO_x (kg/MJ)	7.86E−05	1.35E−05	2.79E−06	2.59E−06
CH_4 (kg/MJ)	7.22E−05	1.03E−04	3.61E−06	4.34E−06
CO_2 (kg/MJ)	0.11	0.09	0.05	0.06
N_2O (kg/MJ)	3.24E−05	4.30E−06	4.49E−07	1.43E−06
NO_2 (kg/MJ)	0	0	0	0
SO_2 (kg/MJ)	0	0	0	0
BC (kg/MJ)	6.36E−07	5.66E−07	1.34E−08	2.75E−07
H_2 (kg/MJ)	0	0	0	0
POC (kg/MJ)	1.10E−06	1.07E−06	4.62E−08	5.47E−07
CO_2_Biogenic (kg/MJ)	−0.07	0	−0.05	0
CO_2_LandUseChange (kg/MJ)	0.01	4.92E−04	0	0
GHG-100 (kg/MJ)	0.05	0.09		
BC_TBW (kg/MJ)	2.21E−07	1.67E−07	1.54E−07	1.17E−07
POC_TBW (kg/MJ)	2.76E−07	2.08E−07	1.93E−07	1.46E−07
PM10_TBW (kg/MJ)	6.04E−06	4.56E−06	4.23E−06	3.19E−06
PM2.5_TBW (kg/MJ)	1.54E−06	1.17E−06	1.08E−06	8.16E−07
VOC_evap (kg/MJ)	1.40E−05	2.64E−05	9.80E−06	1.85E−05

shows the resource use for other resources with a particular focus on the water and energy resources use for the production of dry-grind corn ethanol. In particular, it is insightful to see that majority of the water is used in irrigation while the use of water during processing is minimal. However, it is important to recognize that the irrigation water can be supplied from rainwater also and is usually distributed over large geographic areas. On the other hand, the water used for processing plants is obtained for a single water source and thus may have outsized local impacts compared to the irrigation water use. Comparing the resource use for Gasoline (E10) with Ethanol (E100) pathway shows the while the liquid fossil fuel usage is lower for E100 compared to E10, the usage of other forms of fossil fuels such as Coal and natural gas are actually higher for the E100 pathway compared to the gasoline (E10) pathway. This demonstrates the tradeoffs (Table 9.10).

Table 9.9 Life cycle resource use (well to pump) for the corn ethanol pathway.

Resources	Quantity	Units	Groups	Quantity	Units
Water Total	1465.73	L	Non Fossil Fuel	1259	MJ
Crude Oil	40	MJ	Renewable	1248	MJ
Natural Gas	466	MJ	Biomass	1239	MJ
Coal Average	33	MJ	Fossil Fuel	551	MJ
Forest Residue	648.2	kJ	Natural Gas Fuel	466	MJ
Pet Coke	44.45	kJ	Petroleum Fuel	52	MJ
Corn	1381	MJ	Coal Fuel	33	MJ
Soybeans	−143	MJ	Nuclear	11	MJ
Renewable, Other	247.86	kJ	Renewable Natural Gas	41.24	kJ
Renewable Natural Gas	41.24	kJ			
Uranium Ore	0.1	g	Water	1465.73	L
Hydroelectric Power	3814	kJ	Irrigation	1240.37	L
Nuclear Energy	11	MJ	Process water	163.06	L
GeoThermal Power	217.38	kJ	Mining water	30.71	L
Solar	961.83	kJ	Reservoir Evaporation	17.68	L
Wind Power	3910	kJ	Cooling water	13.92	L
Sugarcane	7934.14	J			
Bitumen	3772	kJ			
Shale Oil (Bakken)	3823	kJ			
Shale Oil (Eagle Ford)	4281	kJ			

Table 9.10 Life cycle resource use for the gasoline (E10) and ethanol (E100).

	E100	E10
Total Energy (J/MJ)	1,755,720	1,266,089
Fossil Fuel (J/MJ)	557,910.7	1,180,140
Coal Fuel (J/MJ)	67,067.57	14,825.66
Natural Gas Fuel (J/MJ)	441,438.7	157,444.2
Petroleum Fuel (J/MJ)	49,404.46	1,007,870
Renewable (J/MJ)	1,189,335	82,214.83
Biomass (J/MJ)	1,182,005	79,031.22
Nuclear (J/MJ)	8473.88	3734.27
Non Fossil Fuel (J/MJ)	1,197,809	85,949.1
Water_Reservoir Evaporation (m^3/MJ)	1.40E−05	6.09E−06
Water_Irrigation (m^3/MJ)	0	7.27E−05
Water_Cooling (m^3/MJ)	1.10E−05	4.82E−06
Water_Mining (m^3/MJ)	2.98E−05	6.17E−05
Water_Process (m^3/MJ)	1.54E−04	3.22E−05

9.2.2 Cellulosic ethanol in the United States

Background: Cellulosic ethanol has been the focus of much research and development, government, and industry focus in the past two decades. Cellulosic ethanol is thought to be a lower impactful method for producing ethanol especially from agricultural residues that are already produced during the production of valuable food crops. Therefore agricultural residues such as corn stover, wheat straw, and barley straw among others are considered as potential feedstocks for producing second generation cellulosic ethanol. Among the agricultural residues, the corn stover is produced in the largest amount and is most suitable for any commercial exploitation due to the concentration of the production areas in the corn belt of Midwest United States with very good infrastructure in place for harvesting, collection, and transportation of agricultural commodities.

Goal and scope:

The goal of this analysis is to:

1. Determine the levels of ethanol that can be produced from corn stover and other agricultural residues without any increase in cultivated area.
2. Understand the technical feasibility, economics, environmental impacts, use of resources, environmental risks, and policy support for ethanol from agricultural residues with a special focus on corn stover.

The scope of the analysis will be limited to the operations between the farm to burning of ethanol in an E10 mix in a regular sedan passenger vehicle. Emissions resulting from use in other types of vehicles or at higher blends will not be part of this analysis. The impact of weather, soil properties, management practices on soil emissions will not be considered nor will be modeled using detailed biogeochemistry models. Agricultural emissions will be estimated using the IPCC factors.

9.2.2.1 Technical feasibility analysis of cellulosic ethanol in the United States

Problem Statement:
- How much ethanol can be produced from corn stover produced in the United States?
- What percentage of the US gasoline needs can this meet?

Data:
- Projected corn stover at <$80/ton is 112.2 million tons and 129 million tons of all agricultural residues (Table 4.7 in [10])
- Corn Stover Composition (dry basis) (Table 9.11).

Assumptions:
- Basis: 1 dry ton.
- Pretreatment and hydrolysis efficiency is 80%
- Inhibitors generation: 1% from hexoses and 2% from pentoses
- Fermentation efficiency: Hexose and pentose fermentation efficiency are 98 and 60%, respectively.
- Distillation efficiency is 99%.

Table 9.11 Composition (% dry basis) of corn stover (Table 4 in [11])

Glucan	Xylan	Lignin	Ash	Acetate	Protein
35.05	19.53	15.76	4.93	1.81	3.1
Extractives	Arabinan	Galactan	Mannan	Sucrose	
14.65	2.38	1.43	0.6	0.77	

PC, purchased cost of equipment; Installation = 0.5 × PC; DC, PC + Installation+A+B+C+D+E+F+G+UI+UEI; DFC, DC+IC+OC.

Calculations:
1. Pretreatment and hydrolysis:
   Hexoses = Biomass* Glucan* $\eta_{hydrolysis}$*Hydrolytic gain*(1-inhibitor production)
   Hexoses = 1000* 0.3505*0.8*1.11*(1 − 0.01) = 308.13 kg
   Inhibitors (HMF) = 1000* 0.3505*0.8*1.11*(0.01) = 3.08 kg
   Pentoses = Biomass* Xylans* $\eta_{hydrolysis}$*Hydrolytic gain*(1-inhibitor production)
   Pentoses = 1000* 0.1953*0.8*1.136*(1 − 0.02) = 173.94 kg
   Inhibitors (F) = 1000* 0.1953*0.8*1.136*(0.02) = 3.55 kg
2. Fermentation
   Ethanol = Hexoses * ethanol yield * hexose fermentation efficiency + Pentoses * ethanol yield * pentose fermentation efficiency
   Ethanol = 308.13*0.511*.98 + 173.94*0.511*0.6 = 207.63 Kg
3. Distillation:
   Ethanol (L) = Ethanol (Kg) * distillation efficiency / ethanol density (kg/L)
   = 207.63 *0.99 /0.789 = 260.5 L
   Overall Ethanol yield = 260.5 L/1000 kg = 0.2605 L/Kg
   Total inhibitors = 3.08 + 3.55 = 6.63 kg
   Residual lignin coproduct: 1000*0.4543 = 454.3 Kg
4. Total corn ethanol = Average US corn stover production × yield = 112.2 million MT × 260.5 L/MT = 29.23 billion L/year (7.71 billion gal/year).
5. Based on the total US gasoline needs of 541.735 billion L/year, the corn stover ethanol represents a 5.4 % of the total gasoline needs.
6. Assuming all the agricultural residues are used for cellulosic ethanol production and the yields per ton remain the same as for con stover, the total ethanol production from all agricultural residues will be 33.6 billion L/year represents a 6.2% of the total United States annual gasoline needs.

Results:
- The ethanol production from corn stover or the available agricultural residues represents only 5.54%–6.2% of the US gasoline needs. The primary advantage of this scenario is that the agricultural residues are already produced and may therefore not incur any *additional* environmental and resource burdens.

Table 9.12 Capital cost adjustments.

Direct cost (DC)		Indirect cost (IC)	
Piping (A)	0.35	Engineering	0.25
Instrumentation (B)	0.4	Construction	0.35
Insulation (C)	0.03	Other Cost (OC)	
Electrical Facilities (D)	0.1	Contractors Fees	0.05
Buildings (E)	0.15	Contingency	0.1
Yard Improvement (F)	0.15	Unlisted Equipment (UE)	0.05
Auxillary Facilities (G)	0.2	Unlisted equipment Installation (UEI)	0.5

9.2.2.2 Techno-economic analysis of cellulosic ethanol in the United States

Problem Statement: Calculate the Payback period, NPV, IRR, uncertainty, and sensitivity analysis of a million L/year (40 million gal/year) corn ethanol plant.

Data: The default corn stover ethanol process model in the Superpro v8.5 was used for conducting techno-economic analysis

Assumptions: The project evaluation parameters were the same as the case for dry grind corn ethanol process and are summarized in Table 9.2. The capital cost adjustments for are more complex for the cellulosic ethanol plant and hence are specified in more detail (Table 9.12) rather than a lumped composite factor. Adjusted Basic rate of 57.5$/h for all operators and a lumped rate of 20 $/hour was assumed for all other labor requirements in the plant.

Calculations: The simulations were performed in SuperPro V 8.5 by Intelligen Inc.

Results: The plant uses about 749,074 MT of corn/year as feedstock (Table 9.13) and produces 176 million L/year ethanol 52.15 million KWh/year electricity (Table 9.14). The capital and operating costs for the 176 million L/year corn ethanol plant are $150.55 million and $108.43 million per year, respectively (Table 9.14). The ROI is 13.24% while the payback period is 7.53 years. The internal rate of return (IRR) is 8.83% before taxes and 416% after taxes. The NPV is $20.23 million ($i$ = 7.0%), −$1646 ($i$ = 9.0%) and −$18,737 ($i$ = 11.0%) (Table 9.15).

Conclusion: The economic analysis of the cellulosic ethanol process as assessed by the NPV under various discount rate scenarios indicates that the process is marginally economically feasible. Especially under high discount rates (9% and 11%), the process has a

Table 9.13 Stream report.

Material	MT/yr
Air	2,450,987
Amm. Sulfate	651
HP Steam	237,600
Hydrolase	2123
RO Water	1,496,019
Stover	749,074
Water	1,334,542
TOTAL	6,270,995

Table 9.14 Economic analysis.

10. Profitability analysis (2020 prices)

			Units
A.	Direct Fixed Capital	138,836,000	$
B.	Working Capital	4,771,000	$
C.	Startup Cost	6,942,000	$
D.	Up-Front R&D	0	$
E.	Up-Front Royalties	0	$
F.	Total Investment (A+B+C+D+E)	150,549,000	$
G.	Investment Charged to This Project	150,549,000	$
H.	Revenue/Savings Rates		
	Ethanol (Main Revenue)	176,232,566	L(STP) UPRF/yr
	Electricity (Revenue)	52,147,231	kW-h/yr
I.	Revenue/Savings Price		
	Product (Main Revenue)	0.66	$/L(STP)
	P-29(Revenue)	0.05	$/kW-h
J.	Revenues/Savings		
	Product (Main Revenue)	116,279,100	$/yr
	P-29(Revenue)	2,607,362	$/yr
1	Total Revenues	118,886,461	$/yr
2	Total Savings	0	$/yr
K.	Annual Operating Cost (AOC)		
1	Actual AOC	108,428,000	$/yr
2	Net AOC (K1-J2)	108,428,000	$/yr
L.	Unit Production Cost /Revenue		
	Unit Production Cost	0.615	$/L(STP) UPRF
	Net Unit Production Cost	0.615	$/L(STP) UPRF
	Unit Production Revenue	0.675	$/L(STP) UPRF
M.	Gross Profit (J-K)	10,458,000	$/yr
N.	Taxes (35%)	3,660,000	$/yr
O.	Net Profit (M-N + Depreciation)	19,987,000	$/yr
	Gross Margin	8.80	%
	Return On Investment	13.28	%
	Payback Time	7.53	years

UPRF = total flow of stream "Product"

negative NPV indicating that the process may not be economically feasible if the investors perceive the process to be of high risk and demand a higher discount rates. Under the positive NPV scenario (discount rate = 7%), the payback period of 7.53 years is still higher than the corn ethanol process (Payback period of 2.08 years) indicates that the corn ethanol technology has higher economic viability compared to cellulosic ethanol process.

9.2.2.3 Environmental impact assessment of cellulosic ethanol in the United States

The Environmental Impact Assessment was performed using GREET [9] which is a tool recommended by the US EPA for conducting life cycle emission calculations for transportation fuels in the United States. The standard GREET pathway for the US corn stover to ethanol (Fig. 9.4) was used for analysis.

Table 9.15 Cash flow analysis (thousand $).

Year	Capital investment	Debt finance	Sales revenues	Operating cost	Gross profit	Loan payments	Depreciation	Taxable income	Taxes	Net profit	Net cash flow
1	−41,651	0	0	0	0	0	0	0	0	0	−41,651
2	−55,534	0	0	0	0	0	0	0	0	0	−55,534
3	−46,422	0	19,814	43,583	−23,769	0	13,189	0	0	−10,579	−57,002
4	0	0	118,886	108,428	10,458	0	13,189	10,458	0	23,648	23,648
5	0	0	118,886	108,428	10,458	0	13,189	10,458	3660	19,987	19,987
6	0	0	118,886	108,428	10,458	0	13,189	10,458	3660	19,987	19,987
7	0	0	118,886	108,428	10,458	0	13,189	10,458	3660	19,987	19,987
8	0	0	118,886	108,428	10,458	0	13,189	10,458	3660	19,987	19,987
9	0	0	118,886	108,428	10,458	0	13,189	10,458	3660	19,987	19,987
10	0	0	118,886	108,428	10,458	0	13,189	10,458	3660	19,987	19,987
11	0	0	118,886	108,428	10,458	0	13,189	10,458	3660	19,987	19,987
12	0	0	118,886	108,428	10,458	0	13,189	10,458	3660	19,987	19,987
13	0	0	118,886	95,239	23,648	0	0	23,648	8277	15,371	15,371
14	0	0	118,886	95,239	23,648	0	0	23,648	8277	15,371	15,371
15	0	0	118,886	95,239	23,648	0	0	23,648	8277	15,371	15,371
16	0	0	118,886	95,239	23,648	0	0	23,648	8277	15,371	15,371
17	0	0	118,886	95,239	23,648	0	0	23,648	8277	15,371	15,371
18	0	0	118,886	95,239	23,648	0	0	23,648	8277	15,371	15,371
19	0	0	118,886	95,239	23,648	0	0	23,648	8277	15,371	15,371
20	11,713	0	118,886	95,239	23,648	0	0	23,648	8277	15,371	27,084

IRR before taxes	11.95 %	
IRR after taxes	8.83 %	

IRR/NPV SUMMARY

Interest %	7.00	9.00	11.00
NPV	20,234.00	−1,646.00	−18,737.00

IRR, internal rate of return.

A systems analysis of first- and second-generation ethanol in the United States 169

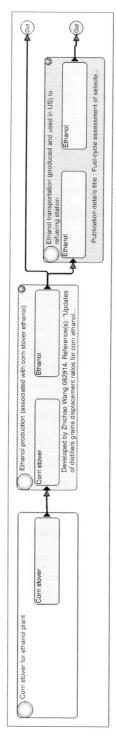

Fig. 9.4 GREET corn stover to ethanol production pathway system.

Goal and scope: The environmental impact assessment of the corn stover to ethanol process was conducted in the GREET. The goal of the analysis is to determine the six life cycle emissions for cellulosic ethanol production. The scope of the analysis is to use the results for assessing the life cycle impacts of corn stover to ethanol process for determining the limited environmental impacts that are regulated by the US EPA for transportation fuels. The system boundary consists of all processes included in the GREET pathway for the conversion of corn stover into ethanol has corn stover collection, transportation, ethanol production in a biorefinery, and ethanol transportation and blending (Fig. 9.4).

Life cycle Inventory and emissions: The inventory for the system described above is presented in Table 9.16 for the two main processes (corn stover collection and cellulosic ethanol production processes). The functional unit for the process pathway was 1000 MJ

Table 9.16 Inventory for main processes in the corn stover ethanol pathway (1000 MJ ethanol).

Corn stover collection			Ethanol production		
Inputs	Quantity	Units	Inputs	Quantity	Units
Nitrogen	3182.81	g	Alpha Amylase	0	g
Phosphoric Acid P_2O_5	2273.43	g	Gluco Amylase	0	g
Potassium Oxide	13.64	Kg	Yeast	26.58	g
Energy	235.9	MJ	Corn Stover	10.67	Kg
Diesel for Non Road Applications	1		Cellulase	106.73	g
Gasoline Blendstock	0		Sulfuric Acid	346.23	g
Liquified Petroleum Gas	0		Ammonia	41.55	g
Natural Gas	0		Corn Steep Liquor	131.57	g
Electricity	0		Diammonium Phosphate	13.85	g
Output			Sodium Hydroxide	117.72	g
Corn Stover	907.18	kg	Calcium Oxide	76.17	g
N_2O	−153.17	g	Urea	20.77	g
NO_x	−66.5	g	Water Process	10.22	L
			Energy	1898.95	KJ
			Natural Gas	100%	
			Electricity	0%	
			Outputs		
			Ethanol	3.789	L
			Electricity	331.004	KJ
			VOC	2.24	g
			PM10	0.86	g
			PM2.5	0.15	g
			CO_2	−14.02	g
			CO_2 Biogenic	−36.12	g

Table 9.17 Life cycle emissions for the corn stover ethanol pathway.

Emissions	Total emissions	Unit	Urban emissions	Unit
CO_2 Total	19.85	kg	1.73	kg
CO_2	19.89	kg	1.73	kg
CO_2_Biogenic	−0.04	kg	−4.64E-04	kg
VOC	60.48	g	12.94	g
CO	33.14	g	1.1	g
NO_x	56.68	g	2.96	g
PM10	15.68	g	0.17	g
PM2.5	6.11	g	0.14	g
SO_x	0.14	kg	1.97	g
CH_4	35.58	g	1.44	g
N_2O	0	kg	44.16	mg
NO_2	0	kg	0	kg
SO_2	0	kg	0	kg
H_2	0	kg	0	kg
BC	1.71	g	14.86	mg
POC	1.08	g	48.48	mg
Groups				
GHG-100	20.16	kg	1.83	kg
Biogenic carbon mass ratio	100	%		

of ethanol produced and made available at the pump. The category emissions for the inventory in Table 9.16 are shown in Table 9.17.

The CO_2 emissions from the production processes account for over 100% of the total CO_2 emissions while the biogenic carbon sequestration mitigates the impacts through negative emission (i.e., sequestration of carbon). The total CO_2 emissions for the process are about 48% lower than the total CO_2 emissions for the corn ethanol pathway indicating that the cellulosic ethanol pathways emit overall less CO_2 compared to the first-generation corn ethanol pathway.

The urban emissions due to dry-grind corn ethanol are also summarized in Table 9.17.

9.2.2.4 Resource use for cellulosic ethanol in the United States

Table 9.18 shows the resource use for other resources with a particular focus on the water and energy resources use for the production of corn stover ethanol. Just as it was for the corn ethanol pathway, the majority of the water is used in irrigation. It is interesting to note that the water use during processing is comparable to the corn ethanol process but the overall water use is only 35% of the water use in corn ethanol process.

Table 9.18 Life cycle resource use for the corn stover ethanol pathway.

Resources	Quantity	Unit	Groups	Quantity	Unit
Water Total	518.97	L	Non Fossil Fuel	2493	MJ
Crude Oil	72	MJ	Renewable	2487	MJ
Natural Gas	186	MJ	Biomass	2483	MJ
Coal Average	23	MJ	Fossil Fuel	303	MJ
Corn Stover	2269	MJ	Natural Gas Fuel	186	MJ
Forest Residue	393.05	kJ	Petroleum Fuel	94	MJ
Pet Coke	80.15	kJ	Coal Fuel	23	MJ
Corn	208	MJ	Nuclear	5533	kJ
Soybeans	4323	kJ	Renewable Natural Gas	1551.69	J
Renewable, Other	128.39	kJ			
Renewable Natural Gas	1551.69	J			
Uranium Ore	52.42	mg	Water Total	518.97	L
Hydroelectric Power	1975	kJ	Irrigation	306.1	L
Nuclear Energy	5533	kJ	Process	161.04	L
GeoThermal Power	112.6	kJ	Mining	35.47	L
Solar	498.21	kJ	Reservoir Evaporation	9.16	L
Wind Power	2025	kJ	Cooling	7.21	L
Sugarcane	262.63	kJ			
Bitumen	6802	kJ			
Shale Oil (Bakken)	6893	kJ			
Shale Oil (Eagle Ford)	7720	kJ			

9.3 Conclusions and perspectives

This chapter focused on two case studies for first-generation biofuels represented by the dry-grind corn ethanol process and the second-generation biofuels represented by the corn stover to ethanol process. The technical feasibility study indicated that both dry-grind corn ethanol and corn stover to ethanol processes cannot replace the US gasoline needs completely. Only a partial replacement of about 10% and 6% of the US gasoline needs even when using all the agricultural residues available in the United States. In general, it is apparent from the techno-economic and life cycle assessment that the second generation ethanol processes have much lower environmental impacts compared to the corn dry-grind ethanol process but the economic viability is better for the dry-grind corn ethanol processes.

References

[1] US Corn yields. 2018. https://www.nass.usda.gov/Statistics_by_Subject/result.php?B10EB84A-5970-335E-A882-5B42E549DF66&sector=CROPS&group=FIELD%20CROPS&comm=CORN. [Accessed October 5, 2018].

[2] EIA petroleum and other liquids consumption. 2018. https://www.eia.gov/dnav/pet/pet_cons_psup_dc_nus_mbbl_a.htm. [Accessed October 5, 2018].

[3] Alternative Fuels Data Center. 2018. https://www.afdc.energy.gov/fuels/fuel_comparison_chart.pdf. [Accessed October 5, 2018].
[4] FAOSTAT. 2018. http://www.fao.org/faostat/en/#data/RL. [Accessed October 5, 2018].
[5] Renewable Fuels Association. GREET. 2018. https://ethanolrfa.org/how-ethanol-is-made/[Accessed October 5, 2018].
[6] USDA NASS. 2018. https://usda.mannlib.cornell.edu/usda/current/USCatSup/USCatSup-06-24-2016.pdf. [Accessed October 5, 2018].
[7] University of Georgia Extension Publication. 2018. http://extension.uga.edu/publications/detail.html?number=B1482&title=Using%20Distillers%20Grains%20in%20Beef%20Cattle%20Diets. [Accessed October 5, 2018].
[8] J.R. Kwiatkowski, A.J. McAloon, F. Taylor, D.B. Johnston, Modeling the process and costs of fuel ethanol production by the corn dry-grind process, Ind. Crops Products 23 (2006) 288–296.
[9] Greenhouse gases and regulated emissions and energy use in transportation. 2020. Argonne National Laboratory. DOI: 10.11578/GREET-Net-2020/dc.20200913.1.
[10] M.H. Langholtz, B.J. Stokes, L.M. Eaton (Leads), U.S. Department of Energy, 2016 Billion-Ton Report: Advancing Domestic Resources for a Thriving Bioeconomy. In: Volume 1: Economic Availability of Feedstocks., ORNL/TM-2016/160, Oak Ridge National Laboratory, Oak Ridge, TN, 2016, p. 448. doi:10.2172/1271651.
[11] Humbird, D., Davis, R., Tao, L., Kinchin, C., Hsu, D., and Aden, A. 2011. Process design and economics for biochemical conversion of lignocellulosic biomass to ethanol dilute-acid pretreatment and enzymatic hydrolysis of corn stover. National Renewable Energy Laboratory, Report NREL/TP-5100-47764: https://www.nrel.gov/docs/fy11osti/47764.pdf. [Accessed October 20, 2020].

CHAPTER TEN

Solar energy in India

Ganti S. Murthy[a,b]

[a]Biological and Ecological Engineering, Oregon State University, Corvallis, OR, United States, [b]Biosciences and Biomedical Engineering, Indian Institute of Technology, Indore, India

10.1 Introduction

Since independence from colonial rule in 1947, India has grown tremendously and is now the third-largest economy of the world. Concomitant with the growth of its economy and prosperity, the installed capacity for electricity has grown from 1.36 GW in 1947 to 370 GW in 2020 (Fig. 10.1). Similarly, the per capita electricity consumption has grown from 16 kWh to 1208 kWh in 2020 [1]. Domestic/imported coal and mostly imported oil are the major sources of energy inflows while the industrial sector is the major consumer (56%) of the energy (Fig. 10.2). With increasing economic growth, the electricity consumption by both the industrial sector and the residential consumers is increasing significantly [2]. Despite being a developing nation with little contribution to historical greenhouse gas emissions, India made a voluntary Intended Nationally Determined Contribution (INDC) commitment to reduce the carbon intensity of its GDP by 33%–35% and 40% cumulative nonfossil fuel-based energy resources from the 2005 baseline year by 2030. Since electricity generation from coal is the highest contributor to GHG emissions, replacing and substituting the future growth of coal-based electricity with renewable energy alternatives has been one of the major ways in which India is on track to exceed its INDC commitments. India's location in the tropics and high potential for renewable energy production has led to a natural focus on solar photovoltaic and wind energy for electricity production (Table 10.1). The estimated potential for solar and wind energy are 68% and 27.6%, respectively, out of the total potential to produce 1096 GWh of renewable electricity in India [3]. The installed electricity production is expected to grow to 118 GW by 2047 and the share of renewable energy is expected to increase to 26% from 10% in 2020. Installed solar and wind energy are expected to be 197 and 222 GW, respectively [2].

Solar energy incident over India is estimated to be about 5000 trillion MWh/year with an average of 4–7 kWh/m^2-day in most parts of India. Using just 3% of the wasteland area in India for solar photovoltaic electricity production can generate up to 748 GW of electricity which is multiple times the currently installed electricity generation capacity [4]. Given the tremendous potential of solar energy for electricity production, India started renewable energy research and development and implementation in the

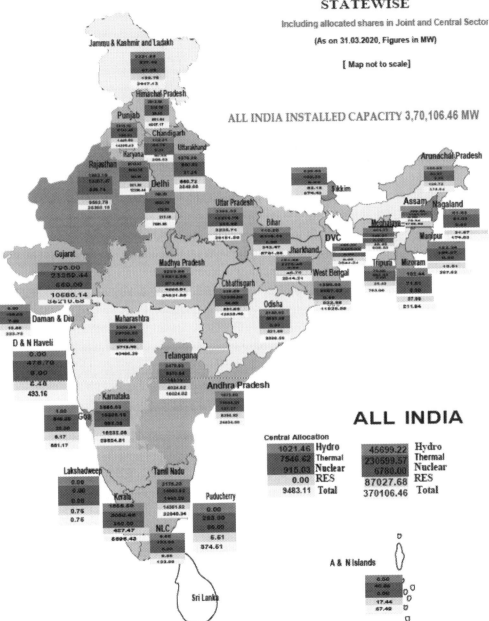

Fig. 10.1 Installed electricity capacity in India (figure from [1]).

Solar energy in India

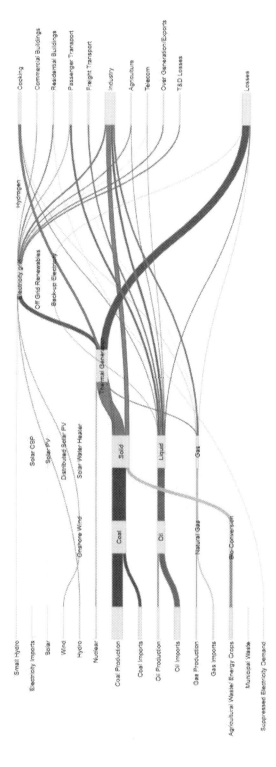

Fig. 10.2 Energy flows (TWhr) of India in 2017 (figure from [2]).

Table 10.1 Renewable energy potential from various resources (Table 1.3 from [3]; all numbers in MW).

States/UTs	Wind power @ 100 m	Small hydro power	Biomass power	Cogeneration-bagasse	Waste to energy	Solar energy	Total estimated reserves	Distribution (%)
Andhra Pradesh	44,229	978	578	300	123	38,440	84,648	7.72
Arunachal Pradesh		1341	8			8650	10,000	0.91
Assam		239	212		8	13,760	14,218	1.30
Bihar		223	619	300	73	11,200	12,415	1.13
Chhattisgarh	77	1107	236		24	18,270	19,714	1.80
Goa	1	7	26			880	913	0.08
Gujarat	84,431	202	1221	350	112	35,770	122,086	11.14
Haryana		110	1333	350	24	4560	6377	0.58
Himachal Pradesh		2398	142		2	33,840	36,382	3.32
Jammu & Kashmir and Ladakh		1431	43			111,050	112,523	10.27
Jharkhand		209	90		10	18,180	18,489	1.69
Karnataka	55,857	4141	1131	450		24,700	86,279	7.87
Kerala	1700	704	1044		36	6110	9595	0.88
Madhya Pradesh	10,484	820	1364		78	61,660	74,406	6.79
Maharashtra	45,394	794	1887	1250	287	64,320	113,933	10.39
Manipur		109	13		2	10,630	10,755	0.98
Meghalaya		230	11		2	5860	6103	0.56
Mizoram		169	1		2	9090	9261	0.84
Nagaland		197	10			7290	7497	0.68
Odisha	3093	295	246		22	25,780	29437	2.69
Punjab		441	3172	300	45	2810	6768	0.62
Rajasthan	18,770	57	1039		62	142,310	162,238	14.80
Sikkim		267	2			4940	5209	0.48

Solar energy in India

Tamil Nadu	33,800	660	1070	450	151	17,670	53,800	4.91
Telangana	4244					20,410	24,654	2.25
Tripura		47	3		2	2080	2131	0.19
Uttar Pradesh		461	1617	1250	176	22,830	26,333	2.40
Uttarakhand		1708	24		5	16,800	18,537	1.69
West Bengal	2	396	396		148	6260	7202	0.66
Andaman & Nicobar	8	8				0	16	0.00
Chandigarh					6	0	6	0.00
Dadar & Nagar Haveli						0	0	0.00
Daman & Diu						0	0	0.00
Delhi					131	2050	2181	0.20
Lakshadweep	8					0	8	0.00
Puducherry	153				3	0	156	0.01
Others★					1022	790	1812	0.17
All India Total	302,251	19,749	17,536	5000	2554	748,990	1,096,081	100.00
Distribution (%)	27.58	1.80	1.60	0.46	0.23	68.33	100.00	

★Industrial Waste

early 1960s. The focus on renewable energy has increase even more after India's INDC commitment. With the achievement of grid parity in terms of the generation costs, the installed solar energy capacity has grown over 11 times from 2.6 GW in 2014 to 34.6 GW in 2020 making India among the top five countries in the world in solar power deployment [4].

10.2 Development of solar energy in India

Historically, although the Council for Scientific and Industrial Research and other research organizations started research into solar cells in the 1950s, the interest of policy makers in renewable energy sources was limited to large and small hydropower projects until the 1980s. The earlier decades saw the development of research programs in various academic and scientific institutions in India. In the 1980s, small solar photovoltaic companies started emerging due to various incentives from government The Commission for Additional Sources of Energy (CASE) formed in 1981 was the first government body to encourage research and development in renewable sources of energy as response to the oil shocks and volatility in the energy markets in the previous decade. A new department of non-conventional Energy sources was formed in 1982 absorbing CASE and extending the mandate to implementational aspects. The department was converted into a full-fledged Ministry of Non-conventional Energy Sources in 1992, which was a first in the world. Reflecting the emerging international consensus around global climate change and the necessity to focus on renewable energy source, the ministry was renamed the Ministry of New and Renewable Energy in 2006. The ministry was formed with a mission to increase energy security, increase the share of clean power, promote energy availability, access affordability and energy equity in India. It was only after 2000 that the renewable energy sources composed mostly of wind, and solar photovoltaic (solar PV) installations saw a significant increase from 1 GW in 2000 to 92.5 GW in January 2021 [5]. Internationally, India and France provided leadership for the formation of International Solar Alliance in 2015 with the partnership of 121 countries rich in solar energy resources. The International Solar Alliance aims to deploy 1000 GW of solar energy globally by mobilizing investments of up to $1000 trillion by 2030.

Many government programs were initiated to increase solar PV utilization in the country (Table 10.2). One of the first programs to promote solar PV in a significant way was the national solar mission (NSM) launched in January 2010. In this initiative, the Government of India set a target of 20GW solar PV installation by 2022 to be achieved in three phases. However, the progress in this program was significantly better than envisaged initially and number of favorable trends including reducing cost of solar PV panels internationally, India's commitment to reduce the GHG intensity, various policies

Table 10.2 Govt policies and programs to support solar and renewable energy in India ([6]).

Name of policy	Features
Electricity act of 2003	1. The electricity act of 2003 modernized the electricity laws of India by consolidating laws related to generation, transmission, and distribution of electricity throughout the country 2. Allowed development of standalone systems 3. Encouraged private sector participation in generation, transmission, and distribution of electricity 4. Introduced the feed in tariff (FIT) and Renewable purchase obligation (RPO) for the first time
Tariff Policy of 2006	1. Goal was to improve the quality, efficiency, and the price of electricity supply 2. Provided special tariffs for renewable energy to encourage private investments 3. This policy was continuation of the Electricity act of 2003 and the national energy policy of 2005
Integrated Energy policy of 2006	1. Required power regulators to provide incentive structures for renewable energy sources 2. Required power regulators to purchase mandated solar energy as per the Electricity act of 2003
National Action Plan on Climate Change (2008)	1. The goal of the NAPCC was to address the climate change concerns. 2. Schemes introduced for electrifying remote areas with local solar PV supported electricity grids 3. NSM was part of the NAPCC
Generation based incentives for Solae (2009)	1. Scheme targeted small grid solar projects below 33 KV 2. Aimed at increasing private sector participation 3. Prime motive was to decrease the gap between the base tariff of INR 5.5 and the tariff f Central Electricity Regulatory Commission
Jawaharlal Nehru National Solar Mission (2010)	1. First major initiative with goals to install 22 GW capacity by 2022 2. Objective was to establish India as an international leader in solar energy 3. The NSM consisted of three phases: • Phase 1 (2010–2013): 1.1 GW grid connected utility, 200 MW off grid solar applications: successfully completed • Phase 2 (2013–2017): 4–10 GW grid connected utility, 100 MW off grid systems, 25 solar parks and ultra-mega solar power projects. Focus to rooftop applications: successfully completed • Phase 3 (2017–2022): 20 GW grid connected utility and 2 GW off grid systems. The scope of the NSM was increased to 100GW by 2022: continuing, 37GW installed in grid connected and off grid systems

to promote manufacturing facilities, the solar PV especially the grid-connected, and solar park solar PV facilities led to a revision of the goals for the NSM from 20 GW in 2022 to 100 GW by 2022 [6].

The policy initiatives were both informed and supported by the emerging trends in the international markets such as the expansion of the solar PV programs around the world, sharp reductions in the solar panel manufacturing costs dominated by the growth in the Chinese solar PV manufacturing industry. This was particularly the case after 2010 when the solar panel cost fell from about INR 200/Wp in 2010 to ~INR 35/Wp in 2020.

In India, the solar PV industry started with the Enertech group which started manufacturing the UPS systems in 1990. Other companies such as EMMVEE founded in 1992, Kotak Urja Pvt. Ltd. Started in 1997 and ICOMM Tele Ltd., Mosar Baer Solar Ltd. entered the solar energy industry. The status of the industry significantly expanded after 2000 with the establishment of Tata power solar system Ltd., Delta India, Toshiba Mitsubishi-Electric Industrial systems, Indosolar Ltd., between 2000 and 2010 [6]. India does not currently have any manufacturing capacity for Polysilicon/wafer/ingots. Manufacturing industries in India have capacities of about 3GW/year of solar PV cells and 10 GW/year of solar PV module manufacturing capacity [4]. There are several recent government incentives to increase the manufacturing capacity through assured takeoff in the form of power purchase agreements (PPAs) for solar power plants.

Several schemes were formulated by the MNRE to encourage both grid-connected solar and off-grid applications (Table 10.3). In addition to the schemes mentioned here, there were many other initiatives which are completed to encourage solar PV installations in the country. As a result of these schemes and policy initiatives, the solar PV sector in India has seen a tremendous amount of growth in recent years. Currently, the installed solar PV capacity in India is 37.9 GW as of March 2020 and represents just 5% of the total solar potential of the country (Table 10.4). The grid-connected solar facilities account for about 28.2 GW capacity, while the standalone off-grid applications constitute 212 MW capacity in 2019. The off-grid applications include solar lanterns, home and streetlights, solar-powered irrigation pumps and the standalone off-grid systems.

10.3 Challenges to solar energy in India

Although the installed capacity of solar PV systems has increased several folds in the last decade, this represents only a small fraction of the total potential and 95% of which is yet to be realized. Several technological, policy, financial, governance, and human resource challenges have been identified that are impeding the growth of the solar PV sector in India.

Industrial and Technological Barriers: most of the solar PV installations in India use the crystalline silicon solar PV cells. One of the biggest technological barriers is the lack of adequate manufacturing capacity in India [7]. The manufacturing capacity for

Table 10.3 Details of the central government schemes for promotion of solar energy active in 2021 [4].

Scheme Name	Features
	Grid connected solar PV projects
Development of solar parks and ultra-mega solar power projects	1. The goal of this scheme is to facilitate the setting up of large (>500 MW) solar parks in a plug and play model 2. Government facilitates the process by developing the sites, installing the transmission lines, providing the roads, and other infrastructure necessary for setting up the solar project 3. The central government provides the central financial assistance (CFA) to states for implementing the scheme in collaboration with state government agencies, CPSU, and private industry 4. The CFA of INR 2 million/MW is provided of which 60% is to be used for site infrastructure development and the 40% for external transmission system
Central Public Sector Undertaking (CPSU) Scheme Phase II	1. Provides assistance to government organizations with up to 70% viability gap funding 2. The power generated by the solar PV project can be used for the CPSUs self-user other Government Organizations through DisComs 3. Requires domestically manufactures solar PV cells and modules
Grid connected Solar Rooftop Programme	1. Objective is to achieve 40GW rooftop solar projects by 2022 2. Scheme consists of three components and must be impleme5.3nted through power distributing companies 3. Component A: Provides CFA to residential sector up to 4GW; 40% CFA for up to 3 KWp, 20% for beyond 3KWp. 4. Component B: provides progressive incentives of up to 10% for achieving installed capacity targets more than 15% compared to the baseline capacity in the previous financial year. The incentives are provided up to initial 18 GW installed capacity.
	Solar Off grid projects
Pradhan Mantri Kisan Urja Surasha evam Utthan Mahabhiyan (PM KUSUM)	1. The scheme aims to add 25.75 GW renewable energy capacity by 2022 and consists of three components. • Component A: installing individual plants of up to 2 MW size for a total installed capacity of 10 GW in decentralized ground mounted grid connected renewable power plants. Discoms will receive incentives of up to IN 0.40/KWh for a period of 5 years • Component B: installation of 1.75 million stand-alone solar powered agricultural pumps with up to 7.5 hp capacity. CFA of 30% and 30% from the state government will be provided. The farmer is responsible for remaining 40% of which 30% can be obtained as a bank loan • Component C: solarization of 1 million grid connected agriculture pumps of individual pump capacity u to 7.5 HP. Similar financial incentives as for component B
Atal Jyoti Yojana (AJAY): Phase II	1. Objective of this scheme is to install over 300,000 solar streetlights 2. States targeted for the implementation are Assam, Bihar, Himachal Pradesh, Jammu and Kashmir, Jharkhand, Odisha, Uttar Pradesh, Uttarakhand, NE states, Andaman & Nicobar, and Lakshadweep

Table 10.4 Potential versus Installed solar capacities in GW (data from [4]).

States/ UTs	Solar energy[a]	Installed capacity[a]	Percentage capacity installed[a]	Lanterns/ lamp nos.[b]	Home lights nos.[b]	Streetlights nos.[b]	Pumps nos.[b]	Stand alone (KWp)[b]
Andhra Pradesh	38,440	3610.02	9.4%	77,803	22,972	8992	34,045	3815.595
Arunachal Pradesh	8650	3620.75	41.9%	18,551	35,065	5008	22	963.2
Assam	13,760	41.23	0.3%	498,361	46,879	9547	45	1605
Bihar	11,200	151.57	1.4%	1,258,294	12,303	29,858	2107	6770
Chhattisgarh	18,270	231.35	1.3%	3311	42,232	2042	61,970	31,249.9
Goa	880	4.78	0.5%	1093	393	707	15	32.72
Gujarat	35,770	2948.37	8.2%	31,603	9253	2004	11,522	13,576.6
Haryana	4560	252.14	5.5%	93,853	56,727	34,625	1293	2321.25
Himachal Pradesh	33,840	32.93	0.1%	33,909	22,592	78,000	6	1905.5
Jammu & Kashmir and Ladakh	111,050	19.3	0.0%	51,224	144,316	14,156	39	8129.85
Jharkhand	18,180	38.4	0.2%	747,295	9450	10,301	3857	3769.9
Karnataka	24,700	7277.93	29.5%	7781	52,638	2694	6343	7754.01
Kerala	6110	142.23	2.3%	54,367	41,912	1735	818	15,825.39
Madhya Pradesh	61,660	2258.46	3.7%	529,101	7920	10,833	17,813	3654
Maharashtra	64,320	1801.8	2.8%	239,297	3497	10,420	4315	3857.7
Manipur	10,630	5.16	0.0%	9058	24,583	11,205	40	1580.5
Meghalaya	5860	0.12	0.0%	40,750	14,874	5800	19	2004
Mizoram	9090	1.52	0.0%	10,512	12,060	5325	37	2955.6
Nagaland	7290	1	0.0%	6766	1045	6235	3	1506
Odisha	25,780	397.84	1.5%	99,843	5274	14,567	9327	567.515
Punjab	2810	947.1	33.7%	17,495	8626	42,758	3857	2066
Rajasthan	142,310	5137.91	3.6%	22,5851	187,968	6852	48,175	30,349
Sikkim	4940	0.07	0.0%	23,300	15,059	504	0	850
Tamil Nadu	17,670	3915.88	22.2%	16,818	290,376	39,419	4984	12,752.6

Telangana	20,410	3620.75	17.7%	0	0	1103	424	7450
Tripura	2080	9.41	0.5%	64,282	32,723	1199	151	867
Uttar Pradesh	22,830	1095.1	4.8%	1,336,733	235,909	258,863	20,465	10,638.31
Uttarakhand	16,800	0	0.0%	163,386	91,595	22,119	26	3145.03
West Bengal	6260	114.46	1.8%	17,662	145,332	8726	653	1730
Andaman & Nicobar	12	12.19	100.0%	6296	468	390	5	167
Chandigarh	41	40.5	100.0%	1675	275	898	12	730
Dadar & Nagar Haveli	5	5.46	100.0%	–	–	–	–	–
Daman & Diu	20	19.86	100.0%	–	–	–	–	–
Delhi	2050	165.16	8.1%	4807	0	301	90	1269
Lakshadweep	0	0	0.0%	5289	600	2465	0	2190
Puducherry	0	0	0.0%	1637	25	417	21	121
Others[a]	790		0.0%	125,797	140,273	9150	4621	23,885
Total	**748990**	37,920.75	5.1%	5,823,800	2E+06	65,9218	237,120	212,054.17

[a] As on 31.3.2020
[b] For year 2018–2019

the solar PV cell and module manufacturing is 3GW/year and 10 GW/year, respectively, which is a fraction of >600 GW of world manufacturing capacity of which the majority is concentrated in China. More critically, there are no foundries in India to manufacture the silicon ingots used for solar cell manufacture [4]. Such import dependence on critical manufacturing base required for large-scale expansion of the solar PV is a critical barrier.

One of the attractive features of the solar PV is its long life and relatively very low maintenance costs. Studies on the long-term performance of the solar PV modules under Indian conditions indicate that the performance of the solar PV modules varied widely and the annual linear degradation rates for crystalline silicon modules varied from 0.6% to 5%/year [8]. The average performance of good sites (mostly large installations with >100 kW) had ~0.89% annual degradation rate was comparable to the international reported averages. However, the performance of the second group of sites (mostly smaller installations) was much worse with annual linear degradation rates >2.21%/year. Unsurprisingly, the locations in hotter climates had higher degradation rates. The modules installed <5 years ago had higher degradation rates compared to the older modules indicating a worrying trend of poor-quality materials in the recent installations and/or possible overrating of the panels by unethical manufacturers [8]. Some of the issues also pointed to the poorer quality materials, improper or hasty installations.

The output from solar PV modules is strongly affected by dust and could lead to a loss of up to 30% power under typical Indian weather conditions [9]. Different regions of India require different methods that are applied at various frequencies for optimum cleaning and maintenance of the PV module power output. For example, the cold climate of Ladakh requires a set cleaning once every six months, while the coastal and southern regions of India with warm and humid climate require monthly cleaning. The composite climate zones of central and northern India require cleaning every three months while the hot and dry climatic regions of western Indian in Rajasthan, Gujarat, and Maharashtra require a weekly cleaning frequency to maintain the power output [9].

With high fraction of renewable energy, the grid balancing often poses a critical implementational problem and often needs to curtailment of the renewable energy sources. A detailed study of India's 175 GW renewable energy by 2022 goal found that the grid balancing can be done at 15 min intervals with this level of renewable energy and minimal renewable energy curtailment [10]. However, as the fraction of the renewable electricity from solar and wind increase in the grid, older coal power plants are replaced and the nature of the consumer loads change in the future, the grid balancing can become a barrier for the growth of renewable energy in India.

Off-grid systems face unique challenges compared to the Gird based solar PV systems. Often the real-time information about the status of installations and operational is lacking. This is a very severe problem for off-grid systems which are typically installed, and but not properly maintained. Solar tracking systems can increase annual power availability by up to 40% but is not practiced widely. Most of the tracking systems are not appropriately designed to withstand the winds in coastal areas.

Resource Barriers: the land, materials used for solar PV module manufacturing, and water are three critical natural resources for manufacturing solar PV modules. About 4–5 acres of land is required for each MW of installed solar capacity. The available wasteland in India is about 557,000 km^2 [11] of which 108,000 km^2 is more likely to be suitable for solar PV installations. Thus, the available wasteland is more than sufficient to meet the land requirements for the solar PV systems. Additional policy measures such as the creation of the solar park will remove the

In addition, the solar panel cleaning requires about 75 L water/MW installed capacity [12]. Since most of the states (Rajasthan, Gujarat, Andhra Pradesh, Madhya Pradesh, Maharashtra, and Karnataka) where the large-scale solar PV installations have taken place are already water-stressed, providing ~18.7 million L water/year mostly from groundwater sources at industrial tariffs for water can pose a huge challenge to the environment and to the operational costs.

A typical solar PV module is made of 75% glass, 10% polymers, 8% aluminum, 5% silicon, 1% copper, and small quantities of silver, tin, lead, and other metals [13]. With the first of the large-scale solar PV installations coming to their end of their projected life, the amount of waste produced from the solar PV plants will increase dramatically and will mirror the increase in the solar installations of the last decade. Currently, there is a very limited or no organized recycling industry for the solar PV modules in India. Establishing a cost effective, environmentally friendly recycling industry for the solar PV modules will be very critical for developing a self-sustained industry.

Policy, governance and financial barriers: a stable and consistent policy environment is required for gaining the confidence of investors to attract the investments into the renewable energy sector [6]. Although there has been a consistent policy regarding the renewable energy sector by the Government of India at the central level, the policy implementation, and consistency at the state level have been less satisfactory. In particular, the clarity required for incentivizing the investments into renewable energy sector through land acquisition, providing transmission lines, timely permissions, clear and transparent bureaucracy is often challenging [7]. Additionally, there is lack of enthusiasm among discoms for regarding the renewable energy due to their prior lock-in agreements with coal-based power plants. This results in discoms being noncooperative and meeting only the bare minimum standards as mandated by the laws. In addition to these challenges, the governance issues such as the lack of transparency and corruption often plague the implementation of even good policies. Solar PV systems are capital cost-intensive and require huge upfront outlays of capital. Additionally, the cost of capital is much higher in India due to very high-interest rates compared to the international markets.

Human resource barriers: trained manpower is critical for executing any project. While the availability of trained manpower to install and operate is a big challenge for the rapidly expanding solar PV industry sector. The solution is more acute for off-grid solution which is typically installed in remote areas and the distributed nature of such solutions make it difficult for scaling up solutions for regular maintenance of these systems. To add to the challenge funding agencies are mostly involved until the installation of the system with very few structured follow-up mechanisms resulting in many poorly maintained or defunct systems. Surya Mitra scheme of the government is aimed to train 50,000 skilled and employable workforce to meet the servicing needs of this rapidly increasing sector.

Nexus barriers: energy-water nexus issues are the biggest cause of many issues faced by the Solar PV systems in agricultural sector. A front-end subsidy on the solar PV systems has resulted in the availability of very cheap reliable power to farmers. In the face of no disincentives, this has led to overexploitation and misuse of groundwater resources. One of the failures of the JNNSM is that the encouragement of the installation of solar irrigation pumps in all parts of the country without the concomitant requirement for optimal utilization of the groundwater. The government has also lost the only lever (electricity supply to farms) available to control the GW use on farms, leading to groundwater depletion in many states in particular northern Karnataka. There is a lack of coordination between the government agencies focusing on implementing solar PV systems and those that are focusing on increasing water use efficiency through drip irrigation systems. Ideally, a coordinated effort between different arms of the government would have resulted in a policy that ties the solar irrigation pump systems with micro-irrigation system installation to minimize the usage of groundwater.

10.4 Innovative responses to the challenges

Industrial mega solar parks: the solar PV systems require 4–5 acres of land per MW of installed capacity. The land acquisition for large projects is a very big challenge in a densely populated country like India. Land acquisition is often the slowest and most expensive part of the solar PV project. Recognizing this, the Government of India started the creation of industrial-scale solar parks in collaboration with the state government (Table 10.3). The land acquisition for these parks is done by the state governments. And they are responsible for creating the basic infrastructure such as site development, access roads, and the development of transmission lines for connection to the state and central electric grid for efficient power evacuation. The program incentivizes the state governments through central funding assistance and the creation of such facilities could help the state discoms improve their profitability by reducing the price of purchased electric power and switching to a cleaner and renewable alternative. The advantage to the potential investors is that the solar parks free the investors from the challenges related to land acquisition and development, and regulatory issues, enabling them to focus on the core technologies for implementation of the solar PV systems. Since the introduction of this measure, the total capacity in solar parks is 25 GW forming a significant part of the overall installed solar PV capacity.

Canal-based solar plants: another innovation that addresses the huge land acquisitions costs of solar PV systems is the introduction of canal top solar plants [4]. India has the largest irrigation canal network in the world. Installing solar PV systems on irrigation canal stretches can eliminate the need for additional land. Solar panels shade the water in the canals, thus reducing the evaporative losses while the cooler temperatures above the water surface increase the efficiency of the solar panels. The water evaporative losses are reduced, efficiency of solar power conversion is increased, reduced soiling losses, maintenance and cleaning of the solar panels is easier due to access to water, and the land requirements are eliminated in canal solar PV systems. However, a more robust support structures for the solar PV panels increasing the cost of the PV systems and the effects of long-term shading on the aquatic life in the water is not well understood. A similar concept is the floating solar PV systems on water bodies to provide similar benefits of eliminating the land requirements, reducing the evaporative losses and increasing the efficiency of the solar PV systems. The claimed benefits for reducing the evaporative losses on canals and large water bodies has not been clearly established and must be viewed with caution. The challenges in the floating solar PV systems, called "flovoltaics," is the need for an effective design of the tether systems, preventing the movement of the floating solar panels due to winds and mitigating the impact of the shading on the aquatic life in the water bodies. Fluctuating water levels in the waterbodies poses additional challenges for effective mooring and anchoring. More than 18,000 km^2 of water bodies in India represent a potential for installing 280 GW of floating solar PV capacity [14]. However, the cost of a floating solar PV system is also higher (~INR 35–54/Wp between 2016 and 2018) and 30%–37% of the systems costs are related to the floating platform, mooring and anchoring costs. Currently, India has 2.7 MW floating solar PV systems already installed and over 1.7 GW capacity projects are in development [14].

Agrivoltaics: Agrivoltaics refers to the colocation of the solar PV panels on agricultural land for simultaneous production of agricultural crops and renewable electricity [15]. The microclimate underneath the solar panels reduces the evapotranspiration, increase the efficiency of the solar panels due to lower temperatures in the underside of the solar panels, and increase the crop productivity (Fig. 10.3). In fact, the potential of agrivoltaics has been shown to be highest on the crop lands [16]. Currently, there are few experimental agrivoltaics around the world including India where the various crops especially the labor-intensive crops such as vegetable crops are grown under the panels. Vast tracts of arable land in India are suitable for agrivoltaics and have the potential to address the renewable energy production without additional land, increasing agricultural productivity, reducing the water usage and increasing the income to farmers through an additional income streams through sale of electricity.

Demand reduction programs: one of the ways to reduce the carbon intensity of the per capita GDP is through elimination of inefficient devices on the consumers side thus reducing the demand. Lighting is one of the basic services provided by the electricity and

Fig. 10.3 Agrivoltaic systems at Oregon State University (Photograph courtesy of Mr. Kyle Proctor).

until the last decade, most of the homes in India used inefficient incandescent bulbs and fluorescent tube lights. In 2015, government launched the Unnat Jyoti by Affordable LEDs for all (UJALA) scheme to replace all the inefficient appliances with efficient appliances were supplied at subsidized prices. This was the world's largest scheme for the distribution of efficient appliances and was expanded to UK in 2017. The scheme was hugely successful and has resulted in 9545 MW of avoided peak demand, savings of 47.6 billion KWh/year of electricity, cost savings of INR 1907 billion, and 38.6 million tons of CO_2 emission reduction/year [17]. Although not implemented yet, similar possibilities for huge energy savings exist in the agricultural sector by increasing the efficiency of the irrigation pumps.

Peak power balancing with pumped storage: with the proposed huge increase in renewable energy, the grid balancing becomes even more challenging. Although the studies have shown that this may not be a concern in the near term, the proposed expansion to 450 GW of renewable energy contributing 60% of the electricity production from renewable sources will pose huge challenges for grind balancing. Several schemes such as using the pumped storage, large lift-irrigation projects and river interlinking projects for utilizing the excess power have been proposed. In addition, new technologies for the production of ammonia and hydrogen from the excess electricity from the wind and solar PV installations can contribute to the flattening of renewable energy generation and aid in balancing the grid.

10.5 Overall scenario

India's solar PV program has been a very big success with the solar PV electricity prices reducing 84% in the last decade and the large-scale increase in the installed capacities. However, there are some headwinds to the expansion of this sector in the last few years with 90 projects for 39.4 GW of new capacity facing various delays. Despite the tremendous increase in the renewable electricity production, India's electricity sector remains critically dependent on thermal energy sector which consists of 88% coal and lignite-based generation contributing to 75% of the actual generation [18]. This legacy of thermal power-based electricity has resulted in many long-term expensive PPAs that are difficult to service by many cash-starved discoms which make it difficult for them to abandon the long term PPAs in favor of the renewable alternatives. Legacy of the PPA contracts by discoms often exceed the peak demand and thus locking the financial positions in few states leading to financially distressed discoms which lack the financial muscle to invest in new renewable electricity alternatives. Although the government has initiated measures to shutdown older plants, the political pressures from the coal industry, workers in the mining sectors and many vested interests make it a long and arduous road of implementation. Lack of enthusiasm for renewable energy sources and apathy among the majority of the political class is cited as a fundamental reason for the lack of necessary push from the political and policy-making side especially at the state level. Government launched Ujwal DISCOM Assurance Yojana (UDAY) scheme in 2015 to transfer 75% of the debts of the state discoms to their respective states to improve the financial positions of the discoms. The scheme was partly successful in narrowing the gap between the cost of electricity and the revenues for discoms thus improving their financial health.

A critical emerging issue in the distributed off-grid solar PV systems is the nexus between the energy and water. This severity of issue is only expected to increase many times with proposed large scale roll out of the solar irrigation pumps. This requires concerted efforts among various ministries to develop a coherent and consistent set of policies to incentivize both energy production and conservation of the already scarce water resources. This calls for a rationalization of subsidies for different areas of the country. For example, the front-end capital subsidy could be provided in areas in Gangetic plains that are rich in groundwater. In areas with stressed groundwater resources such as the western and southern states, the subsidy must be tied to the optimal use of the system with incentives for exporting electricity to grid and minimal utilization of the system for irrigation. Karnataka's Surya Raitha program which incentivizes farmers to generate more electricity by feed-in tariffs could be a model program to reduce overexploitation of the groundwater resources [19]. However, preventing the misuse of net metering by the unscrupulous players requires use of reliable internet-of-things technologies for real-time monitoring of the systems.

India is already on track to exceed the 33%–35% reduction in the carbon intensity by 2030 promised to the international community at the Paris Climate Accord in 2015. Setting the policy directions for the next decade, the Prime Minister of India announced in 2019 G20 meeting that India will try to achieve 450 GW by 2030 [20]. India aims to produce 60% of its grid electricity through renewable sources by 2030 and will shift to the production of "green ammonia" and hydrogen [21]. These goals will be achieved by adoption of floating solar, wind-solar hybrid, peaking power, and round-the-clock procurements, innovative storage solutions, streamlining the regulations, reducing the price of renewable alternatives by increasing efficiencies of solar PV systems, increasing domestic manufacturing bases, encouraging innovation, and enhancing affordability and access to energy.

10.6 Conclusions and perspectives

Solar renewable energy from solar PV systems have been successful in reducing the dependence on polluting coal-based electricity and reducing the per capita carbon intensity of India. The growth from 1.0 GW in 2000 to 92.5 GW in January 2021 represents a success for this industry. However, with the proposed increases in capacities of up to 450 GW by 2030, the sector is facing several headwinds which can be addressed through innovative strategies. Increasing the manufacturing base for the solar PV systems, decreasing the policy uncertainty, reducing the regulatory barriers, effective implementation of the purchase agreements, increase access to low-cost financing, concerted efforts to address the energy-water nexus issues are some of the possible solutions to make India's electricity sector a model for the world.

References

[1] Central Electric Authority, Govt. of India Electricity consumption in India. 2021. Growth of electricity sector in India from 1947-2020. https://cea.nic.in/wp-content/uploads/pdm/2020/12/growth_2020.pdf. [Accessed February 13, 2021].

[2] IESS. 2020. India energy security scenarios. http://iess2047.gov.in. [Accessed February 13, 2021].

[3] Energy statistics. 2019. Energy Statistics 2019 (Twenty Sixth Issue). http://mospi.nic.in/sites/default/files/publication_reports/Energy%20Statistics%202019-finall.pdf. [Accessed February 13, 2021].

[4] MNRE. Solar energy. 2021. https://mnre.gov.in/solar/current-status/. [Accessed February 13, 2021].

[5] National Power Portal. 2021. Growth of installed renewable energy capacity. https://npp.gov.in/dashBoard/cp-map-dashboard. [Accessed February 13, 2021].

[6] G. Raina, S. Sinha, Outlook on the Indian scenario of solar energy strategies: policies and challenges, Energy Strat. Rev. 24 (2019) 331–341.

[7] M. Irfan, Z.-Y. Zhao, M. Ikram, N.G. Gilal, H. Li, A. Rehman, Assessment of India's energy dynamics: prospects of solar energy, J. Renew. Sustain. Energy 12 (2020) 053701.

[8] Palchak, D., Cochran, J., Ehlen, A., McBennett, B., Milligan, M., Chernyakhovskiy, I., Deshmukh, R., Abhyankar, N., Soonee, S.K., Narasimhan, S.R., Joshi, M., Sreedharan, P., 2017. Greening the grid: pathways to Integrate 175 gigawatts of renewable energy into India's electric grid, Vol. I – National Study (No. NREL/TP-6A20-68530, 1369138).

[9] A. Ghosh, Soiling losses: a barrier for India's energy security dependency from photovoltaic power, Challenges 11 (2020) 9.

[10] Chattopadhyay, S., Dubey, R., Kuthanazhi, V., Zachariah, S., Bhaduri, S., Mahapatra, C., Rambabu, S., Ansari, F., Chindarkar, A., Sinha, A., Singh, H.K., Shiradkar, N., Arora, B.M., Kottantharayil, A., Narasimhan, K.L., Sabnis, S., Vasi, J., 2016. All-India Survey of photovoltaic module reliability: https://nise.res.in/wp-content/uploads/2018/09/Report-on-All-India-Survey-of-Photovoltaic-Module-Reliability-2016.pdf. [Accessed February 13, 2021].

[11] Dept. of Land Resources. 2019. Wasteland Atlas of India. https://dolr.gov.in/documents/wasteland-atlas-of-india. [Accessed February 13, 2021].

[12] Saur Energy. 2019. MNRE roots for robots to reduce water usage at solar plants. https://www.saurenergy.com/solar-energy-news/mnre-roots-for-robots-to-reduce-water-usage-at-solar-plants. [Accessed February 13, 2021].

[13] Saur Energy. 2020. Recycling, the coming challenge for solar panels—Saur Energy International. https://www.saurenergy.com/solar-energy-news/recycling-the-coming-challenge-for-solar-panels. [Accessed February 13, 2021].

[14] M. Acharaj, S. Devraj, Floating Solar Photovoltaic (FSPV): A Third Pillar to Solar PV Sector? TERI Discussion Paper: Output of the ETC India Project, The Energy and Resources Institute, New Delhi, 2019.

[15] K.W. Proctor, G.S. Murthy, C.W. Higgins, Agrivoltaics align with green new deal goals while supporting investment in the US' rural economy, Sustainability 13 (2020) 137.

[16] E.H. Adeh, S.P. Good, M. Calaf, C.W. Higgins, Solar PV power potential is greatest over croplands, Sci Rep 9 (2019) 11442.

[17] UJALA. National UJALA dashboard. 2021. http://www.ujala.gov.in/ [Accessed February 13, 2021].

[18] Saur Energy. 2021. The triple challenge dragging down solar In India—Saur Energy International. https://www.saurenergy.com/solar-energy-articles/the-triple-challenge-dragging-down-solar-in-india. [Accessed February 13, 2021].

[19] T. Shah, S. Verma, N. Durga, Karnataka's smart, new solar pump policy for irrigation 5, Econ. Pol. Weekly 48 (2014) 10–14.

[20] Press Information Bureau. 2019. Need, not greed, has been India's guiding principle: says PM. https://pib.gov.in/PressReleasePage.aspx?PRID=1585979. [Accessed February 13, 2021].

[21] Press Information Bureau. 2020. Increasing storage and democratizing RE deployment will be the next frontier for the Government of India: Power Minister Shri R K Singh at valedictory session of 3rdGlobal Re-Invest. https://pib.gov.in/Pressreleaseshare.aspx?PRID=1676863. [Accessed February 13, 2021].

CHAPTER ELEVEN

A systems analysis of solar and wind energy in the United States

Kyle Proctor[a], Ganti S. Murthy[a,b]
[a]Biological and Ecological Engineering, Oregon State University, Corvallis, OR, United States, [b]Biosciences and Biomedical Engineering, Indian Institute of Technology, Indore, India

11.1 Introduction

In recent decades, the focus has increasingly turned toward renewable sources of energy, largely as a result of increasing awareness of climate change, improvements in renewable energy technologies, and shifting economic conditions. Wind and solar have often been at the forefront of these discussions as they are some of the most abundantly available renewable resources in the United States. Although research and investment in these technologies have intensified in recent years, their histories stretch back far longer than a few decades. Using the wind as a power source is a practice that dates back thousands of years. References to windmills as a source of mechanical energy for irrigation can be found in the writings of the Babylonian king Hammurabi in 1700 BCE [1]. By the ninth century CE, wind power was widely used for pumping water and grinding grain [1]. The first windmill used for the production of electric power in the United States was developed in 1888 by Charles F. Brush. However, the current wave of modern wind energy production did not begin until it was sparked by the widespread oil shortages of the 1970s.

This oil crisis led to federal investment in the research and development of large wind turbines and other alternative energy sources [2]. The first US wind plants began installation in 1980 but after an initial burst of development, government incentives, and political will be dried up and there was little growth in the US wind sector from the mid-1980s to the turn of the century [3]. From 2000 onwards, generation from wind resources has grown rapidly from 6.7 TWh in the year 2000–300.1 TWh in 2019 [4]. As of 2019, wind energy accounts for 7.29% of the United States total electricity production and is the largest source of renewable energy in the country, surpassing hydroelectric power [4]. By the end of 2019, 19 states had more than 1000 MW of installed capacity. The five largest wind-producing states, Texas, Iowa, Oklahoma, Kansas, and California account for over half of the nation's total wind energy generation [5]. At this point, onshore wind energy production is considered a mature technology, indicating that it has been in use for long enough for many of the initial challenges and drawbacks to be overcome. Offshore wind turbine technologies are less mature but show substantial potential [3].

Similar to wind power, the history of humans harnessing solar energy dates back thousands of years to when various forms of reflective and magnifying lenses were used for lighting torches. Solar energy is the largest and most evenly distributed renewable energy source in the United States. The United States land surface receives as much energy from the sun in a few hours as the United States population consumes each year, about 4000 TWh [4]. This energy is generally harvested using either photovoltaic (PV) solar cells, or concentrating solar power, each method has its own benefits and drawbacks. Solar PVs function by converting light energy directly into electricity via the photoelectric effect. CSP uses mirrors or lenses to concentrate a large area of sunlight onto a much smaller area, producing heat, this heat is then used to drive a heat engine that powers a turbine that produces electrical energy.

The modern history of solar energy can be traced back to the late 19th century when the first solar cells were developed by American inventor Charles Fritts using selenium wafers. In 1954, the first silicon PV cells were developed at Bell Labs in the United States, these silicon solar cells had a 4% efficiency and were the first cells capable of converting enough energy to power contemporary electrical equipment [6]. In the decades since this initial development, solar panel efficiencies have improved substantially with most modern commercial PV panels having efficiencies between 15% and 20% [3]. Solar energy generation in the United States has also increased substantially in recent decades, going from 0.54 TWh in 2000 to 72.2 TWh in 2019 [4].

11.2 Technical feasibility analysis

11.2.1 Can we generate 100% electricity with solar and wind technologies?

As solar and wind are the most abundant of the United States' renewable resources, it is crucial to assess what role these resources will play in reaching goals of 100% renewable energy. In order to assess this potential, it is first important to consider the ways in which wind and solar energy differ from traditional energy sources. "Dispatchability" refers to the capability for electricity to be produced on demand. Traditional fossil fuel energy sources are highly dispatchable, some renewable energy sources such as hydropower, geothermal, biopower, and concentrating solar with storage are also dispatchable, on the relevant time scales. Wind and solar PV, however, are substantially less dispatchable. This lack of dispatchability is linked with the fact that both wind and solar energy generation include significant levels of variability and uncertainty as a result of their dependence on weather conditions. Energy output can be decreased but cannot be increased if no wind or sunshine is available.

Furthermore, individual solar PV systems, with power output varying on the scale of seconds to minutes as a result of passing clouds, are more variable than wind systems. However, this variability is lessened significantly with the use of large PV systems (multi-MW) or with many small PV systems distributed over a wider area [3]. Still, even if the weather could be predicted 100% accurately, these energy sources would require planning and coordination with other energy generation resources or storage to ensure adequate power to meet required demand at all times.

11.2.2 Renewable electricity futures study

An in depth analysis of the technical feasibility of powering the United States with renewable energy was conducted by the National Renewable Energy Laboratory (NREL) in their Renewable Electricity Futures Study (RE-Futures Study) [7]. The study assessed an array of more than two dozen scenarios with renewable electricity generation ranging from 30% to 90% of the US' total generation, including varying simulations of how demand and renewable technology development might progress through 2050. The study placed an emphasis on the scenario where 80% of US generation was produced from renewable sources. Within this scenario, simulations were considered where energy demand increased substantially, where transmission was constrained, where grid flexibility was constrained, and at varying levels of technological advancement.

The study concluded that using current, commercially available energy technologies and a more flexible electric system it is possible to supply 80% of the total United States electricity generation in 2050 using renewables, while consistently meeting electricity demand on an hourly basis in every region of the United States. In this scenario, wind and solar technologies could account for nearly 50% of the US' total electricity generation.

In the RE-Futures study's 80% renewable energy scenario, wind generation is expected to account for between 32% and 43 % (386–756 GW capacity) of total generation. Within this scenario, wind energy is expected to account for more than 30% of total generation in all simulations, with the bulk of this energy coming from onshore wind turbines. Solar was expected to account for between 13% and 22% (225 GW–325 GW capacity) of the total generation by 2050. The study predicts the highest solar energy penetration in the "Constrained resources" simulation where environmental concerns have placed constraints on the developable potential for many of the technologies, in this simulation 13.9% of generation comes from CSP while 7.9 % comes from PV.

This study gives a broad overview of the potential energy mix penetrations of wind and solar, however, it does not account for technologies that are not yet commercially available but may have substantial impacts, such as improved offshore wind turbines [8] and next-generation PV technologies such as dye-sensitized PV cells [9] and PV nanostructures such as quantum dots [10].

11.2.3 Storage

A fundamental aspect of the electrical grid is the need to balance generation and demand at all times. Given the variable, intermittent, and uncertain nature of wind and solar technologies, it seems clear that these two resources, although abundant, will not have the potential to meet 100% of the country's electricity needs. This is to be expected, as an energy mix which combines a wide range of renewable resource technologies will be more robust. Still, there are approaches that can allow for greater penetration of wind and solar technologies into the energy mix including improved grid flexibility and improved storage.

Energy storage improves the viability of highly variable technologies like wind and solar by allowing energy to be shifted from periods of high generation and low demand to periods of low generation and high demand. Storage also allows for improved overall stability by evening out short-scale fluctuations in a generation. Generally, energy storage technologies can fulfill three different primary functions that are highly dependent on the technologies discharge time: (1) power quality and regulation [seconds to minutes], (2) bridging power [minutes to ~1 h], and (3) energy management [hours] [3]. As energy mix penetration of highly variable technologies like wind and solar increase, the importance of operating reserves also increases.

The largest quantity of energy storage in the United States comes in the form of Pump-Stored Hydro (PSH) [11]. This is a longer time scale discharge technology useful for energy management and load leveling. Pump storage hydro facilities push water uphill to an elevated reservoir during periods of surplus energy production to convert the electrical energy to potential energy. During periods of high electricity demand, the stored water flows downward through turbines and the power is converted back to electricity using the same technology as conventional hydropower facilities [11].

The capacity of bulk energy storage in the United States grew sharply in the early 1970s to a capacity of approximately 10 GWs as a result of the oil shortages mentioned in Section 11.2. Since then, bulk storage capacity growth has tapered off, remaining relatively stable around 20 GW from the late 1980s through today. As of 2018, PSH accounts for 97% of utility-scale energy storage in the US with a total capacity of 23 GW [11].

To deal with the short-term variability of wind and solar, storage technologies with shorter discharge times are necessary. High-energy batteries, including lithium-ion and sodium-sulfur batteries, are one such technology with growing interest and investment. US utility-scale battery storage capacity has grown rapidly in recent years from 214 MW of capacity in 2014 to nearly 900 MW in 2019 [12]. This rapid growth is expected to continue with estimates of 2500 MW capacity by 2023 [13]. Although it seems clear that wind and solar will not meet all of the US energy needs, these energy sources are likely to play an increasingly important role in the countries energy mix in the coming years.

11.3 Environmental Impact assessment
11.3.1 Wind

Wind power technologies have a significantly lower environmental impact than traditional, fossil fuel based, electricity generation technologies. Wind turbines have the second smallest global warming potential per unit of energy generated, with only hydropower providing more energy per CO_2 equivalent emitted [14]. The Intergovernmental Panel on Climate Change (IPCC) 2014 report found the lifecycle emissions of onshore wind power to be between 7 and 55 gCO_{2eq}/kWh with a median life cycle emission of 11 gCO_{2eq}/kWh [15]. These values are substantially less than traditional energy technologies (see Table 11.1). Although wind energy has no direct emissions of GHGs or other air pollutants during operation and requires only minimal amounts of water for periodic blade cleaning, it still has negative impacts in other stages of the life cycle. There are also concerns related to the total land use and ecological impact of wind turbines [3].

Generally, the life cycle of energy production technologies can be separated into three distinct stages: (1) upstream processes, (2) operational processes, and (3) downstream processes. For wind energy generation, these three stages can be further broken down into five phases: (1) raw material acquisition and manufacturing, (2) transportation, (3) installation, (4) operation and maintenance, (5) end-of-life (Fig. 11.1).

When considering the environmental impact of wind power in the United States, the first phase, raw material acquisition and manufacturing, contributes the largest environmental impact, accounting for more than 60% of impacts in all TRACI impact categories, and more than 90% of impacts in most categories. Installation ranked 2nd, followed

Table 11.1 Lifecycle emissions for energy technologies (adapted from IPCC 2014 [15] and NREL [3]).

Energy technology	IPCC [15] Direct emissions Min.	Median	Max.	IPCC [15] Lifecycle emission (including albedo effect) Min.	Median	Max.	NREL [3] Lifecycle emission Min.	Media	Max.
Coal—PC	670	60	870	740	820	910	725	990	1690
Gas—combined cycle	350	370	490	410	490	650	340	450	690
Hydropower	0			1	24	2200	1	5	160
Nuclear	0			3.7	12	110	2	13	110
Concentrated Solar Power	0			8.8	27	63	11	27	81
Solar PV—utility	0			18	48	180	25	45	220
Wind onshore	0			7	11	56	2	11	44
Wind offshore	0			8	12	35	8	11	22

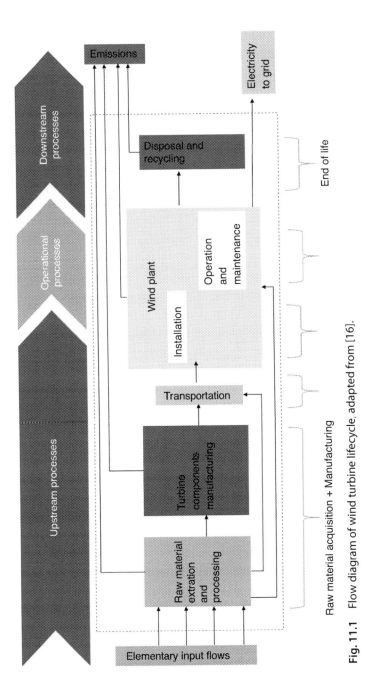

Fig. 11.1 Flow diagram of wind turbine lifecycle, adapted from [16].

by operation and maintenance, Transportation, and finally End-of-life [16]. When life cycle analyses of wind power were conducted in other countries, Manufacturing consistently ranked first and End-of-life consistently ranked last, the ranking of the other three phases varied between studies and locations [16]. This indicates that the manufacturing phase provides the largest room for reduction in GHG emissions, that end-of-life treatment generally has relatively small impacts, and that the importance of transportation, installation, and O&M for overall environmental impact are all highly context dependent.

11.3.2 Solar

Similar to wind power, solar energy is known to have much lower GHG emissions than traditional energy generation technologies. The IPCC 2014 report estimated lifecycle GHG emissions to be between 18 and 180 gCO2eq/kWh with a median value of 48 gCO_{2eq}/kWh for utility-scale PV [15] (Table 11.1). Concentrated solar power showed slightly lower lifecycle emissions with a median value of 27 gCO_{2eq}/kWh.

The NREL conducted an LCA harmonization study that reviewed over 400 solar PV LCA articles and attempted to harmonize the results to achieve consistent values. The study found that between 60% and 70% of all GHG emissions in the life cycle of solar PV result from upstream processes such as raw material extraction, materials production, and module manufacture. Operational processes accounted for between 20% and 26% of GHG emissions and downstream processes accounted for between 5% and 20% of GHG emissions [17]. The study reported no significant differences in life cycle GHG emissions between monocrystalline and polycrystalline silicon technologies and no significant life cycle GHG emissions between ground-mounted and roof-mounted systems.

11.4 Resource sustainability analysis
11.4.1 Wind energy resource sustainability

A wind turbine is generally composed of a foundation, a rotor, a nacelle, and a tower. The rotor has a central hub and three rotor blades. The nacelle contains major components such as the gearbox and the generator. The first two components are housed on top of the tower. Wind turbines are composed primarily of steel (71%–79%), fiberglass, resin, or plastic (11%–16%), iron (5%–17%), copper (1%), aluminum (0%–2%) [18]. Availability of these principal materials is not expected to limit the development of wind power technologies [3]. Some modern turbines use the rare-earth alloy neodymium-iron-boron to make the permanent magnets used in the turbine generators. There is currently uncertainty regarding whether availability of this resource may pose a supply chain risk due to over-dependence on the China-based supply chains [3].

11.4.1.1 Wind energy land and water use

Wind turbines require substantial amounts of land. For efficient energy, production turbines must be well spaced to allow for an uninterrupted flow of air. The exact requirements of the spacing are dependent primarily on the rotor diameter but are also impacted by local topography and predominant wind direction. It is recommended that turbines have a spacing of 3–5 rotor diameters between turbines aligned in a row, and 10–12 rotor diameters between rows of turbines [3]. It estimated that for turbines with a 2–3 MW capacity (80–95 m rotor diameter) a single turbine can require between 28 and 52 ha [3]. It is important to keep in mind that the actual turbine itself occupies a very small portion of this land, allowing for dual-use scenarios with livestock or agriculture. The water used for wind energy production is effectively zero. No water is required for the operation of wind turbines, water is only used for periodic blade cleaning.

11.4.1.2 Wind turbine end of life

The weight contribution of the wind turbines foundation, tower, nacelle, and rotor is 78%, 13.5%, and 4%, respectively [19]. The foundation is mostly made of concrete and steel is destroyed or left in situ after the end of the life. While the steel can be recycled, the concrete in the foundation is not recyclable. The metals such as steel, aluminum, and copper in the tower, nacelle, and rotor represent 94% of their mass and can be completely recycled. Remaining 6% weight mostly consists of plastics, rubber, and fiber-reinforced composites used in the wind turbine blades. Possible end-of-life technologies for the wind turbine blades are reuse, incineration, and mechanical grinding. Much research has indicated that the turbine blades are still usable even at the end of the projected 20-year life of a wind turbine. The reuse of the turbines is the most environmentally friendly and economically effective. In the incineration process, the turbine blades are combusted at high temperatures but the glass fibers present in the composites pose difficulties for efficient combustion and result in fly ash production which poses additional difficulties in the disposal. Some of the new thermal recycling technologies using pyrolysis and fluidized bed being developed can allow recycling of fibers but are still in research stages. The mechanical grinding of the turbine blades results in shredded composite mixtures which can be used in other industries as a filler and reinforcement of composites and cement concrete. Even with the developments in the technologies discussed int his section, there are currently no good mechanisms to recycle the turbine blades other than reuse of the turbine blades in the near future [19].

11.4.2 Solar energy resource sustainability

A PV solar energy system is generally composed of a solar panel, a solar controller, an inverter, and potentially some form of storage. A variety of semiconductor materials have been used for the production of PV panels and the panels are generally categorized into one of three generations.

11.4.2.1 First-generation solar panel resource requirements

The first generation of solar PV technologies also referred to as conventional, traditional, or wafer-based solar cells are constructed with crystalline silicon (c-Si), either monocrystalline silicon (mono-Si), or polycrystalline silicon (poly-Si). First-generation panels are the most dominant solar PV technology, accounting for about 95% of the total production in 2017. Polycrystalline silicon panels account for the largest share of the first generation panels, accounting for 65% of total production in 2017 [20].

Primary components of crystalline silicon solar panels include:
1. Solar PV cells:
   -Silica sand, (SiO2) which is then refined into elemental silicon.
2. Toughened Glass.
3. Extruded Aluminum Frame.
4. Encapsulation—ethylene vinyl acetate (EVA) film layers.
5. Polymer rear back sheet.
6. Junction box.

Silica sand, also referred to as quartz, is the second most abundant mineral in the Earth's continental crust. It makes up approximately 12% of the Earths land surface and 20% of the earth's crust [21]. Thus, material feedstocks for crystalline silicon PV are abundant and unlikely to limit growth. The materials for the glass, aluminum frame, EVL encapsulation, and polymer rear sheet are also relatively abundant materials [3].

11.4.2.2 Second-generation solar panel resource requirements

Second-generation solar cells are referred to as thin film (TF) solar cells. They are created by depositing a thin layer, or film, of PV material onto a substrate, commonly used substrates include metal, glass, and plastic. Semiconductor materials for TF cells include cadmium telluride (CdTe), copper indium gallium diselenide (CIGS), and amorphous thin-film silicon (a-Si, TF-Si) [22]. As the name suggests, TF cells are substantially thinner than first-generation cells, TF cell thickness varies from a few nanometers to a few micrometers while traditional silicon wafers can be up to 200 micrometers thick [22]. This reduction in materials required and manufacturing costs allows producers to sell TF cells at a much lower price than traditional panels, the trade-off to this low cost is that TF cells generally have lower efficiency [22]. In 2017, TF solar cells accounted for 5% of total global PV production with the majority of that production coming from CdTE and CIGS solar cells [20].

Second-generation cells are more likely to have their development limited as a result of resource availability. About 60–90 MT of tellurium is required per GW of CdTe PV cells. The global supply of Tellurium was estimated to be 630 MT/year in 2011 [23]. This supply may increase over time as Tellurium is primarily a byproduct of electrolytic copper refining, and global copper demand is expected to increase [3]. Each GW capacity of CIGS requires between 25 and 50 MT of indium, global supply of indium was estimated to be 1300 MT/year in 2011 [3,20]. Indium is primarily used for the production of transparent conductive oxidative coatings; this demand leads to projected increases in global supply but will also lead to competition between PV and non-PV applications of the material [3].

11.4.2.3 Third-generation solar panel resource requirements

The third generation of PV cells defines a broad range of technologies often described as "emerging PVs" which have the potential to overcome the Schockley–Quisser efficiency limit that exists for single bandgap solar cells (~33.7%) [24]. Third-generation technologies include quantum dot solar cells, perovskite solar cells, organic solar cells, dye-sensitized solar cells, and copper zinc tin sulfide solar cells [24]. As third-generation PV solar cells draw from such a large range of technologies while not yet having large-scale deployment, they will not be considered in this resource sustainability analysis.

11.4.2.4 Concentrating solar energy resource requirements

CSP systems generally require relatively abundant construction elements such as steel, aluminum, glass, concrete, and salts which are not expected to limit development [3].

11.4.2.5 Solar energy land use

The land area suitable for PV and CSP within the United States is orders of magnitudes larger than what would be required to supply the country's energy needs with solar panels. On average, the area required per MW capacity ranges between 2.5 and 3.32 ha/MWac for utility-scale PV and between 1.88 and 4 ha/MWac for CSP technologies [25]. Note that area requirements will be highly dependent on the location and the technology being used (for example tracking panels vs fixed). Taking the average value given for large scale PV (1.36 ha/GWh/year) and the total electricity consumption in 2018 (3970 TWH) an estimated total area of 54390 km^2 (0.55% of the US's total land area) would be required to meet the total US electricity demand. When looking at this reduced order estimate it is critical to keep in mind the discussion in Section 11.3 of this chapter about the infeasibility of powering the country with a non-dispatchable energy source.

There are current strategies being developed to mitigate the large land requirements by creating dual-use systems which combine solar PV with livestock and/or agricultural production. These dual-use systems, often referred to as "Agrivoltaic" systems have the potential to increase overall land-use efficiency while simultaneously reducing crop water demand and increasing panel efficiency as a result of the cooler panel temperatures that result from crop latent heat exchange [26–28].

11.4.2.6 Solar energy water use

Water used in Solar PV systems is minimal. No water is required directly for energy generation; however, water is used for periodic washing of the panels to remove dust which can lower panel efficiencies. The water required for washing PV panels has been estimated to be <19 L/MWh produced [3]. Water use for wet-cooled CSP technologies has been estimated to be in the range of 284–3790 L/MWh produced [29].

11.4.2.7 Solar energy end of life

Typical solar PV panel is made of a silicon PV cell sandwiched between EVA (ethylene/vinyl acetate copolymer) layers supported by a backboard made of Topotecan Hydrochloride and a tempered glass in the front framed in an aluminum alloy frame.

Mass composition of a typical solar panel is glass (54.7%), aluminum (12.7%), adhesive sealant (10%), silicon (3.1%), and other materials (19.5%). Silicon, cadmium, selenium, tellurium, gallium, molybdenum, indium and other relatively rare, and expensive metals are all located in the silicon PV cell layer. Based on the installed capacities of the solar panels, silicon-based (C-si), CIGS, CdTE are the most important types of the solar panels. The techniques for recycling solar PV panels are based on repairing old panels, separation of the modules into individual components, and complete recycling of the solar panels into individual raw materials for production of new solar PV panels [30]. Solar world company developed a thermal processing-based method for recycling solar PV panels which involve, thermal decomposition of the EVA and polymer layers by incineration at 600C in a process that protects the semiconductors from oxidation, recovery of the tempered glass, and aluminum components and re-etching of the silicon cells for use in new panels. A wet extraction process for recycling of TF solar PV panels was developed by the First Solar company in the United States. The process involves comminuation of the panels into 5–6 mm particles, leaching of the semiconductor layers in a stainless steel drum with acid and hydrogen peroxide. The leached solution is further processed to recover the semiconductor materials. The solids are separated into the glass and the plastic laminate materials. The metal-rich liquid can be precipitated at various pH to obtain high purity unrefined metal-rich slurries that can be used for production of new solar PV panels [31]. In a comparative analysis, Maani et al. [32], concluded that thermal methods of recycling are more ecofriendly compared to the mechanical and chemical methods. Recovery of Ag, Al, S, and glass should be prioritized in c-Si PVs while the recovery of tellurium, coper, and glass should be prioritized for CdTe PV cells.

11.5 Policy, governance, and social impact analysis

As described earlier, solar PV and wind energy technologies have significant positive environmental impacts and reduce the use of fossil fuels. These technologies are therefore thought of as credible alternatives for fossil fuels for electricity generation.

The United States has developed renewable portfolio standards (RPS) at both federal and state levels since the 1980s which were applicable to 56% of the total retail US electricity sales in 2016 [33]. The RPS policies vary from setting targets for both the amount and time frames for renewable electricity production, phase-out of older technologies, contracting requirements, cost caps, and other policy instruments. These RPS policies have contributed to over 50% of total US renewable energy generation growth since 2000 with most of the growth shifting from wind energy in the earlier decades to solar energy in recent years (Fig. 11.2).

As a response to the Paris Climate Accords [34] countries around the world have begun assessing the feasibility of meeting 100% of their energy needs using renewable sources. Although currently, only approximately 11% of US energy comes from renewables [4], individual states and cities have begun looking toward the possibility of 100% renewable electricity. In late 2018, California signed Senate Bill 100 (SB100) which commits to converting the state to 100% clean energy by

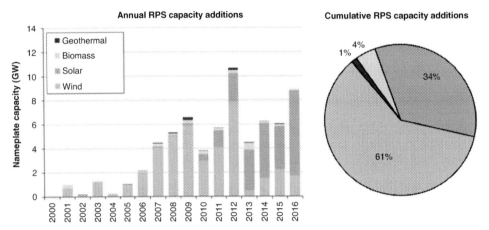

Fig. 11.2 Annual and cumulative RPS capacity additions by technology [33]. *RPS, renewable portfolio standards.*

2045, the bill breaks this into 60% renewable power and 40% energy from "zero carbon sources" [35]. Since then, nine other states (Arizona, Connecticut, Nevada, New Jersey, New Mexico, New York, Virginia, Washington, and Wisconsin) have all pledged to reach 100% clean energy by 2045. Additionally, four states (Hawaii, Maine, Minnesota, Rhode Island) and the territory of Puerto Rico have taken this pledge further and committed to providing 100% of their electricity from renewable sources with timelines ranging from 2030 to 2050 [36]. At a federal level, the SunShot program of the US Government was designed to increase the basic research to bring new technologies to the market along with many other initiatives for reducing the financial barriers for the adoption of solar PV technologies.

While there was a significant policy push for renewable energy through RPS, it is to be noted that there was greater support for centralized renewable energy schemes from the large utility companies rather than the distributed solar energy production [37]. The advocates of distributed renewable energy systems (mostly distributed rooftop solar PV with net metering) insist that the consumers must be able to connect to the grid and freely export excess electricity to the grid as it is an environmentally friendly alternative to other forms of electricity, the utilities put forth the arguments that the net metering results in consumers benefitting from the existing transmission grid connections and thus is unfair to the customers who do not have the capability for net metering. Grid connection fees and reduced rates for electricity produced from the distributed solar producers are the two strategies employed by the utility companies they have not always been successful in limiting the growth of the distributed solar PV production [37].

11.6 Conclusions and perspectives

Significant progress has been made in the past few decades in the adoption of renewable energy systems for electricity production has dramatically increased. Wind energy systems were adopted in the earlier decades and in the last one decade, the solar PV systems have become cost-competitive and their rate of installations has overtaken the wind energy sources. The wind and solar PV systems have been shown to be environmentally better alternatives compared to the fossil-based electricity generation systems. Many policy initiatives and Govt. incentives to support the adoption of wind and solar PV systems have been formulated both at the federal and state level in the United States. There has been greater support for larger centralized schemes rather than distributed systems from the utilities and power companies. Looking into the future, it will be expected that the advances in solar cell technologies, manufacturing, end of life technologies for recycle, and recovery of valuable materials will further contribute to increasing economic viability, positive environmental impacts, and more extensive adoption of solar PV and wind energy technologies.

References

[1] R. Gasch, J. Twele, "Historical Development of Windmills". In: R. Gasch, J. Twele, (Eds.), Wind Power Plants: Fundamentals, Design, Construction and Operation. Springer Berlin Heidelberg, Berlin, Heidelberg, 2012, pp. 15–45.
[2] Energy Information Administration, "History of Wind Power," Wind explained, 2020. https://www.eia.gov/energyexplained/wind/history-of-wind-power.php. [Accessed 16th May, 2021].
[3] C. Augustine, et al., Volume 2: Renewable Electricity Generation and Storage Technologies. In: Renewable Electricity Futures Study, NREL/TP-6A, National Renewable Energy Laboratory, Golden, CO, USA, 2012.
[4] Energy Information Administration, "EIA Electricity Data Browser," 2020. https://www.eia.gov/electricity/data/browser/. [Accessed 16th May, 2021].
[5] American Wind Energy Association, "AWEA 3rd quarter 2019 Public Market Report," 2019. https://cleanpower.org/resources/. [Accessed: 25 May, 2020].
[6] Energy Efficiency and Renewable Energy, "The history of solar," US Department of Energy. https://www1.eere.energy.gov/solar/pdfs/solar_timeline.pdf. [Accessed: 16th May, 2021].
[7] M.M. Hand and D. Baldwin S.; E. DeMeo; J.M. Reilly; T. Mai; D. Arent; G. Porro; M. Meshek; D. Sandor, "Renewable electricity futures study (entire report)," Golden, CO, 2012.
[8] P. Beiter, W. Musial, L. Kilcher, M. Maness, and A. Smith, 2017. An Assessment of the Economic Potential of Offshore Wind in the United States from 2015 to 2030 (No. NREL/TP--6A20-67675, 1349721). https://www.nrel.gov/docs/fy17osti/67675.pdf?xid=PS_smithsonian. [Accesed 16th May, 2021].
[9] S. Sharma, B. Siwach, S.K. Ghoshal, D. Mohan, Dye sensitized solar cells: from genesis to recent drifts, Renew. Sustain. Energy Rev. 70 (2017) 529–537.
[10] M.V. Kovalenko, Opportunities and challenges for quantum dot photovoltaics, Nat. Nanotechnol. 10 (12) (2015) 994.
[11] National Hydropower Association, "2018 pumped storage report," https://www.hydro.org/wp-content/uploads/2018/04/2018-NHA-Pumped-Storage-Report.pdf. [Accessed 16 May, 2021].
[12] Energy Information Administration, "Annual electric generator report," 2019. https://www.eia.gov/electricity/data/eia860/. [Accessed: 16 May, 2021].
[13] Energy Information Administration, "Preliminary monthly electric generator inventory," 2019. https://www.eia.gov/electricity/data/eia860m/. [Accessed: 16 May, 2021].
[14] B. Guezuraga, R. Zauner, W. Pölz, Life cycle assessment of two different 2 MW class wind turbines, Renew. Energy 37 (1) (2012) 37–44.

[15] IPCC, 2014: Climate Change 2014: Synthesis Report. Contribution of Working Groups I, II and III to the Fifth Assessment Report of the Intergovernmental Panel on Climate Change [Core Writing Team, R.K. Pachauri and L.A. Meyer (eds.)]. IPCC, Geneva, Switzerland, p 151. https://www.ipcc.ch/site/assets/uploads/2018/05/SYR_AR5_FINAL_full_wcover.pdf. [Accessed 16 May, 2021].

[16] A. Alsaleh, M. Sattler, Comprehensive life cycle assessment of large wind turbines in the US, Clean Technol. Environ. Policy 21 (4) (2019) 887–903.

[17] National Renewable Energy Laboratory, "Life cycle greenhouse gas emissions from solar photovoltaics," 2012.

[18] C. Mone, M. Hand, 2017. 2015 Cost of Wind Energy Review. Renewable Energy 115. NREL/TP-6A20-66861 National Renewable Enery Laboratory Golden, CO. USA. https://www.nrel.gov/docs/fy17osti/66861.pdf. [Accessed: 16 May, 2021].

[19] K. Borch, N.E. Clausen, G. Ellis, Environmental and social impacts of wind energy, in: H. Hvidtfeldt Larsen, L. Sønderberg Petersen (Eds.), DTU International Energy Report 2014: Wind energy—Drivers and Barriers for Higher Shares of Wind in the Global Power Generation Mix, Technical University of Denmark, 2014, pp. 86–90. https://backend.orbit.dtu.dk/ws/files/128071350/Wind_Turbine_Blades.pdf. [Accessed June 19, 2020].

[20] Fraunhofer Institute for Solar Energy Systems, "Photovoltaics report," 2020. https://www.ise.fraunhofer.de/content/dam/ise/de/documents/publications/studies/Photovoltaics-Report.pdf. [Accessed 16 May, 2021]

[21] S.P. Anderson, RS. Anderson, Geomorphology: The mechanics and Chemistry of Landscapes, Cambridge University Press, Cambridge, UK, 2010.

[22] K. Chopra, P. Paulson, V. Dutta, Thin-film solar cells: an overview, Prog. Photovoltaics 12 (Mar. 2004) 69–92.

[23] DOE (US Department of Energy), "Critical materials strategy," 2011. https://www.energy.gov/sites/prod/files/DOE_CMS2011_FINAL_Full.pdf. [Accessed 16 May, 2021]

[24] G. Conibeer, Third-generation photovoltaics, Mater. Today 10 (11) (2007) 42–50.

[25] S. Ong, C. Campbell, P. Denholm, R. Margolis, and G. Heath, "Land-use requirements for solar power plants in the United States," 2013. National Renewale Energy Laboratory. NREL/TP-6A20-56290 Golden, CO, USA. https://www.nrel.gov/docs/fy13osti/56290.pdf. [Accessed 16 May, 2021].

[26] H. Dinesh, J. Pearce, The potential of agrivoltaic systems, Renew. Sustain. Energy Rev. 54 (2016) 299–308.

[27] E.H. Adeh, J.S. Selker, and C.W. Higgins, "Remarkable agrivoltaic influence on soil moisture, micrometeorology and water-use efficiency," 2018. PLOS ONE 13, e0203256.

[28] G. Barron-Gafford, et al., Agrivoltaics provide mutual benefits across the food–energy–water nexus in drylands, Nat. Sustain. 2 (2019).

[29] C.F. Turchi, C.S. Wagner, M.J. Kutscher, "Water use in parabolic trough power plants: summary results from Worley Parsons' analyses," 2010. NREL/TP-5500-49468 Golden, CO, USA. https://www.nrel.gov/docs/fy11osti/49468.pdf. [Accessed 16 May, 2021].

[30] Y. Xu, L. Jinhui, Q. Tan, A.L. Peters, C. Yang, Global status of recycling waste solar panels: a review, Waste Manag. 75 (2018) 450–458.

[31] United Nations Treaty Collection, "Paris agreement," 2015. https://unfccc.int/files/meetings/paris_nov_2015/application/pdf/paris_agreement_english_.pdf. [Accessed 16 May, 2021].

[32] Public Utilities Code, SB-100 California Renewables Portfolio Standard Program: emissions of greenhouse gases, no. 100. California Senate, 2018.

[33] G. Barbose, U.S. Renwewables Portfolio Standards: 2017 Annual status updates, 2017. https://eta-publications.lbl.gov/sites/default/files/2017-annual-rps-summary-report.pdf. [Accessed 15 June, 2021].

[34] J. Tao, S. Yu, Review on feasible recylcing pathways and technologies of solar photovoltaic modules, Solar Energy Mat. Solar Cells 141 (2018) 108–124.

[35] T. Maani, I. Celik, M.J. Heben, R.J. Ellingson, D. Apul, Environmental impacts of recycling crystalline silicon (C-SI) and cadmium telluride (CDTE) solar panels, Sci. Total Env. 735 (2020) 138827.

[36] J. Heeter, B. Speer, B.M. Glick, International Best Practices for Renewable Portfolio Standard (RPS) Policies, National Renewable Energy Laboratory, Golden, CO, 2019. NREL/TP-6A20-72798. https://www.nrel.gov/docs/fy19osti72798.pdf. [Accessed June 19, 2020].

[37] D. Hess, The politics of niche-regime conflicts: Distributed solar energy in the United States, Env. Innov. Soc. Transform. 19 (2016) 42–50.

CHAPTER TWELVE

Biofuels and bioproducts in India

Ganti S. Murthy[a,b]
[a]Biological and Ecological Engineering, Oregon State University, Corvallis, OR, United States, [b]Biosciences and Biomedical Engineering, Indian Institute of Technology, Indore, India

12.1 Introduction

India has the largest percentage of arable land in the world and its rich agriculture supports the second largest population in the world and Indian subcontinent accounts for 12% of the world's biodiversity despite having only a 5% landmass of the world. India's historical emission contribution to greenhouse gases is <3% while it is 25% for the United States, 23% for the European Union, and 6.8% for China. India also imports over 80% of its fossil fuels which support its bustling economy which is moving millions of people from extreme poverty to the middle class. India is the third-largest user of energy in the world and has much lower per capita emissions compared to other developed economies. Still, India has made a voluntary commitment to reduce its per capita greenhouse gas emissions by 33%–35 % below 2005 levels by 2030 under the Paris Agreement. Meeting this commitment requires large-scale introduction of renewable energy sources to meet the sustainable development goals while protecting the environment. Due to the large agricultural base and multiple cropping systems, over 600 MMT of agricultural residues are produced every year. This biomass is used extensively in India for domestic cooking, animal fodder, and bedding, and as manure. The use of biomass for transportation fuels is relatively new in India and has been experimentally tried in the last two decades starting with 2% sugarcane ethanol blending pilot programs in 2001. The biofuels policies were formulated in 2009 which formulated the policies for the use and expansion of bioethanol and biodiesel. This policy was revised again in 2018 as national biofuels policy 2018 sets a goal of blending 20% ethanol and biodiesel in petrol and diesel fuels, respectively, by 2030. India currently uses 232 billion L of petrol every year.

12.2 Systems analysis of biofuel technologies

There have been many flip-flops in the policies and nonachievement of targets set in earlier biofuel programs and policies. In that context, it becomes necessary to estimate the realistic biofuel potential from agricultural residues in India. The system analysis approach will be used to assess the feasibility of producing 10%, 20%, and 50% of the Indian liquid transportation fuel requirements from agricultural residues. Additionally, the potential for producing biodiesel from oilseeds, and other valuable bioproducts will be assessed. It is important to understand how much biofuels can be produced, what are the potential feedstocks, selection of technologies for production of biofuels based on feedstocks, the geographical distribution of biofuel feedstocks and the economic viability. The techno-economic feasibility, environmental, and resource impacts will be assessed using the methods discussed in earlier chapters.

The goal of this analysis is to:
1. Determine the levels of ethanol that can be produced from agricultural residues without any increase in cultivated areas.
2. Understand the technical feasibility, economics, environmental impacts, use of resources, environmental risks, and policy support for ethanol from agricultural residues and biodiesel from nonedible oilseed crops.

The scope of the analysis will be focused on the resource sustainability assessment with limited analysis of the technologies and the environmental impacts. Currently, the resource sustainability issues are the most fundamental issues compared to relatively well-known technology challenges which will have similar strengths and weaknesses as seen in other countries and analyzed thoroughly in Chapter 9. The operations between the farm to the burning of ethanol in an E10 mix in a regular sedan passenger vehicle. Emissions resulting from use in other types of vehicles or at higher blends will not be part of this analysis.

12.3 Resource assessment for bioethanol from agricultural residues

Agricultural crops are produced in three seasons (Kharif, rabi, and zaid) in India. Of these, Kharif and Rabi are the main cropping seasons, and over 85% of the agricultural residues are produced during these two seasons. Top agricultural crops in India are rice, wheat, sugarcane, cotton, pulses, and soybean. A comprehensive assessment of biomass availability and its potential use for bioethanol/bioproducts is presented here with a focus on the second-generation biofuels from lignocellulosic feedstocks. Biomass availability study [1] will be used as the data source for biomass availability. The average ethanol yield from biomass is assumed to be 260 L/ton (using the calculations from Chapter 9). The analysis is aimed at answering the following questions:
- What are the feedstocks available and what is their geographical distribution?
- What is the largest size of bioethanol plant possible in each state of India?
- What percentage of India's petrol needs can this biomass meet?

The biomass generation, consumption, surplus agricultural residues are reported in Tables 12.1–12.3 respectively. Surplus biomass density is reported in Table 12.4. The estimated ethanol potential is reported in Table 12.5. The total generation of biomass from all types of agricultural residues is over 600 MMT (Table 12.1). Top agricultural

Table 12.1 Annual agricultural residue generation (MMT/year) in India.

State	Total	Rice straw	Wheat straw	Sugarcane bagasse	Sugarcane tops	Oilseecs various	Maize stover	Rice husk	Pulses various	Cotton stalk	Jowar stalk	Water hyacinth	Bajra stalk
Total	604	112.0	109.9	101.3	97.8	57.7	22.7	22.4	18.9	18.9	15.6	15.0	12.2
Uttar Pradesh	138	13.3	36.3	38.2	36.8	2.4	1.7	2.7	2.6		0.5	1.0	1.9
Maharashtra	81	3.1	2.4	22.4	21.6	8.8	1.7	0.6	3.1	6.4	8.2	0.7	1.5
Andhra Pradesh	43	14.2		6.2	6.0	3.2	3.7	2.8	1.8	2.0	0.9	1.8	0.1
Punjab	41	12.2	21.2	1.7	1.7	0.2	0.7	2.4		1.2			
Tamil Nadu	39	7.9		11.7	11.3	2.6	1.1	1.6	0.4	0.3	0.6	1.7	0.1
Karnataka	35	4.1	0.3	8.2	7.9	2.7	4.1	0.8	1.2	0.8	2.9	1.4	0.3
Madhya Pradesh	37	1.6	10.6	0.8	0.8	13.8	1.3	0.3	4.3	1.3	1.3	0.9	0.4
Rajasthan	33		10.2			12.3	1.7		2.0	0.7	0.8	0.7	5.0
Gujarat	31	1.7	4.4	4.5	4.3	6.1	0.5	0.3	0.8	4.9	0.2	1.5	1.5
Haryana	30	4.0	14.6	2.7	2.6	2.0		0.8	0.2	1.1	0.1		1.5
West Bengal	26	17.7	1.2	0.4	0.3	1.5	0.4	3.5	0.2			1.2	
Bihar	20	6.0	5.7	1.7	1.6	0.4	2.6	1.2	0.6			0.3	

Table 12.2 Annual agricultural residue consumption (MMT/year) in India.

State	Total	Rice straw	Wheat straw	Bagasse	Oilseeds (various)	Rice husk	Maize stover	Sugarcane tops	Jowar stover	Pulse (various)	Bajra stalk	Cotton stalk	Ragi stalk
Total	461	103.56	100.87	96.26	40.36	21.96	21.52	19.55	14.00	13.23	11.01	7.43	4.12
Uttar Pradesh	100	12.28	33.39	36.27	1.71	2.62	1.65	7.37	0.47	1.84	1.69	0.00	0.00
Maharashtra	53	2.84	2.17	21.27	6.18	0.60	1.64	4.32	7.38	2.14	1.39	2.57	0.24
Punjab	37	11.56	19.48	1.63	0.13	2.38	0.68	0.33	0.00	0.00	0.00	0.48	0.00
Andhra Pradesh	33	13.10	0.00	5.87	2.26	2.79	3.51	1.19	0.80	1.26	0.07	0.67	0.13
Madhya Pradesh	29	1.51	9.78	0.76	9.65	0.32	1.20	0.15	1.18	2.98	0.33	0.52	0.00
Karnataka	28	3.81	0.28	7.76	1.88	0.81	3.88	1.58	2.57	0.83	0.25	0.30	2.71
Tamil Nadu	27	7.30	0.00	11.13	1.79	1.55	1.08	2.26	0.57	0.27	0.13	0.11	0.44
Rajasthan	27	0.00	9.21	0.00	8.59	0.00	1.60	0.00	0.72	1.38	4.47	0.29	0.00
Haryana	25	3.84	13.42	2.59	1.38	0.79	0.00	0.53	0.06	0.13	1.33	0.44	0.00
West Bengal	23	16.28	1.07	0.34	1.08	3.47	0.36	0.07	0.00	0.14	0.00	0.00	0.04
Gujarat	20	1.53	4.00	4.23	4.27	0.33	0.51	0.86	0.20	0.55	1.33	1.97	0.03
Bihar	17	5.51	5.22	1.61	0.25	1.17	2.45	0.33	0.00	0.41	0.00	0.00	0.02
Orissa	11	7.53	0.00	0.34	0.30	1.60	0.00	0.07	0.02	0.33	0.00	0.00	0.09

Table 12.3 Annual agricultural residue surplus (MMT/year) in India.

State	Total	Sugarcane tops	Oilseeds various	Water hyacinth	Cotton stalk	Wheat straw	Rice straw	Sugarcane bagasse	Pulses various	Bamboo	Maize cobs	Jowar stalk
Total	163.3	79.5	17.3	14.0	11.4	9.1	8.5	6.4	5.7	3.3	1.7	1.6
Uttar Pradesh	38.4	29.5	0.7	0.9		2.9	1.1	1.9	0.8		0.1	0.1
Maharashtra	28.5	17.3	2.6	0.7	3.9	0.2	0.2	1.1	0.9	0.2	0.1	0.8
Tamil Nadu	13.4	9.0	0.8	1.7	0.2		0.6	0.6	0.1		0.1	0.1
Andhra Pradesh	11.7	4.8	1.0	1.7	1.3		1.1	0.3	0.5	0.1	0.3	0.1
Karnataka	11.5	6.3	0.8	1.3	0.5	0.0	0.3	0.4	0.4	0.1	0.3	0.3
Gujarat	10.9	3.4	1.8	1.4	3.0	0.3	0.1	0.2	0.2		0.0	0.0
Madhya Pradesh	9.3	0.6	4.1	0.8	0.8	0.9	0.1	0.0	1.3	0.2	0.1	0.1
Rajasthan	7.2		3.7	0.6	0.4	1.0			0.6		0.1	0.1
West Bengal	6.3	1.6	0.5	1.1		0.1	1.4	1.4	0.1		0.0	
Haryana	5.1	2.1	0.6		0.7	1.2	0.2	0.1	0.1			0.0
Punjab	4.7	1.3	0.1		0.7	1.7	0.6	0.1			0.1	
Orissa	3.5	0.3	0.1	2.0			0.7	0.0	0.1	0.2		
Bihar	3.4	1.3	0.1	0.3		0.5	0.5	0.1	0.2		0.2	0.0

Table 12.4 Annual surplus biomass density.

State	Total	Sugarcane tops	Oilseeds various	Water hyacinth	Cotton stalk	Wheat straw	Rice straw	Sugarcane bagasse	Pulses various	Bamboo
Total	49.7	24.2	5.3	4.3	3.5	2.8	2.6	2.0	1.7	1.0
Uttar Pradesh	160.9	123.5	3.1	3.9	0.0	12.2	4.5	8.0	3.3	0.0
Haryana	115.1	47.7	13.4	0.0	14.9	26.4	4.6	3.1	1.3	0.0
Tamil Nadu	102.7	69.6	5.9	12.7	1.3	0.0	4.9	4.5	0.9	0.0
Punjab	92.7	26.3	1.1	0.0	14.4	33.6	12.1	1.7	0.0	0.0
Maharashtra	92.6	56.2	8.6	2.2	12.5	0.6	0.8	3.6	3.0	0.6
West Bengal	71.4	18.1	5.2	12.7	0.0	1.0	16.0	15.7	0.7	0.0
Karnataka	60.1	32.9	4.2	6.8	2.4	0.1	1.7	2.1	1.9	0.5
Gujarat	55.5	17.5	9.3	7.3	15.1	1.8	0.7	1.1	1.2	0.0
Andhra Pradesh	42.5	17.3	3.5	6.0	4.8	0.0	4.1	1.1	2.0	0.3
Bihar	36.1	13.9	1.1	3.4	0.0	4.8	5.1	0.9	1.9	0.0
Tripura	31.5	0.0	0.0	0.0	0.0	0.0	0.0	0.0	0.0	31.5
Madhya Pradesh	30.2	2.0	13.4	2.7	2.5	2.8	0.4	0.1	4.1	0.7
Uttarakhand	29.6	25.1	0.0	0.0	0.0	2.2	0.0	1.6	0.0	0.0
Orissa	22.6	1.8	0.8	12.6	0.0	0.0	4.2	0.1	0.9	1.6
Rajasthan	21.1	0.0	10.8	1.8	1.3	3.0	0.0	0.0	1.7	0.0
Nagaland	20.0	0.0	0.0	0.0	0.0	0.0	0.0	0.0	0.0	20.0

Biofuels and bioproducts in India

Table 12.5 State-wise cellulosic ethanol potential of India.

State	Total	Sugarcane tops	Oilseeds various	Water hyacinth	Cotton stalk	Wheat straw	Rice straw	Sugarcane bagasse	Pulses various	Bamboo	Maize cobs	Jowar stalk
Total	42.5	20.7	4.5	3.6	3.0	2.4	2.2	1.7	1.5	0.9	0.4	0.4
Uttar Pradesh	10.0	7.7	0.2	0.2	0.0	0.8	0.3	0.5	0.2	0.0	0.0	0.0
Maharashtra	7.4	4.5	0.7	0.2	1.0	0.0	0.1	0.3	0.2	0.0	0.0	0.2
Madhya Pradesh	3.5	2.4	0.2	0.4	0.0	0.0	0.2	0.2	0.0	0.0	0.0	0.0
Rajasthan	3.0	1.2	0.3	0.4	0.3	0.0	0.3	0.1	0.1	0.0	0.1	0.0
Punjab	3.0	1.6	0.2	0.3	0.1	0.0	0.1	0.1	0.1	0.0	0.1	0.1
Tamil Nadu	2.8	0.9	0.5	0.4	0.8	0.1	0.0	0.1	0.1	0.0	0.0	0.0
Karnataka	2.4	0.2	1.1	0.2	0.2	0.2	0.0	0.0	0.3	0.1	0.0	0.0
West Bengal	1.9	0.0	1.0	0.2	0.1	0.3	0.0	0.0	0.2	0.0	0.0	0.0
Haryana	1.6	0.4	0.1	0.3	0.0	0.0	0.4	0.4	0.0	0.0	0.0	0.0
Orissa	1.3	0.5	0.2	0.0	0.2	0.3	0.1	0.0	0.0	0.0	0.0	0.0
Gujarat	1.2	0.3	0.0	0.0	0.2	0.4	0.2	0.0	0.0	0.0	0.0	0.0
Others	0.9	0.1	0.0	0.5	0.0	0.0	0.2	0.0	0.0	0.1	0.0	0.0
Andhra Pradesh	0.9	0.3	0.0	0.1	0.0	0.1	0.1	0.0	0.0	0.0	0.0	0.0

residues (>90 MMT/year) are rice straw (112 MMT/Year), wheat straw (109.9 MMT/year), sugarcane bagasse, and sugarcane tops. However, the consumption of the residues for various uses including cooking, cattle feed, and bedding requires more than 75% of the total generated biomass (Table 12.2). For example, sugarcane bagasse is almost completely utilized for energy production in sugar mills, and ~101 MMT/year wheat straw out of 110 MMT/year is used for various applications. This leaves very little biomass for the production of biofuel/bioproducts production. Therefore, the surplus availability of the biomass is an important factor to determine the available biomass for biorefineries (Table 12.3). Surprisingly, the sugarcane tops and water hyacinth are the two major sources of available biomass. Wheat straw, sugarcane bagasse, and rice straw although produced in large quantities (>100 MMT/year), their actual surplus availability is much less (<10 MMT/year). An important factor in the establishment of any biorefinery is the available biomass density. High biomass density is preferred as it reduced the footprint of the biorefinery and minimizes the biomass transportation costs while reducing the emissions associated with the transportation. Although the overall surplus biomass density for India is just 50 MT/km^2, some of the states such as Uttar Pradesh, Haryana, Punjab, and Tamil Nadu have relatively higher biomass densities (Fig 12.1D). A more careful look at the available surplus biomass indicates that most of the surplus biomass can be primarily attributed to sugarcane tops, water hyacinth, and wheat straw. Based on the surplus biomass from agricultural residues, a total ethanol potential of 42.5 billion L/year with sugarcane tops, oilseed crop residues, water hyacinth, cotton stalk, and wheat straw being the top five crop residues for ethanol production. The annual potential of 42.5 billion L represents about 18% of the annual petroleum needs. The primary advantage of bioethanol production from agricultural residues is that they are already produced, and many times burned in the field for lack of better economic opportunities and creates air pollution, health hazards for the general population. Therefore, utilizing this surplus biomass may help achieve multiple goals even though the total amount of ethanol that could be produced is <20% of the annual needs.

12.4 Techno-economic analysis

Agricultural residues produced in India are varied and among them, rice straw has been studied comprehensively for bioethanol production. Rice straw is an established agricultural residue that is produced in large quantities. Therefore, the economic analysis is focused on utilizing rice straw as a model feedstock for bioethanol production from agricultural residues. It is expected that the overall results will be similar for other feedstocks.

Overall surplus biomass density in India is ~50 MT/km^2 (Table 12.4) and the top five states with surplus agricultural residues have a surplus biomass density close to 100 MT/km^2. Assuming a maximum transport distance of 20 km from the field to

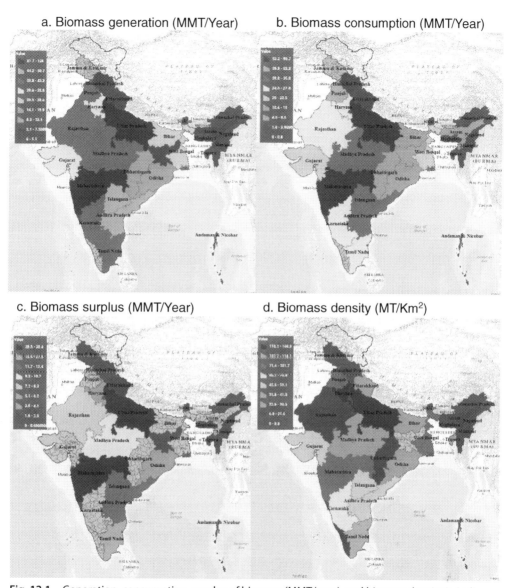

Fig. 12.1 Generation, consumption, surplus of biomass (MMT/year), and biomass density in India.

the plant, each plant can therefore have access to biomass in a 1257 km² area. For the average Indian scenario, this translates to a total of 62,800 MT/year (i.e., 172 MT/day) which can be approximated to 150 MT/day availability considering other factors that may limit availability. A similar analysis for the top five producers indicates an annual and daily availability of biomass as 125,600 MT/year and 300 MT/day respectively. Therefore, the two scenarios that will be examined in the techno-economic analysis are for bioethanol plants with 300 MT/day and 150 MT/day, respectively, representing the average biomass availability in the top five biomass producing states and the average scenario for India. Based on this background information, the payback period, NPV, and internal rate of return (IRR) of a 300 and 150 MT/day bioethanol plant using rice straw as the model feedstock were calculated.

A SuperPro model reported by Rajendran and Murthy [2] was used as the base model. The actual parameters in the model were modified for the Indian conditions. The composition of the rice straw and the fuel used for the harvesting and transportation was obtained from Sreekumar et al. [3].

The rice straw price was assumed to be $74/MT. The selling price of ethanol at the producer plant was $0.67/L ethanol, electricity was sold to the grid at $0.17/KWh and gypsum was priced at $30/MT. Adjusted basic rate of 3.0 $/h resulting in a total cost of $5.1/h including all benefits was assumed for all labor requirements in the plant reflecting the scenarios specific for India. The assumptions for the techno-economic analysis are summarized in Table 12.6.

The simulations were performed in SuperPro V 8.5 by Intelligen Inc. Most of the materials cost for the plant was contributed from the feedstock (~60%) and the enzymes (~30%) in line with the observations made by other studies (Table 12.7). The economic analysis summary is presented in Table 12.8.

The plant uses about 99,000 MT of rice straw/year as feedstock and produces 28.5 million L /year ethanol, 22 million KW of electricity, and 2836 MT/year of gypsum as coproducts (Tables 12.7 and 12.8). The total equipment cost for the 300 MT/Day plant was $18.65 million. The total capital costs were $53.3 million and the annual operating cost was $15.9 million. The return of investment is 9.19 % while the payback period is 10.89 years. The IRR is 6.95% before taxes and 4.61% after taxes. The NPV is $2.0 million ($i$ = 4.0%), -4.4 million (i = 6.0%), and −9.3 million (i = 8.0%) (Table 12.9). The economic analysis summary for t 150 MT/day plant (Table 12.10) indicates that the minimum performance criteria (NPV > 0, unit production price of ethanol < selling price, etc.) are not met and the 150MT/day plant is an economically unviable venture. Further analysis of the performance of even the 300 MT/day plant indicates that the plant is only

Table 12.6 Project economic evaluation parameters.

Year of analysis	2020		NPV interest		
Construction start	2020		Low	4	%
Construction period	30	Months	Medium	6	%
Startup period	4	Months	High	7	%
Project life time	20	Years	Depreciation	10	Years
Inflation	4	%	Salvage value	5	%

Table 12.7 Material input costs for 300 MT/day bioethanol plant.

Bulk material	Unit cost ($)	Annual amount	Unit	Annual cost ($)	%
Water	0.002	184,227,639	kg	295,870	4.87
gasoline	0.800	109,612	kg	87,689	1.44
Calcium hydroxide	0.100	629,022	kg	62,902	1.04
Air	0.000	215,267,784	kg	0	0.00
Boiler water	0.000	180,426,146	kg	63,871	1.05
Sulfuric acid	0.035	918,117	kg	32,134	0.53
Rice_straw	74.000	49,500	MT	3,663,000	60.31
Cellulase	0.517	3,427,465	kg	1,771,999	29.17
DAP	0.210	19,800	kg	4158	0.07
YeastMx	1.860	49,500	kg	92,070	1.52
TOTAL				6,073,693	100.00

Table 12.8 Economic analysis summary for 300 MT/day bioethanol plant.

A.	Direct fixed capital	52,003,000	$
B.	Working capital	1,267,000	$
C.	Startup cost	2,600,000	$
D.	Up-front R&D	0	$
E.	Up-front royalties	0	$
F.	Total investment (A+B+C+D+E)	55,870,000	$
G.	Investment charged to this project	55,870,000	$
H.	Revenue Rates		
	Gypsum (revenue)	2,836,812	kg /year
	Ethanol (main revenue)	28,529,305	L(STP) /year
	P-33(revenue)	22,046,934	kW-h/year
I.	Revenue price		
	Gypsum (revenue)	30.00	$/1000 kg
	Ethanol (main revenue)	0.67	$/L(STP)
	P-33(revenue)	0.17	$/kW-h
J.	Revenues		
	Gypsum (revenue)	85,000	$/year
	Ethanol (main revenue)	19,115,000	$/year
	P-33(revenue)	3,748,000	$/year
	Total revenues	22,948,000	$/year
K.	Annual operating cost (AOC)		
	AOC	22,627,000	$/year
L.	Unit production cost/revenue		
	Unit production cost	980.54	$/MT MP
	Unit production revenue	994.44	$/MT MP
M.	Gross profit (J–K)	321,000	$/year
N.	Taxes (40%)	128,000	$/year
O.	Net profit (M-N + depreciation)	5,133,000	$/year
	Gross margin	1.40	%
	Return on investment	9.19	%
	Payback time	10.89	years

Table 12.9 Cash flow analysis (Thousand $) for 300 MT/day bioethanol plant.

Year	Capital investment	Debt finance	Sales revenues	Operating cost	Gross profit	Loan payments	Depreciation	Taxable income	Taxes	Net profit	Net cash flow
1	−10,401	0	0	0	0	0	0	0	0	0	−10,401
2	−10,401	0	0	0	0	0	0	0	0	0	−10,401
3	−11,667	0	11,474	15,905	−4431	0	4940	0	0	509	−11,158
4	−10,401	0	22,948	22,627	321	0	4940	321	128	5133	−5268
5	−10,401	0	22,948	22,627	321	0	4940	321	128	5133	−5268
6	0	0	22,948	22,627	321	0	4940	321	128	5133	5133
7	0	0	22,948	22,627	321	0	4940	321	128	5133	5133
8	0	0	22,948	22,627	321	0	4940	321	128	5133	5133
9	0	0	22,948	22,627	321	0	4940	321	128	5133	5133
10	0	0	22,948	22,627	321	0	4940	321	128	5133	5133
11	0	0	22,948	22,627	321	0	4940	321	128	5133	5133
12	0	0	22,948	22,627	321	0	4940	321	128	5133	5133
13	0	0	22,948	17,687	5261	0	0	5261	2104	3157	3157
14	0	0	22,948	17,687	5261	0	0	5261	2104	3157	3157
15	0	0	22,948	17,687	5261	0	0	5261	2104	3157	3157
16	0	0	22,948	17,687	5261	0	0	5261	2104	3157	3157
17	0	0	22,948	17,687	5261	0	0	5261	2104	3157	3157
18	0	0	22,948	17,687	5261	0	0	5261	2104	3157	3157
19	0	0	22,948	17,687	5261	0	0	5261	2104	3157	3157
20	3867	0	22,948	17,687	5261	0	0	5261	2104	3157	7023

IRR/NPV summary
IRR before taxes 6.95 % Interest % 4.00 6.00 8.00
IRR after taxes 4.61 % NPV 2071.00 −4439.00 −9333.00

Table 12.10 Economic analysis summary for 150 MT/day bioethanol plant.

A.	Direct fixed capital	50,086,000	$
B.	Working capital	686,000	$
C.	Startup cost	2,504,000	$
D.	Up-front R&D	0	$
E.	Up-front royalties	0	$
F.	Total investment (A+B+C+D+E)	53,276,000	$
G.	Investment charged to this project	53,276,000	$
H.	Revenue Rates		
	Gypsum (revenue)	1,419,087	kg/year
	Ethanol (main revenue)	14,264,659	L(STP)/year
	P-33(revenue)	10,163,138	kW-h/year
I.	Revenue price		
	Gypsum (revenue)	30.00	$/1000 kg
	Ethanol (main revenue)	0.67	$/L(STP)
	P-33(revenue)	0.17	$/kW-h
J.	Revenues		
	Gypsum (revenue)	43,000	$/year
	Ethanol (main revenue)	9,557,000	$/year
	P-33(revenue)	1,728,000	$/year
	Total revenues	11,328,000	$/year
K.	Annual operating cost (AOC)		
	AOC	15,893,000	$/year
L.	Unit production cost/revenue		
	Unit production cost	1377.48	$/MT MP
	Unit production revenue	981.76	$/MT MP
M.	Gross profit (J−K)	−4,566,000	$/year
N.	Taxes (40%)	0	$/year
O.	Net profit (M−N + depreciation)	192,000	$/year
	Gross margin	−40.31	%
	Return on investment	0.36	%
	Payback time	276.97	years

MP = Total flow of stream "Ethanol"

marginally profit under the most optimistic scenario ($i = 4\%$ for NPV calculations). Commercializing technologies with marginal economic performance often needs support from the government in the form of incentives.

Conclusion: the economic analysis of the agricultural residue-based bioethanol production indicates that the process may not be economically feasible if the investors perceive the process to be of high risk and demand a higher discount rate. Even under the positive NPV scenario (discount rate = 4%), the payback period of 10.89 years is still unacceptably high for most investors in the top five surplus biomass states of India.

12.5 Environmental impact assessment

Goal and scope: the goal of this life cycle assessment is to evaluate the environmental impacts of the rice straw ethanol process in India. The scope of the analysis includes all the processes from rice straw collection, transportation, and processing of the biomass to bioethanol in a biorefinery. The purpose of the analysis is to quantify the environmental impacts of agricultural residues in general and rice straw ethanol in particular in the Indian context. The functional unit of analysis is 1000 MJ of ethanol at the biorefinery gate. Open LCA software V1.10.3 with Ecoinvent database 3.3 will be used for all analyses.

Life cycle inventory: the rice straw collection is assumed to be taking place from a 20 km radius around the plant. The rice straw is collected in the field and consolidated in the collection centers near the field. The total fuel requirements for the collection and transportation of biomass to the biorefinery in 6-ton capacity tractors requires 5.76 L diesel/MT biomass (.e 4.78 Kg diesel/MT biomass) as reported by Sreekumar et al. [3].

The biorefinery process was similar to the one reported by Rajendran and Murthy [2]. Briefly, the process begins with eh arrival of the biomass in trucks which is washed and shredded in a knife mill. The shredded straw is mixed with the process and fresh waters and pretreated using the dilute acid pretreatment method. The solids and liquids are separated after the pretreatment and the liquid fraction is detoxified with the addition of calcium hydroxide. The resulting gypsum and other particulate matter are recovered as a coproduct. The solid fraction and detoxified liquid fraction are combined and fermented after adjustment of pH, the addition of cellulases and yeast. The fermented broth is distilled to separate ethanol which is further processed using molecular sieves in a pressure swing adsorption process to recover anhydrous ethanol, the main product of the process. The remaining solids are dewatered and sent to oiler for the production of steam. Any steam not used in the process is used to produce electricity, the major coproduct for the process. Process waters are treated in an anaerobic digestion process followed by aerobic bio-oxidation to reduce the COD of the wastewaters and the resulting methane is used to as a supplemental fuel in the steam production process. The water is also treated in multiple-effect evaporator to recover clean process water that is recycled back to the front of the process. The syrup from the multiple effect evaporators is used as a supplemental fuel for steam production.

Detailed inputs and outputs for the process are presented in Table 12.11. The inputs for the processing of biomass were obtained from the detailed techno-economic analysis model described in the above section.

Life cycle impact assessment: the ReCipE 2016 (H) was used as the impact assessment for evaluating the environmental impacts. The heirarchist (H) impact assessment model is a consensus model indicating the scientific consensus on the values of the

Table 12.11 Life cycle inventory of rice straw ethanol process.

Flow	Quantity	Reference unit
Inputs		
Rice straw	99,000	Metric ton
Diesel for transportation and collection	570,900	L
Cellulase	6,854,930	Kg
Sulfuric acid	1,83,274	Kg
Calcium hydroxide (as lime)	1,257,398	Kg
Phosphate as diammonium phosphate	19,800	Kg
Yeast	49,500	Kg
Water	307,95,139	m3
Outputs		
Gypsum	2,836,812	Kg
Electricity	22,046,934	Kwh
Ethanol	28,529,305	L (STP)

reported environmental impacts. The method has 18 midpoint indicators and three endpoint indicators. The midpoint indicators for the process are shown in Table 12.12.

Uncertainty analysis: LCA often is incomplete and has large uncertainties. Therefore, an uncertainty analysis must be conducted to understand the uncertainties in the environmental impacts. The ethanol production efficiency is dependent on multiple factors such as composition, pentose sugar fermentation efficiency, etc., Considering a range of variations in the moisture content (10%–15%), cellulose

Table 12.12 Life cycle impacts of rice straw ethanol using ReCiPe 2016 (H) method.

Impact category	Result	Reference unit
Fine particulate matter formation	3.473E−02	kg PM2.5 eq
Fossil resource scarcity	1.446E+00	kg oil eq
Freshwater ecotoxicity	1.141E−01	kg 1,4-DCB
Freshwater eutrophication	7.030E−04	kg P eq
Global warming	2.659E+01	kg CO_2 eq
Human carcinogenic toxicity	6.055E−02	kg 1,4-DCB
Human noncarcinogenic toxicity	3.733E+00	kg 1,4-DCB
Ionizing radiation	7.747E−02	kBq Co-60 eq
Land use	1.963E−02	m2a crop eq
Marine ecotoxicity	1.629E−01	kg 1,4-DCB
Marine eutrophication	3.778E−05	kg N eq
Mineral resource scarcity	1.877E−02	kg Cu eq
Ozone formation, human health	5.687E−03	kg NOx eq
Ozone formation, terrestrial ecosystems	5.871E−03	kg NOx eq
Stratospheric ozone depletion	1.218E−06	kg CFC11 eq
Terrestrial acidification	1.167E−01	kg SO2 eq
Terrestrial ecotoxicity	9.114E+00	kg 1,4-DCB
Water consumption	5.512E+00	m3

content (30%–40%), ash (0.5%–5%), and xylose fermentation efficiency (40%–90%), the range of ethanol productivity was estimated as 293.0–362.4 L/Kg biomass. The cellulase enzyme production environmental impacts are highly uncertain and a range of 1.0–10.0 Kg/Kg cellulase with a mode value of 2.264 Kg/Kg cellulase as reported by Kumar and Murthy [4] has been used. All other inputs were assumed to have a normal distribution with a standard deviation equal to 10% of the mean value. A detailed Monte–Carlo Analysis was performed, and the results are presented in Table 12.13. The results demonstrate that there can be significant uncertainties in the environmental impacts as can be seen from the values for the difference between the mean values calculated for the base case scenarios and the values obtained through uncertainty analysis (Fig. 12.2). The wide distribution of the uncertainties for some categories such as global warming compared to other categories such as freshwater eutrophication demonstrates that the propagation of uncertainties in the inputs to output environmental impact categories is not always uniform across various impact categories. In such contexts, the sensitivity analysis will indicate which input uncertainties have the largest impacts for various impact categories and can be used to prioritize data collection and validation for the most sensitive inputs. This will ensure optimum use of available resources for conducting life cycle assessments.

12.6 Policy and social aspects of biofuels in India

India's biofuels programs started in 2001 with a 5% ethanol blending program. The program was expanded in 2003 to include biodiesel and an aim to achieve a 20% blending of biodiesel by 2011–2012. The 5% ethanol blending was made mandatory in 2007. First national policy on biofuels (NPBs) were promulgated in 2009 with an aspirational goal of 20% blending of ethanol and biodiesel by 2017. Many policy flips-flops, insufficient capacity, and lack of translation of laboratory results to the field, and lack of economic viability resulted in the nonachievement of the aspirational goals set forth in the NPB 2009 [5]. Much of the focus of the NPB 2009 was on addressing the supply side challenges of the biofuels. These and other deficiencies were addressed in a comprehensive. A revised NPB 2018 was formulated with a mandatory goal of 20% ethanol bending and 5% biodiesel bending by 2030. In a departure from the NPB 2009, the NPB 2018 allows the production of first-generation biofuels from starch and sugar-rich feedstocks such as sugar beet, sweet sorghum, corn, and cassava and damaged cereals like wheat and broken rice. Additionally, surplus food grains are also allowed for bioethanol production with prior approval of the National Biofuel Coordination Committee.

Despite a severe shortage of animal fodder, agricultural residues in many parts of India are burned in the field. Lack of adequate manpower, use of mechanical harvesting, and short inter-cropping season encourages farmers to burn the crop residues in

Table 12.13 Range of environmental impacts based on uncertainty analysis.

Impact category	Reference unit	Mean	Standard deviation	Median	5% Percentile	95% Percentile
Fine particulate matter formation	kg PM2.5 eq	3.357E−02	3.339E−03	3.356E−02	2.819E−02	3.914E−02
Fossil resource scarcity	kg oil eq	1.393E+00	9.896E−02	1.394E+00	1.234E+00	1.555E+00
Freshwater ecotoxicity	kg 1,4-DCB	2.003E−01	1.596E−01	1.572E−01	7.037E−02	4.715E−01
Freshwater eutrophication	kg P eq	1.167E−03	1.211E−03	8.726E−04	4.920E−04	2.604E−03
Global warming	kg CO_2 eq	3.936E+01	2.685E+01	3.305E+01	7.205E+00	9.490E+01
Human carcinogenic toxicity	kg 1,4-DCB	2.084E−01	3.138E−01	1.258E−01	5.274E−02	6.004E−01
Human noncarcinogenic toxicity	kg 1,4-DCB	6.138E+00	5.431E+00	4.556E+00	1.783E+00	1.583E+01
Ionizing radiation	kBq Co-60 eq	1.431E−01	1.762E−01	9.479E−02	3.715E−02	4.011E−01
Land use	m2a crop eq	2.157E−02	3.154E−03	2.127E−02	1.712E−02	2.706E−02
Marine ecotoxicity	kg 1,4-DCB	2.827E−01	2.189E−01	2.227E−01	1.019E−01	6.608E−01
Marine eutrophication	kg N eq	5.019E−05	2.191E−05	4.565E−05	3.272E−05	7.821E−05
Mineral resource scarcity	kg Cu eq	2.177E−02	7.067E−03	2.005E−02	1.425E−02	3.465E−02
Ozone formation, human health	kg NOx eq	6.425E−03	1.502E−03	6.119E−03	4.869E−03	8.873E−03
Ozone formation, terrestrial ecosystems	kg NOx eq	6.628E−03	1.513E−03	6.326E−03	5.047E−03	9.095E−03
Stratospheric ozone depletion	kg CFC11 eq	1.473E−06	4.689E−07	1.386E−06	9.142E−07	2.339E−06
Terrestrial acidification	kg SO2 eq	1.122E−01	1.132E−02	1.122E−01	9.399E−02	1.312E−01
Terrestrial ecotoxicity	kg 1,4-DCB	1.331E+01	5.516E+00	1.196E+01	7.181E+00	2.443E+01
Water consumption	m3	5.819E+00	9.907E−01	5.652E+00	4.602E+00	7.575E+00

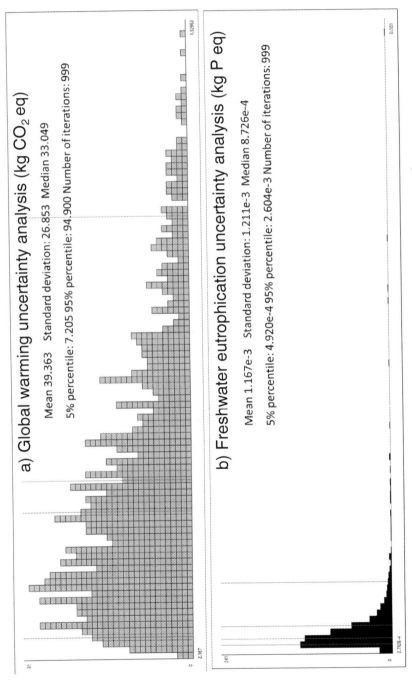

Fig. 12.2 Uncertainty analysis of the global warming and freshwater eutrophication impact categories.

the field to prepare the land for next planting season. Rice straw stubble is burned in many Indian states and the problem is much more prevalent in Punjab, Haryana, and Uttar Pradesh. Even among these states, Punjab accounts for 79% of the stubble burning. Stubble burning has become a national issue in the last one decade and is primarily attributed to the banning of early planting of rice in May by the State Government of Punjab [6]. The ban on early planting of rice was due to an effort to prevent groundwater depletion in the Punjab region. Similarly, the sugarcane crop although covering only 4% of the total cropped area consumes over 70% of the irrigation water and thus represents a huge stress on water resources. The government think tank NITI Aayog has suggested reducing the area under sugarcane plantation by 300,000 ha. NITI Aayog has also proposed to provide up to Rs. 15,000/ha financial incentives to farmers for adopting crops that require less water [7].

However, in recent years, India's sugar production has exceeded the domestic demand of 26 million MT/year by 4 million MT/year in 2018. Most of the ethanol in India is currently produced from molasses, a byproduct of sugarcane processing. In late 2020, the Indian government has increased the support prices for the bioethanol produced from B-heavy and C-heavy molasses to Rs. 54.27 and Rs. 45.69, respectively. Ethanol directly produced from sugarcane is priced at Rs. 62.65 to incentivize diversification of sugarcane markets [8,9]. The government is also promoting ethanol production from damaged food grains as an alternative market for the flood and rotten food grains.

Currently, India produced 3.25 billion L of ethanol mostly produced from sugarcane molasses and other waste cereal grains is used to meet the blending requirements. The policy goal of 10% ethanol blending by 2022 which will increase to 15% by 2026 and 20% by 2030 may be achieved ahead of the timelines. This is especially true if the production of the second-generation biofuels supplements the first-generation biofuels from sugarcane and waste food grains.

Sustainable production of biofuels at scale requires addressing the resource sustainability constraints, techno-economic viability and environmental issues, and policy imperatives.

It is clear from the resource assessment that the complete replacement of fossil fuels even for the transportation sector is a nonstarter for India. Growing dedicated energy crops for biofuels production is not possible due to inadequate arable land resources and the already severe water stress will become unsustainable with large-scale cultivation of new bioenergy crops. Therefore, the best possible solution for the Indian context is the use of agricultural residues which are already produced in large quantities from the annual agricultural production activities. With large demands for the use of agricultural residues for fodder, animal bedding, and other uses even use of all the agricultural residues is infeasible. This is indicated in the major differences between the generated and surplus biomass. Realistically, a 20%–30% replacement of the petrol demand is possible, and it is formulated in the latest NPBs.

Examination of the techno-economic viability of the second-generation biofuels that are being promoted in the NPB 2018 indicates that the second-generation biofuels are marginally viable in the most optimistic scenarios with low discount rates (<4%) and long payback periods. This indicates any second-generation biofuels ventures will require incentivization through policy support throughout the supply chain without which the establishment of a viable economic enterprise by commercial enterprises will be challenging.

The environmental assessment of the second-generation biofuels indicates that it has many positive environmental benefits such as lower greenhouse gas emissions but also suffers from some negative impacts, such as high-water footprint.

The policy environment analysis indicates that while the policy makers recognize many of the challenges in the technological and economic domains as indicated by the various financial and tax incentives for advanced biofuels and the support for pilot ventures for technology demonstration, the resource sustainability issues have not received adequate attention. For example, the largest surplus biomass available is sugarcane tops that are difficult to collect from the fields when sugarcane is harvested manually. Therefore, there should be promotion of mechanical harvesting of sugarcane to enable the collection of the sugarcane tops from the fields.

Socially, there is much less awareness about the global impacts of climate change in much of the population although there is a great sensitivity to local pollution issues. India remains a very price-sensitive market with very less scope for premium pricing of environmentally responsible biofuels as in most parts of the world. With increasing incomes and reducing poverty, the awareness of the greater social issues is expected to increase and will constitute an additional pull for the biofuels from the environmentally conscious consumers in future.

12.7 Conclusions and perspectives

This chapter focused on a comprehensive analysis of second-generation biofuels in India. The resource sustainability analysis indicated that there is a possibility of replacing about 20%–30% of India's petrol requirements using agricultural residues. The techno-economic analysis indicated that the process remains economically viable only under some of the most optimistic scenarios and thus the profitability will be highly susceptible to market variations. The environmental impacts are lower compared to fossil fuels although with a higher water footprint. The policy analysis indicates that some of these issues are recognized by policy makers and are actively formulating policies to promote biofuels with dual goals of encouraging the use of agricultural residues and reduce imports of fossil fuels.

References

[1] IMRB International, Sustainable availability of potent biomass resources for bioethanol production, Business and Industrial Research Division (BIRD) of IMRB International, 2009 Report no: IMRB/RM/VS/SS/RK/2009/AUG.

[2] K. Rajendran, G.S. Murthy, How does technology pathway choice influence economic viability and environmental impacts of lignocellulosic biorefineries? Biotechnol. Biofuels. 10 (2017) 268.

[3] A. Sreekumar, Y. Shastri, P. Wadekar, M. Patil, A. Lali, Life cycle assessment of ethanol production in a rice-straw-based biorefinery in India, Clean Technol. Environ. Policy 22 (2) (2020) 409–422. https://doi.org/10.1007/s10098-019-01791-0.

[4] D. Kumar, G.S. Murthy, Impact of pretreatment and downstream processing technologies on economics, energy and water use in cellulosic ethanol production, Biotechnol. Biofuels 4 (27) (2011), doi:10.1186/1754-6834-4-27.

[5] S. Das, The national policy on biofuels in India-A perspective, Energy Policy 143 (2020) 111595.

[6] M.R. Subramani, Power from biomass, bio-energy, compost: how other countries manage the issue of stubble burning a, Swarajya, 2020. https://swarajyamag.com/ideas/power-from-biomass-bio-energy-compost-how-other-countries-manage-the-issue-of-stubble-burning. [Accessed 22 May, 2021].

[7] M.R. Subramani, Explained: pros and cons of NITI Aayog recommendation to lower sugarcane production b, Swarajya, 2020. https://swarajyamag.com/business/explained-pros-and-cons-of-niti-aayog-recommendation-to-lower-sugarcane-production. [Accessed 22 May, 2021].

[8] Press Information Bureau, Cabinet approves modified scheme to enhance ethanol distillation capacity in the country, Ministry of Consumer Affairs, Food & Public Distribution, 2020. pib.gov.in/Pressreleaseshare.aspx?PRID=1684627. [Accessed January 7, 2021].

[9] M.R. Subramani, How sugarcane farmers stand to benefit from centre's modified scheme to hike ethanol distillation capacity, Swarajya, 2021. https://swarajyamag.com/economy/how-sugarcane-farmers-stand-to-benefit-from-centres-modified-scheme-to-hike-ethanol-distillation-capacity. [Accessed 22 May, 2021].

CHAPTER THIRTEEN

A case study on integrated systems analysis for biomethane use

Sarath C. Gowd[a,b], Deepak Kumar[c], Karthik Rajendran[a]

[a]Department of Environmental Science, SRM University-AP, Amaravati, India, [b]School of Engineering and Applied Sciences, SRM University-AP, Amaravati, India, [c]Department of Chemical Engineering, State University of New York College of Environmental Science and Forestry, Syracuse, NY, United States

13.1 Introduction

Biomethane, an upgraded form of biogas (by removing carbon dioxide), is a renewable and one of the most efficient energy-efficient sources among biofuels [1]. Biomethane replaces natural gas, a fossil energy source; it could be used as diverse energy vectors including heat, electricity, and transportation [2]. Biogas is produced by anaerobic digestion (AD) of organic materials, such as food waste, animal manure, municipal solid waste, agricultural wastes, and other organics [3]. The use of biomethane as an alternative to natural gas addresses the concerns of climate change as it is carbon-neutral fuel, apart from being a renewable energy technology [4].

As of 2019, the average CO_2 emission per capita stands at 4.78 kg/p/a [5]. This alarming rate suggests the need for capturing carbon dioxide from the atmosphere not only to reduce global warming but beyond. Several CO_2 capturing technologies are available including direct air capture, oxy-fuel combustion to name a few. An alias for CO_2 capturing technologies is negative emission technology (NET), wherein they trap the emissions released and convert it to a value-added product or store them underground. Likewise, combining biogas production with carbon capture could function as a NET [6]. The composition of biogas includes 40% carbon dioxide, referred to as biogenic CO_2 (not accounted for in greenhouse gas emissions calculation). However, trapping CO_2 from a biogenic source reduces the stress on other sectors. AD is an attractive technology from several perspectives including waste management, decarbonization, and decentralization of energy systems, and as a NET.

According to the World Bank, global solid waste production has increased to 2-Billion tons as of 2016; by 2050, it might increase by another 70%. Land required to process and landfill such a volume of waste is challenging, especially in developing countries. Managing and treating the waste has become an hazzle in most of the developing countries due to several reasons including population, lack of technology, and economics. The amount of disposable income determines the level of technology that a country can adapt in handling solid waste. Low-income countries prefer economically

Biomass, Biofuels, Biochemicals
DOI: https://doi.org/10.1016/B978-0-12-819242-9.00015-4

cheap disposal methods, such as landfilling or open burning. However, open burning release toxic gases into the atmosphere, and negatively affect the environment.

There are several factors that affect the adoption of biogas including the maturity of the technology, economic feasibility, social acceptance, market for the product, and policy support. Germany had realized AD could be a potential source of energy and waste treatment in the early 2000s; as of date, there are more than 12,000 digesters in operation. As of 2016, the global biomethane market has reached 1.7 Billion USD and it is expected to reach 2.6 Billion USD by 2025 (6). China dominates the small-scale biogas digesters, while Germany is the leader in large-scale systems.

The choice of adopting a technology differs based on usage, requirements, and location. Technology readiness level (TRL) measures the maturity of a technology that has the potential to penetrate the market [7]. A higher TRL indicates a higher maturity of a technology. When compared with fossil energy, most of the biofuel technologies have low TRL. Competing against well-established energy sources such as oil and natural gas needs the intervention of the government, either in the form of policies or subsidies to keep a fair play [8].

Technology implementation in a country is based on in-house research and development or via technology transfer, developed elsewhere. In developing countries, cost plays a key role in the adoption of a technology [9]. Most technologies are expensive in their early phase as the incurred research and development costs are higher. Industries take advantage of early entrants in the market to recover most of their investments. As technology matures, the cost gets cheaper with innovation and competition. Most of the developing countries could not afford initial research and development costs, and hence they tend to obtain from developed countries as technology transfer.

Achieving a global goal such as reducing global warming need some accelerated penetration of technologies to the market. This is where intervention of government through its policies plays a pivotal role [10]. Promotion of a technology through strong policies penetrate the market faster as government helps a technology either through subsidies or regulation mechanism. A strong policy could help biomethane systems to compete against natural gas. Policies could regulate the market in the form of financial aids, subsidies, and guaranteed prices for the product.

Though technology, economics, and policies are in place, social acceptance of a technology is necessary for its final reach out, that is, end consumers [11]. Biomethane from organic wastes requires segregation, which is a social issue, where the responsibility of a citizen is key to the acceptance of a technology. Segregated wastes ease the choice of treatment and the result could be a better efficient process.

As discussed above, adapting a technology, and commercializing it involve several factors and it should complement each other. This chapter presents a case study from Ireland to show the role of integrated systems analysis for biomethane production. The factors discussed include technology, economics, environment, market requirement, social acceptance, and policies (Fig. 13.1).

Fig. 13.1 Factors involved in a technology commercialization.

13.2 Dimensions of systems analysis
13.2.1 Technology

Technology plays a primary role in implementing new inventions to take over conventional systems. Developed countries through extensive inhouse research and development have access to latest and advanced technologies, while in developing countries access to such technologies are minimal [12]. For example, biomethane systems in Germany use advanced AD processes, including high solids digestion, nutrient recovery after energy recovery, recycling, and reuse of process water [13,14]. In developing countries, lagoon, fixed dome, or floating drum models are available for manure treatment.

However, developing an innovative technology is not easy as it should meet requirements including meeting the market demand, and easier enough for a wide penetration. Moreover, focus is necessary when developing technology in terms of effectiveness, efficiency, and energy recovery. [15]. Although the capital investments might be higher with the advanced processes, however, these technologies supply advantages of high overall energy recovery.

In developing countries, waste collection, and segregation are challenging aspects to treat food/organic wastes. Contrastingly, in developed countries, segregation of organics, inorganics and metals happens at source than processing centers [2]. Several countries dump wastes in landfills due to lack of awareness and the availability of technology to segregate. Beyond availability, affordability is the main challenge in adopting modern biogas systems in developing countries. Overcoming such issues requires affordable technologies at large. Biomethane systems need to be simple to work, affordable, and scalable, so that commercialization is easier.

Types of biomethane reactors used in a developed country include continuously stirred tank reactors (CSTR), which accounts for most of the digesters commissioned worldwide (Fig. 13.2). In CSTR, continuous stirring helps the microorganisms to degrade the organics effectively. A disadvantage with CSTR system is its inefficiency to distinguish between the fresh and degraded organics, due to contents getting mixed thoroughly always. Beyond CSTR, recent commercialization technologies include plug-flow systems, wherein the material begins flowing from one end to another in a prompt fashion without mixing. The plug flow system does not require a mixing as the inlet material pushes the existing material forward. CSTR systems could be scaled up to 15,000 m^3, while the plug flow system could reach up to 2,000 m^3.

In developing countries, fixed dome, and floating drum models are common. China had developed fixed dome model, while India had designed a floating drum model. Both models could run at a community level and scaling beyond 200 m^3 is difficult. These models do not have temperature control and hence they do not work at their best conditions. Unlike these models, CSTR and plug flow systems offer temperature control hence a better biogas yield [7].

The processing of biogas produced to biomethane determines the level of emissions avoided or saved. The usage of biogas in a developing country is cooking or electricity production using a small generator. However, in developed countries purification of biogas to biomethane happens via upgrading technologies. Unlike developing countries, the end user of biomethane is different in developed countries includes vehicle fuel, electricity, and heat cogeneration. There are different upgrading mechanisms available including water scrubbing, adsorption, absorption, membrane systems, and cryogenic systems. Novel upgrading systems such as the power to gas and microalgae-based upgrading offers carbon capture, which removes biogenic carbon dioxide produced from the biogas [16].

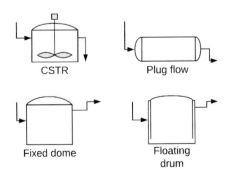

Fig. 13.2 Schematics of typical biogas systems in a developed country (CSTR, plug flow) versus developing countries (fixed dome, floating drum). *CSTR*, continuous stirred tank reactors.

Recently, using the digestate to produce value-added products is gaining importance because of the nutrient resources present in it. Most of the current technologies focus less or no treatment on digestate, where it is either discarded or given away to the farmers for free. An added issue with digestate is its lack of certification to use in agricultural fields. Unlike synthetic fertilizer, digestate from biogas does not have proper certification. This reduces the economic value of bio-based fertilizer produced from AD. Lately, several research focuses on using digestate for struvite production or algal cultivation, which increases the quality of it [17–19]. Digestate reclamation recycles most of the water. Usage of recycled water includes reusing in the process or for secondary purposes such as gardening.

13.2.2 Economics

Economic feasibility is the major barrier for any technology to make it self-sustainable and acceptable by communities. The implementation of promising technologies face hurdles on a full-scale due to its economic feasibility. Capital expenditure (CAPEX) and operational expenditure are two major parameters concerned about the economic viability of any process [20]. Adopting cost-effective technologies according to local conditions (best suitable for local feedstocks and environmental conditions) could enhance productivity and decrease the production cost. However, one technology cannot suit the requirements of all regions of the world. Finding suitable materials and developing technology according to local requirements could enhance economic feasibility.

During the preliminary stages of development, most technologies need subsidies from the government to be economically workable. On one hand, subsidies reduce capital expenditure and the operating cost. On the other hand, giving tax relaxation to the modern technologies intensifies their growth in the competitive industry. Biomethane can receive extra benefits based on the emissions they save, for example, imposing a higher carbon tax on fossil fuels. One good example of technology promotion is tax credit ($1.00 per gallon) on bioethanol production from corn- or cellulosic-based ethanol. Due to the government policies, currently, there are more than 200 commercial corn ethanol plants in the United States with more than 16 billion gallon of annual ethanol production. The subsidies/incentives provided by the government can also attract international companies to make the investment for developing cutting-edge technologies [12]. Alternatively, gate fee on landfilling banned wastes such as organics attracts added revenues. These measures could make the biomethane system sustainable and profitable. It is important to understand that besides economic feasibility, renewable fuel technology needs to match sustainability criteria, which makes it environmentally attractive.

The cost of biogas digesters depends on the feedstock, reactor type, and upgrading method used. The cost of the digester in a developing country might be different from that of a developed country. In addition, the cost varies based on type of installation, that is, for example, the small-scale digesters cost at the rate of 100 $/m^3 of sizes between 1- and 6-m^3 [21]. Households use these digesters for reducing waste and use the biogas for cooking, however, they do not have any temperature control. Typical CSTR operating food waste cost at 400 $/t of waste treated, while lignocellulosics such as grass-based biogas plant costs between 300 $/t [12,16].

The cost of upgrading biogas is around 50% of the running of the plant. Biogas upgrading via conventional upgrading technologies such as scrubbing, absorption, and adsorption costs between 1700 and 2300 $/m^3 biogas/h [22] for a biogas flowrate of 1500 m^3/h. When the flowrate of biogas decreases to say 100 m^3/h, the investment cost exponentially increases in the order of 5500–6000 $/m^3 biogas/h [22]. The cost of a combined heat and power lies in the range between 700 and 900 $/kW.

13.2.3 Environment

An expectation from a modern-day technology is it should offer excessive environmental benefits. Measuring environmental benefits quantitively requires certain parameters such as pollution, emission, and resource exploitation. TRACI is a tool that measures a number of these parameters. According to TRACI (tools for reduction and assessments of chemicals and other environmental impacts) ozone depletion, climate change, acidification, ecotoxicity, human health impacts, etc., are the parameters to be analyzed for any compound/product to know whether it is environmentally safe [23].

Natural gas and biomethane looks similar, however, their production methods is different. A biomethane system uses organic wastes as a feedstock, while fossils end up as natural gas. Given this difference, their emission levels are different though the product is the same. The emission from natural gas is not a part of the biogeochemical cycle, while that of biogas is. Burning of biomethane releases carbon dioxide, a renewed form of atmospheric carbon dioxide, like plants. This carbon dioxide is a part of bio-geochemical cycles and it creates no harm in warming the planet. Producing biomethane makes a positive impact on the environment by reducing the uncontrolled greenhouse gas emissions (methane) by decomposing organic matter to a useful product [24]. A support through proper policy is necessary for environmentally sustainable technologies like biomethane to takeover fossil fuels and reap the environmental benefits.

Natural-gas emits at the rate of 180 g-CO_2/kWh, while petrol and diesel emit at the rate of 240 and 260 g-CO_2/kWh, respectively. Coal has the highest carbon dioxide emissions at 350 g-CO_2/kWh [25]. Carbon dioxide emissions from biogas lie in the range of 100–200 g-CO_2/kWh. In addition, around 80% of these emissions account for biogenic [26]. This comparison shows that biogas offers excellent environmental benefits over natural gas and other fossil-based sources. Europe mentions in renewable-energy directive (Recast) that when biofuels reduce 70% GHG emissions with respect to fossil energy is when it qualifies as advanced biofuels.

13.2.4 Policy

Inducting innovative policies helps modern renewable-energy systems to fight against its antecedents such as fossil-energy. The policy should consist of detailed parameters, which are needed for the sustainable development of technologies [27]. The policy design should account for cost implications throughout its value-chain, starting from raw material to digestate use. A new policy should attract industries by offering higher-level of incentives at the beginning of introducing a technology; lowering or removal of incentives could happen after a brief period of technology maturation. The policy will support the industry to achieve a reasonable price for the product in a profitable manner. A good policy can promote technology to prosper in all stages of production. Nonetheless, policy alone cannot make an impact on technology; the product should have a value in the market.

13.2.5 Market

Even with superior technical advantages, technologies fail at a stage of creating a market for their product. Creating the market to the product is necessary to make it self-driven and sustainable. The government should innovate policies with assured tariffs for biomethane to reach larger sections within the community [28]. Tax allowances and parking fee exemption for biomethane users can create a niche market for biomethane. There are three areas where an AD plant could generate revenue. This includes gate fee cost on the feedstock to treat waste, selling the product of methane, and creating revenue through digestate. The government fixes the gate fee, where revenue generation happens only from municipal or food waste, as landfilling is forbidden. The second key revenue source is methane. Methane being an energy carrier will have a constant market; however, the price fluctuation of natural gas and crude oil hinders it heavily. Therefore, trade agreements are necessary to have a standard revenue from biomethane.

A vulnerable revenue source in biomethane industry is digestate. Currently, digestate do not receive any certification like synthetic fertilizer and hence its market value is low or zero. A proper recognition is necessary for digestate to sold as a commercial fertilizer. This added revenue boosts the profitability and bankability of AD systems. However, the problem with digestate is standardizing the output limits of carbon, nitrogen, and phosphorus. In addition, a retrofitting is necessary to recover nutrient including dewatering, drying, and stabilizing digestate. Sorting this problem could standardize digestate as a revenue source for biogas systems. People's acceptance is necessary for the market to reach consumers, which is a social dimension with a different dynamic.

13.2.6 Social aspects

Majority of the people do not want to change from their conventional lifestyle. A change is difficult to imbibe in a society unless it is self-driven. The lack of awareness of the necessity to reduce our emissions and waste is minimal in our society. No proper channel of communication had created awareness of sea-level rise, global warming, and other potential threat to our planet. Only few developed countries had started the thought to move from fossil to renewable energy. However, this fraction is less. The attitude of people in using a resource, energy, or water needs to change and an understanding is necessary about the limitation of availability of these sources. At present, people have a do not care attitude as they think money can solve their problems. Renewable energy in many countries not yet have a social acceptance, however, it is growing slowly.

Biomethane possess several environmental benefits such as reduction in emission, improved air quality, and less stress on resource usage. However, from a social dimension, people think that it is difficult to segregate or having a digester in their backyard will have unpleasant smell. Technologies such as biomethane require human intervention in the form of segregation, acceptance of produced fuel and fertilizer. Biomethane is an environmentally friendly product with lots of useful features for sustainable development [29]. Creating awareness about sustainable fuels and their important in future among the people can make the change to adopt new fuels like biomethane.

13.3 Case study of Ireland for biomethane use

13.3.1 Background

Ireland imports major share of energy as it is an island that is devoid of natural energy sources. Oil is the primary source of energy in the country, which accounts for 46% of the total energy. The share of renewable was negligible around 8% by 2016 as compared with conventional energy sources. Within renewables, wind energy corresponds to 45% of total energy, while biomass, biogas, hydropower, etc., constitutes rest [30].

However, the state is taking several measures to fight climate change and the decarbonization of energy systems. Ireland have set a target of replacing 16% of total energy consumption by renewables by 2020. It also has a subtarget of 10% in the transportation sector, 40% for renewable electricity, and 12% renewable heat [6]. Meeting the target for transportation and renewable heat is uncertain, while achieving the target for renewable electricity is possible via the scale of wind energy systems installed.

Biomethane can play a significant role in achieving the renewable energy targets in transportation and heat sector. Ireland generates about 820,000 t/year (about 180 kg/person/year) of food waste. Similarly, the grass production and cow manure accounts for 213,188 Mt/year [12]. These three feedstocks (food waste, grass, and cow manure) has the possibility to reduce the emissions of Ireland by 13% via biomethane production [31]. The problem with biomethane commercialization is lack of technology assessment, policy support, and economic feasibility. It is essential to conduct an integrative assessment to understand the potential and realization of biomethane in Ireland.

13.3.2 What is the role of technology and economics in system analysis?

Biogas upgrading has evolved over time. This includes conventional technologies such as water scrubbing, followed by capturing CO_2 and converting it to methane and lower TRL technologies such as microalgae-based upgrading. In microalgae upgrading, CO_2 is captured and converted as algal biomass, which could be processed for further applications or it could function as a platform precursor. Of these technologies, water scrubbing is proven and optimized in industries. Hence, the cost and energy requirement to run such a system are low. However, power to gas or capturing CO_2 to methane technologies have a low TRL, which means its profitability and operation needs to be optimized. When compared with, power to gas, microalgae-based upgrading is a primitive technology, which requires reproducible pilot-scale validations.

Ireland has 6.4 Mt/a food waste, cattle slurry at 28.5 Mt/a, and Grass at 31.3 Mt/a, which if converted to biomethane could potentially reduce 13% of the country's emissions. For a food waste biogas plant running at a capacity of 100,000 t/a, could generate 1.18 MWh/t energy which leads to avoiding emissions at 0.27 tCO_2/t. Constructing and commissioning such a plant needs capital investment in the order of €32M and €50M depending on the upgrading system employed. The levelized cost of energy for a food waste plant was 87 €/MWh, while for plant that uses slurry and grass requires 121 €/MWh. Biogas plants that uses food waste require 13 €/MWh as incentives; while slurry and grass-based biogas plants, require 88 €/MWh for an established water scrubbing technique as an upgrading mechanism. Food waste requires less incentives as it generates income via gate fee at 75 €/t, through the ban of landfilling organic wastes in the EU region. (Table 13.1)

As depicted in the table, power to gas and microalgae upgrading operation costs and energy consumption are higher because of low TRL's [12]. Systems analyses can help in calculating the incentives needed and energy consumption of complex systems. Such an analysis could be used by the policy makers to innovate a constructive policy to promote this technology [12].

13.3.3 How policy influences technology commercialization?

The production cost of biomethane was high as compared to traditional oil like diesel. It is difficult for any technology to kick start without economical support from the government. A policy could be innovative to make biomethane production financially

Table 13.1 Energy consumption and operational expenditure for different biogas upgrading techniques [12].

Upgrading method	Operational expenditure (€/t)	Energy consumption (kWh/m^3)	References
Water scrubbing	62–87	0.13–0.15	[12]
Power to gas	106–166	1.02–1.05	
Microalgae upgrading	70–110	0.25–0.28	

sustainable. There are multiple ways to incentivize biomethane, for example, carbon tax. Adding carbon tax to fossil fuel would bring down the cost of biogas or it helps in competing against fossil fuels. Currently, the carbon tax is €30/tCO$_2$, while increasing it to 50 and 80 €/tCO$_2$ helps biomethane systems to be financially workable. The cost varies depending on which upgrading method is employed. Regular upgrading systems emit carbon dioxide and receives incentives. However, there are carbon dioxide capturing technologies such as power to gas and microalgae upgrading. These upgrading systems avoid more emissions which requires a top-up incentive than regular upgrading mechanisms.

Incentive mechanism is different in different countries. For example, in Norway, electric vehicles have no import tax, capital incentives, no VAT, and free charging. This motivates people to shift toward electric vehicles. Seeing the change on a policy requires a minimum of 2 years once developed and implemented. A typical example to explain this phenomenon is solar energy in Germany, where feed-in tariffs reduction and simultaneous increase in the price of domestic electricity led to a surge in renewable energy production. The initial high feed-in tariffs motivated industries to innovate on cost-cutting technologies. In those periods, the government supported the industries through incentives and regulations. A reduction or removal of incentives shall pertain to when technology gets matured. This helped Germany to be a leader in solar energy systems today. Such a comparative analysis of different policies leads to how all these factors could be combined for a successful technology implementation and adaptation [21]. In the case of Ireland as well, currently, green gas certification program is under progress to implement the decarbonization of energy systems. Proper policy and incentives could make biomethane as a workable possibility to reach renewable energy targets.

13.4 Conclusions and perspectives

Biomethane is a promising technology with high efficiency to compete with conventional fossil fuels. However, it needs intervention from different dimensions, including advancement in technology, economic feasibility, support from the government, and policymakers. A policy intervention from the government helps technology adoption smoother, and the transition faster is a key lesson learned from the Irish case study. Besides, a successful policy will help a technology to get its maturity in a short span of time. The Biomethane industry in Ireland required incentives between 0.13 €/m^3 to 1.03 €/m^3 based upon the type of upgrading and feedstock used.

The Irish case study is, an example, however, the results might not be applicable globally as local conditions are different. Intertwining and estimation of integrative systems analysis are necessary of factors including technology evaluation, economic feasibility, environmental benefits assessment, the role of policy, and how people react.

However, there is no one common method exists to measure them together. A large set of data is necessary like this case study to understand. Based on this, a trained model can predict how technology can cross the valley of death in commercializing it. In addition, diverse research groups should work on a common agenda on understanding the factors mentioned above and their response to implementation. Such studies reduce the risk and improve the chances of commercialization.

References

[1] A.I. Adnan, M.Y. Ong, S. Nomanbhay, K.W. Chew, P.L. Show, Technologies for biogas upgrading to biomethane: a review, Bioengineering 6 (2019) 1–23. https://doi.org/10.3390/bioengineering6040092.

[2] Z. Chu, W. Wang, B. Wang, J. Zhuang, Research on factors influencing municipal household solid waste separate collection: Bayesian belief networks, Sustain 8 (2016). https://doi.org/10.3390/su8020152. [Accessed 5 Nov 2019].

[3] A. Molino, F. Nanna, Y. Ding, B. Bikson, G. Braccio, Biomethane production by anaerobic digestion of organic waste, Fuel 103 (2013) 1003–1009. https://doi.org/10.1016/j.fuel.2012.07.070.

[4] A. Tilche, M. Galatola, The potential of bio-methane as bio-fuel/bio-energy for reducing greenhouse gas emissions: a qualitative assessment for Europe in a life cycle perspective, Water Sci. Technol. 57 (2008) 1683–1692. https://doi.org/10.2166/wst.2008.039.

[5] S. Sahota, G. Shah, P. Ghosh, R. Kapoor, S. Sengupta, P. Singh, et al., Review of trends in biogas upgradation technologies and future perspectives, Bioresour. Technol. Reports 1 (2018) 79–88. https://doi.org/10.1016/j.biteb.2018.01.002.

[6] European Academies Science Advisory Council, Negative emission technologies: What role in meeting Paris Agreement targets? EASAC Policy Rep (2018) 1–45.

[7] K. Rajendran, S. Aslanzadeh, M.J. Taherzadeh Household biogas digesters—a review. vol. 5. 2012. https://doi.org/10.3390/en5082911. [Accessed 20 Nov 2019].

[8] K. Rajendran, B. O'Gallachoir, J.D. Murphy, The combined role of policy and incentives in promoting cost efficient decarbonisation of energy: a case study for biomethane, J. Clean Prod. 219 (2019) 278–290. https://doi.org/10.1016/j.jclepro.2019.01.298.

[9] W.M. Budzianowski, D.A. Budzianowska, Economic analysis of biomethane and bioelectricity generation from biogas using different support schemes and plant configurations, Energy 88 (2015) 658–666. https://doi.org/10.1016/j.energy.2015.05.104.

[10] T. Patterson, S. Esteves, R. Dinsdale, A. Guwy, An evaluation of the policy and techno-economic factors affecting the potential for biogas upgrading for transport fuel use in the UK, Energy Policy 39 (2011) 1806–1816. https://doi.org/10.1016/j.enpol.2011.01.017.

[11] I. D'Adamo, P.M. Falcone, F. Ferella, A socio-economic analysis of biomethane in the transport sector: the case of Italy, Waste Manag. 95 (2019) 102–115. https://doi.org/10.1016/j.wasman.2019.06.005.

[12] K. Rajendran, J.D. Browne, J.D. Murphy, What is the level of incentivisation required for biomethane upgrading technologies with carbon capture and reuse? Renew. Energy 133 (2019) 951–963. https://doi.org/10.1016/j.renene.2018.10.091.

[13] A. Jeihanipour, S. Aslanzadeh, K. Rajendran, G. Balasubramanian, M.J. Taherzadeh, High-rate biogas production from waste textiles using a two-stage process, Renew Energy 52 (2013) 128–135. https://doi.org/10.1016/j.renene.2012.10.042.

[14] S. Aslanzadeh, K. Rajendran, M.J. Taherzadeh, A comparative study between single- and two-stage anaerobic digestion processes: effects of organic loading rate and hydraulic retention time, Int. Biodeterior. Biodegrad. 95 (2014) 181–188. https://doi.org/10.1016/j.ibiod.2014.06.008.

[15] A. Datta, D. Mukherjee, L. Jessup, Understanding commercialization of technological innovation: taking stock and moving forward, R&D Manag. 45 (2015) 215–249. https://doi.org/10.1111/radm.12068.

[16] T.T.Q. Vo, D.M. Wall, D. Ring, K. Rajendran, J.D. Murphy, Techno-economic analysis of biogas upgrading via amine scrubber, carbon capture and ex-situ methanation, Appl. Energy 212 (2018) 1191–1202. https://doi.org/10.1016/j.apenergy.2017.12.099.

[17] T. Liu, X. Zhou, Z. Li, X. Wang, J. Sun, Effects of liquid digestate pretreatment on biogas production for anaerobic digestion of wheat straw, Bioresour. Technol. 280 (2019) 345–351. https://doi.org/10.1016/j.biortech.2019.01.147.

[18] C. Herbes, U. Roth, S. Wulf, J. Dahlin, Economic assessment of different biogas digestate processing technologies: a scenario-based analysis, J. Clean Prod. 255 (2020) 120282. https://doi.org/10.1016/j.jclepro.2020.120282.

[19] W. Peng, F. Lü, L. Hao, H. Zhang, L. Shao, P. He, Digestate management for high-solid anaerobic digestion of organic wastes: a review, Bioresour. Technol. 297 (2020) 122485. https://doi.org/10.1016/j.biortech.2019.122485.

[20] F. Cucchiella, I. D'Adamo, Technical and economic analysis of biomethane: a focus on the role of subsidies, Energy Convers. Manag. 119 (2016) 338–351. https://doi.org/10.1016/j.enconman.2016.04.058.

[21] K. Rajendran, S. Aslanzadeh, F. Johansson, M.J. Taherzadeh, Experimental and economical evaluation of a novel biogas digester, Energy Convers. Manag. 74 (2013) 183–191. https://doi.org/10.1016/j.enconman.2013.05.020.

[22] K. Rajendran, G.S. Murthy, Techno-economic and life cycle assessments of anaerobic digestion —a review, Biocatal. Agric. Biotechnol. 20 (2019) 101207. https://doi.org/10.1016/j.bcab.2019.101207.

[23] J. Bacenetti, A. Fusi, M. Negri, R. Guidetti, M. Fiala, Environmental assessment of two different crop systems in terms of biomethane potential production, Sci. Total Environ. 466–467 (2014) 1066–1077. https://doi.org/10.1016/j.scitotenv.2013.07.109.

[24] G. San Miguel, B. Corona, Hybridizing concentrated solar power (CSP) with biogas and biomethane as an alternative to natural gas: analysis of environmental performance using LCA, Renew. Energy 66 (2014) 580–587. https://doi.org/10.1016/j.renene.2013.12.023.

[25] U.S. Energy Information Administration (EIA), How much carbon dioxide is produced when different fuels are burned? EiaGov. (2019). https://www.eia.gov/tools/faqs/faq.php?id=73&t=11. [Accessed April 28, 2020].

[26] T.T.Q. Vo, K. Rajendran, J.D. Murphy, Can power to methane systems be sustainable and can they improve the carbon intensity of renewable methane when used to upgrade biogas produced from grass and slurry? Appl. Energy 228 (2018) 1046–1056. https://doi.org/10.1016/j.apenergy.2018.06.139.

[27] S. Eker, E. van Daalen, A model-based analysis of biomethane production in the Netherlands and the effectiveness of the subsidization policy under uncertainty, Energy Policy 82 (2015) 178–196. https://doi.org/10.1016/j.enpol.2015.03.019.

[28] C. Da Costa Gomez, Biogas as an energy option, Biogas Handb. Sci. Prod. Appl. (2013) 1–16. https://doi.org/10.1533/9780857097415.1. [Accessed 28 Nov 2019].

[29] R. Selvaggi, G. Pappalardo, G. Chinnici, C.I. Fabbri, Assessing land efficiency of biomethane industry: a case study of Sicily, Energy Policy 119 (2018) 689–695. https://doi.org/10.1016/j.enpol.2018.04.039.

[30] D. Dineen, M. Howley, M. Holland, E. Cotter, Energy Security in Ireland: A Statistical Overview 2016, Sustain. Energy Auth. Irel. 80 (2016).

[31] K. Rajendran, B.Ó. Gallachóir, J.D. Murphy, The Role of Incentivising Biomethane in Ireland Using Anaerobic Digestion, Environ. Prot. Agency Reseach (2019) 1–19.

CHAPTER FOURTEEN

Alternative ammonia production processes and the use of renewables

Gal Hochman[a], Alan Goldman[b], Frank A. Felder[c]

[a]Department of Agriculture, Food & Resource Economics, Rutgers, The State University of New Jersey, New Brunswick, NJ, United States, [b]Department of Chemistry and Chemical Biology, Rutgers, The State University of New Jersey, New Brunswick, NJ, United States, [c]Center for Energy, Economic & Environmental Policy, Edward J. Bloustein School of Planning and Public Policy, Rutgers, The State University of New Jersey, New Brunswick, NJ, United States

14.1 Introduction

In a single day, agriculture produces 19.5 million tons of cereals, roots, tubers, fruits, and vegetables. In a single day, crop production uses 300,000 tons of fertilizers. These amounts are used to feed 7 billion people today, and world population is expected to grow to 9.5 billion by the year 2050 [1]. Population growth is expected to result in agriculture needing to increase global production by 60%, while in developing countries it is likely to increase by 100% because of dietary changes attributed to income changes [2]. To meet this growing demand for food, yields should be maximized, where the intensification of land use for crop production will likely result in significant higher uses of nutrient inputs such as nitrogen.

Nitrogen is one of the essential nutrients for plant growth and usually a limiting nutrient in industrial-scale crop production. Nitrogen is taken up by plants in the form of ammonium compounds and various forms of nitrogen oxides. In nature, atmospheric nitrogen is converted (nitrogen fixation) into ammonium form by nitrogen-fixing bacteria associated with leguminous plants (Eq. 14.1).

$$N_2 + 8H^+ + 8e^- \rightarrow 2NH_3 + H_2 \qquad (14.1)$$

These forms of natural nitrogen cycles are susceptible to losses through leaching from soil to water or loss to the atmosphere; only about 30%–50% of applied nitrogen is taken up by the crop. Thus, naturally fixed nitrogen undermined by losses cannot support current levels of global food production. In fact, synthetic nitrogenous fertilizers fed nearly 45% of the world's population at 2011 [1]. As the world's population increases so will the requirement for nitrogenous fertilizers.

The gap between the requirement for naturally fixed nitrogen and agriculture requirement is filled by the Haber–Bosch (H–B) process which uses elemental nitrogen (N_2) and hydrogen (H_2) to synthesis ammonia (NH_3) (Eq. 14.2).

$$N_2 + 3H_2 \rightarrow 2NH_3 \qquad (14.2)$$

Biomass, Biofuels, Biochemicals
DOI: https://doi.org/10.1016/B978-0-12-819242-9.00007-5

The Haber–Bosch process came into use around 1920. Its share in fertilizer production went from 20% in 1920 to almost 100% in 1990. Fertilizer synthesis using Haber–Bosch process led to multiple-fold increase in crop production in the 20th century. Without a synthetic source for ammonia, crop production would require significantly more land to compensate for the reduced productivity of the soil overtime [1]. However, the Haber–Bosch process is an energy-intensive polluting process; an unsustainable process that is generating about 2% of global greenhouse gas emissions [3]. However, is there a more sustainable alternative to that of ammonia production through Haber–Bosch?

In answering this question, first, the below section describes the current practice of ammonia production. The below sections, then, introduce two alternative processes to the Haber–Bosch and discuss the implications of adopting these technologies. Policy discussion and perspectives are offered in the below section.

14.2 Ammonia production via current practices

The Haber–Bosch is a two-step, energy-intensive process [4]. The first step includes hydrogen production, which then is followed by the Haber–Bosch process that combines the hydrogen with nitrogen (Eq. 14.2) to generate the ammonia (Fig. 14.1).

The hydrogen feed for Haber–Bosch can be produced via various methods, including natural gas reforming (and coal gasification) as well as electrolytic splitting of water (Fig. 14.1). However, natural gas reformation is currently most common [5]. So, any increase in ammonia production through Haber–Bosch means an increase in natural gas consumption and emission of natural gas pollutants to the environment.

The nitrogen feed used in the second step may be either produced as a coproduct of coal gasification and natural gas reforming or extracted from air using an air separation unit (ASU) [3]. The ASU uses a combination of compression, cooling, and expansion to separate the nitrogen, oxygen, and other compounds from the air.

Given the preferred hydrogen production methods (methane steam reforming), next energy is discussed, economic, and emissions implications of implementing these alternatives. To answer these questions, this paper builds on [6].

14.2.1 Energy requirements of Haber–Bosch based on natural gas

The overall energy requirement for ammonia production varies based on the choice of hydrogen and nitrogen production methods, where the Haber–Bosch technology is least energy-intensive among alternative fossil fuels when based on natural gas. Under the natural gas alternative, the net energy input of a typical modern natural-gas-based Haber–Bosch plant is 8.87 MWh/mt-NH_3 which is equivalent to 30.3 MBTU natural gas per mt-NH_3 [2,7].

14.2.2 Economics of the Haber–Bosch process

Following [6] while assuming plant capacity of 2200 mt-NH_3 per day, the construction cost for a natural-gas-based Haber–Bosch production facility is $404,000 per mt-$NH_3$/

Alternative ammonia production processes and the use of renewables

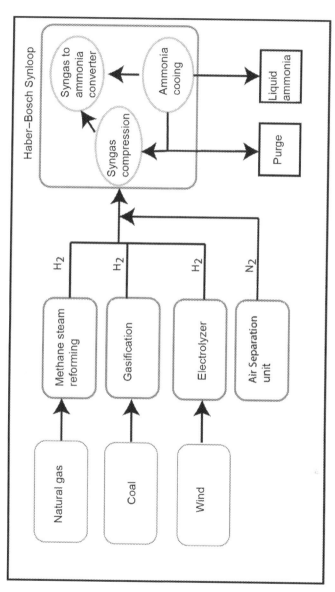

Fig. 14.1 Ammonia process based on Haber–Bosch synthesis using different energy sources.

day capacity. Of this, $235,000 per mt-NH$_3$/day capacity is the cost of the Haber–Bosch process and the ASU without the gas turbine. The cost of the Haber–Bosch process alone is $134,000 per mt-NH$_3$/day capacity. Based on the value of 30.3 MBTU natural gas per mt-NH$_3$, and using the average Henry Hub natural gas price of $3.08/MBTU for January 2017 through December 2018, (with annual averages of $2.99 and $3.17 respectively) [8] the cost for the natural gas required by a plant of the type considered is $93 per mt-NH$_3$ produced.

Using the above construction costs, the capital cost is estimated. The assumptions made in the U.S. Department of Energy H2A Distributed Hydrogen Production Model (Version 3) [9] are assumed and consider a capital cost of 5.00% per year (corresponding to an interest rate of 4.00% per year repaid over 40 years). At this rate the construction cost of $404,000 per mt-NH$_3$/day capacity corresponds to $20,200/year per mt-NH$_3$/day capacity, corresponding to $55/mt-NH$_3$. To approximate operation and maintenance (O&M) costs, the employment values reported for the recently built Yara/BASF ammonia plant (2018 start-up) in Freeport, TX, with a capacity of 2055 mt-NH$_3$/day with 35 full-time employees are used [10]. The DOE H2A estimates of salary, administrative costs, and insurance costs, are used to obtain yearly expenses of $16.6 million, or $22/mt-NH$_3$ [11].

The estimated capital and O&M costs, $55/mt-NH$_3$ and $22/mt-NH$_3$ respectively, combined with the estimated cost of natural gas noted above ($93/mt-NH$_3$) contribute $170/mt-NH$_3$ to the production cost of ammonia for large plants (Table 14.1). Smaller plants, ranging from 90 mt-NH$_3$/day to 550 mt-NH$_3$/day capacity, incur substantially greater per-ton capital and operating costs from [6], with examples shown in Table 14.1.

14.2.3 CO$_2$ emissions from a Haber–Bosch plant

A minimum of 0.97 mt-CO$_2$ emissions per mt-NH$_3$ produced is required to provide the necessary hydrogen from the steam reforming of gas (or from direct reaction of gas with N$_2$), based only on consideration of the stoichiometry of (Eq. 14.3) (3.0 mol CH$_4$

Table 14.1 Estimated costs of ammonia production ($/mt-NH$_3$) via Haber–Bosch, by natural-gas-based Haber–Bosch plants of varying capacity (based on a cost of $3.08/MBTU natural gas).

	Haber–Bosch plant size (mt-NH$_3$/day)		
	Large Haber–Bosch (ca. 2000 mt/day)	Medium Haber–Bosch (545 mt/day)	Small Haber–Bosch (91 mt/day)
Natural gas	$93	$93	$93
Capital	$55	$88	$113
O&M	$22	$62	$133
Total	$170	$243	$339

Source [6].

per mol NH_3). Since (Eq. 14.3) and methane steam reforming are both endothermic, additional combustion of methane is required, thus increasing the minimum CO_2 emissions from a natural-gas-based Haber–Bosch plant.

$$3CH_4(g) + 4N_2(g) + 6H_2O(g) \rightleftharpoons 8NH_3(g) + 3CO_2(g) \quad (14.3)$$

$H°_{450} = 7.10 \text{kcal/mol}, S°_{450} = -106.9 \text{cal/deg} \bullet \text{mol}, G(450°C) = 84.4 \text{kcal/mol}$

$$CH_4(g) + 2O_2(g) \rightleftharpoons CO_2(g) + 2H_2O(g) \quad (14.4)$$

$H°_{450} = -191.2 \text{kcal/mol}, S°_{450} = 0.2 \text{cal/deg} \bullet \text{mol}, G(450°C) = -191.3 \text{kcal/mol}$

$$3.5CH_4(g) + O_2(g) + 4N_2(g) + 5H_2O(g) \rightleftharpoons 8NH_3(g) + 3.5CO_2(g) \quad (14.5)$$

$H°_{450} = -65.4/\text{mol}, S°_{450} = -68.3 \text{cal/deg} \bullet \text{mol}, G(450°C) = -11.3 \text{kcal/mol}$

Since Eq. (14.3) is only modestly endergonic, while methane combustion (Eq. 14.4) is highly exergonic, only a small amount of methane for combustion is thermodynamically required to drive the reaction of Eq. (14.2). (With 0.42 mol CH_4 combusted per 3.0 mol CH_4 required for hydrogen, as per Eq. (14.3), G(450°C) = 0). Thus (Eq. 14.5), in which 0.5 mol methane is combusted per 3 mol methane used as a hydrogen source is highly exothermic and significantly exergonic. This is used as the approximate theoretical lower limit of CO_2 production required for a methane-based Haber–Bosch process, 3.5 mol CO_2 per 8 mol NH_3, or 1.13 kg-CO_2/kg-NH_3. In practice, various analyses of Haber–Bosch plants conclude that CO_2 emissions range from 1.33 mt to 1.69 mt per mt-NH_3 produced [12,13]. (Note that these estimates do not include methane leakages, which may significantly increase greenhouse gas emissions from Haber–Bosch.)

Similar to the analysis above [9], suggesting a ratio of 1.33 ton of CO_2 per 1 ton of anhydrous NH_3 with 99.9% purity. Other factors used in Morgan include air emissions caused during steam reforming include NO_x, SO_2, CO, amines, and particulates. During the Haber–Bosch process, some methane left from the steam-reforming process is released to the atmosphere, which can also cause some amount of ammonia to be released as well. There are also emissions of water pollutants in the form of methanol, liquid ammonia, and other organics [14].

This creates a need for an alternative method of hydrogen production and better management of agroecosystem to reduce nitrogen losses. Electrolysis of water [15] and direct electrochemical nitrogen reduction (ENR) [6] are potential methods of hydrogen synthesis, which can use renewable sources of electricity thus significantly reducing the environmental cost of ammonia production. The distributed sources of renewable electricity like biomass, wind, and solar energy in rural areas can be coupled with the alternative forms of hydrogen synthesis and can further reduce the cost of transportation to agricultural regions.

14.3 Haber–Bosch using electrochemical H$_2$ production (E/H–B)

Our starting point for the electrochemical H$_2$ process is the U.S. Department of Energy H2A Distributed Hydrogen Production Model (Version 3), specifically the modeling of a process for the production of hydrogen from the electrolysis of water using grid-based electricity [8]. The system on which this paper focuses is a standalone grid-powered PEM electrolyzer system with a hydrogen capacity of 50,000 kg (50 mt) H$_2$/day (corresponding to 282 mt-NH$_3$/day). Our baseline model is the generic model, which uses process water and grid electricity. Costs are in 2017 US dollars.

More specifically, the energy of H$_2$ is 143 MJ/kg or 39.7 kWh/kg. The energy efficiency of electrochemical H$_2$ production can be as high as ca. 80% corresponding to 49.6 kWh required per kg-H$_2$. For a reference point, at a price of $0.05/kWh ($50/MWh) this corresponds to a cost for electrical power of $2480/mt-H$_2$. Assuming 100% efficiency for the Haber–Bosch reaction (Eq. 14.2) (0.178 mt-H$_2$/mt-NH$_3$) this corresponds to $441/mt-NH$_3$ for the H$_2$ feed (Table 14.2).

In the analysis, and similar to the Haber–Bosch process, assume that the electrolysis-based ammonia plant operates 330 days, or 7920 operating hours, per year. Then, while the Haber–Bosch through natural gas consumes 8.86 MWh per metric tons of NH$_3$ (in the above section), the electrolysis consumes 11.60 MWh per metric tons of NH$_3$; energy use is 29% higher under the electrolysis process than the natural gas reforming process.

14.4 Direct electrochemical nitrogen reduction

The process of ENR process involves oxidation of water at the anode to yield O$_2$ and H$^+$, and reduction of N$_2$ at the anode, and protonation to yield ammonia (Eq. 14.6) (Fig. 14.2).

$$\text{At the anode: } 3H_2O \rightarrow \frac{3}{2}O_2 + 6e^- + 6H^+ \quad (14.6a)$$

$$\text{At the cathode: } 6e^- + 6e^+ + N_2 \rightarrow 2NH_3 \quad (14.6b)$$

Table 14.2 Estimated costs of ammonia production ($/mt-NH$_3$) via E/H–B, based on H2A model, PEM electrolyzer system with capacity 50 mt-H$_2$/day, with accompanying Haber–Bosch plants of varying capacity, at a fixed benchmark electric power cost of $50/MWh.

Haber–Bosch plant size (mt-NH3/day)		
Large Haber–Bosch (2000 mt/day)	Medium Haber–Bosch (545 mt/day)	Small Haber–Bosch (91 mt/day)
$627	$669	$725

Source [6].

Alternative ammonia production processes and the use of renewables

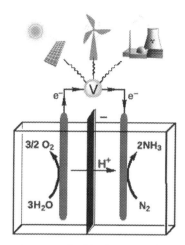

Fig. 14.2 Electrochemical approach to sustainable dinitrogen fixation.

The electrochemical potentials of these processes are as follows:

$$N_2 + 6e^- + 6e^+ \rightarrow 2NH_3(aq); E^0 = +0.092V$$

$$3H_2O \rightarrow 1.5O_2 + 6e^- + 6H^+; E^0 = -1.229V$$

Therefore: $N_2 + 3H_2O \rightarrow 2NH_3 + 1.5O_2; E^0 - -1.137V$

The purely theoretical (thermodynamic) energy requirement for ENR obtained from these values is 5.37 kWh per kg NH_3 produced. By comparison, to produce H_2 needed for E/H–B, the anodic reaction (H_2O oxidation) is the same while the cathodic reaction (reduction of H^+) has a (defined) potential of 0.0 V. The corresponding thermodynamic energy requirement is about 8% greater, at 5.80 kWh/kg. These values of course only represent purely theoretical lower limits of the energy requirements.

Unlike the Haber–Bosch process as currently implemented, ENR or E/H–B would not necessarily involve the use of fossil fuel as a source of hydrogen and commensurate emission of CO_2. Thus, both avoid the production of CO_2 (assuming that carbon-free energy, e.g., wind, solar, or nuclear are used) as well as many of the other gases currently emitted in the course of ammonia production. ENR, however, unlike E/H–B, obviates the need for the highly energy- and capital-intensive Haber–Bosch process. Moreover, the ENR process would have the flexibility to exploit intermittent supplies of renewable electricity to produce NH_3, without the need to store hydrogen. The ENR process thus benefits from the use of electricity during periods of low demand at a much lower energy and capital cost than does E/H–B.

The efficiency of an electrochemical synthesis like ENR is determined by the overpotential required to generate NH_3 (i.e., the operating voltage that is beyond the thermodynamic potential) and the Faradaic efficiency (the fraction of the current that leads to product according to the desired reaction). The values of these parameters are the main determinant of cost, and therefore economic viability, for ENR process. For example [6], it showed that assuming an aspirational 95% Faradaic efficiency, 0.60 V overpotential, and $50 per MWh, yields energy cost per mt-NH$_3$ of $432. However, if the Faradaic efficiency is changed to 60% (as opposed to 95%) then the energy cost per mt-NH$_3$ is $690; reducing the Faradaic efficiency by 37% resulted in energy cost increasing by 60%.

When assessing the economic viability of the ENR plant, the levelized cost of NH$_3$ was used, that is, a measure of the lifetime costs of the project divided by total ammonia produced during that same period was used. The above assumptions then suggest a cost of $508 per metric ton of NH$_3$, where Table 14.3 shows the breakdown of the various cost factors with the feedstock (energy cost) being 85% of the cost per kg of NH$_3$.

Building on [6], parameters determining the cost of electrochemical direct nitrogen reduction were identified and linked these parameters to electricity prices. To establish the economic viability of intermittent production of ammonia, the year 2015 was analyzed for electricity price data taken from the Electric Reliability Council of Texas (ERCOT) region in the United States. ERCOT manages the flow of electricity in most of Texas and performs financial settlement for the competitive wholesale bulk-power market and administers retail switching [16]. The ERCOT price data used below are hourly data. Because the ERCOT price is the wholesale price, the following assumptions were needed to calculate the industrial price. Specifically, assume that the difference between industrial and wholesale prices is fixed and therefore the difference between the industrial and wholesale prices equals the difference between the average annual industrial price and the average wholesale price. EIA annual industrial electricity price data [17], together with ERCOT prices, were used to calculate this difference [18].

Given the industrial price calculated, the estimated number of hours that ammonia plants would operate throughout the year was calculated, where the data suggests electricity prices fluctuate significantly over time, within a day, over a week, and across seasons. In the simulation, assume future prices of electricity can be set a day in advance and that the plant management can decide whether the plant should operate during a given hour. The hourly price density graphs of the ERCOT wholesale prices show that the density peak occurs sufficiently below the price ceiling of $0.04 per kWh for the fall and winter seasons, while it covers a significant portion of the peak in the spring and summer seasons. The density for the various hours' and seasons' wholesale prices per MWh are depicted in Fig. 14.3. The analysis shows the frequency where wholesale

Table 14.3 The baseline specific item cost calculations.

Cost component	Cost contribution ($/kg of NH$_3$)	Percentage of NH$_3$ cost (%)
Capital costs	$0.035	6.9
Fixed O&M	$0.041	8.1
Feedstock costs	$0.432	85.0
Total	$0.508	

Alternative ammonia production processes and the use of renewables 251

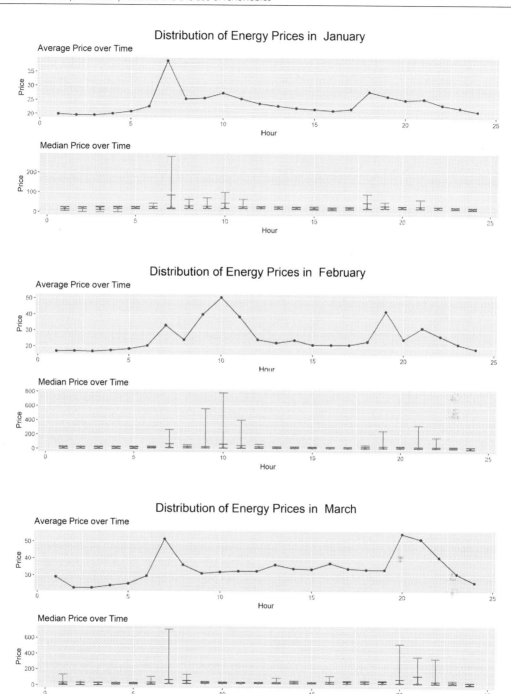

Fig. 14.3 Distribution of wholesale electricity prices at ERCOT. The figure shows the distribution of wholesale electricity prices for each month. The first graph shows the average hourly price while the second shows the median price, with error bars for the 1st to 99th percentile and the 5th–95th percentile. All prices are USD per MWh. *ERCOT*, Electric Reliability Council of Texas.

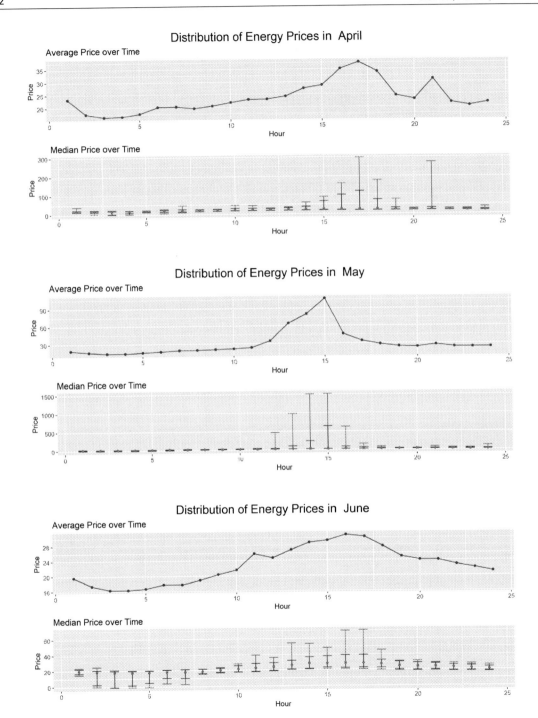

Fig. 14.3 (continued)

Alternative ammonia production processes and the use of renewables

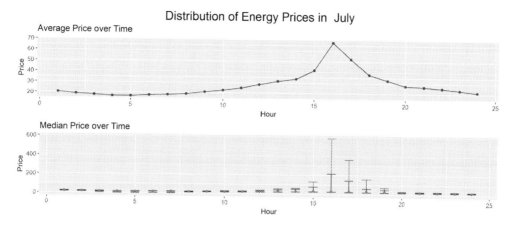

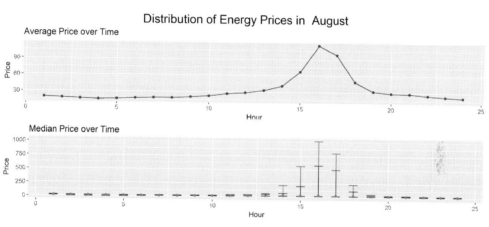

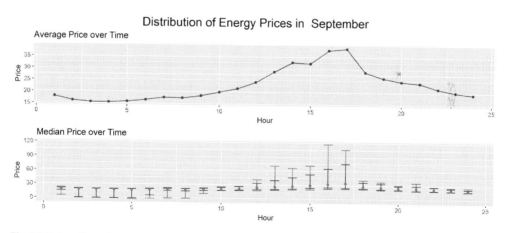

Fig. 14.3 (continued)

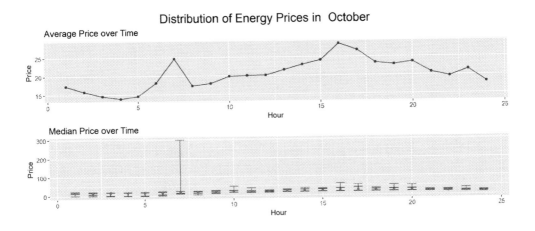

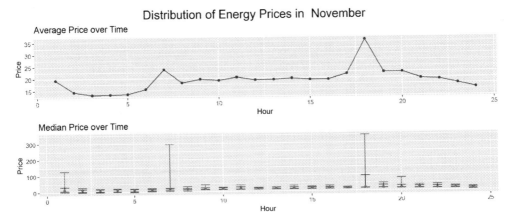

Fig. 14.3 (continued)

ERCOT prices are above the price of $40 per MWh—that price at which it is not economical to produce ammonia. This analysis is done for the four seasons. For all the seasons, most of the time, the wholesale prices are below the price ceiling. The analysis shows that, ex-post, the number of working hours per year in which the price is below $40 per MWh is around 6500 out of 8760 hours in a year. Hence, the analysis suggests that an ammonia plant utilizing ENR technologies can operate about 74% of the year if the price the ammonia plant faces is the ERCOT wholesale price.

However, industrial facilities usually do not have the capacity to utilize high voltage and thus require a supply of low voltage. An ammonia plant can invest in an off-grid solar/wind farm [19], but it might also elect to purchase low-voltage electricity from the grid. When looking at the industrial price, the distribution of the price in the Texas

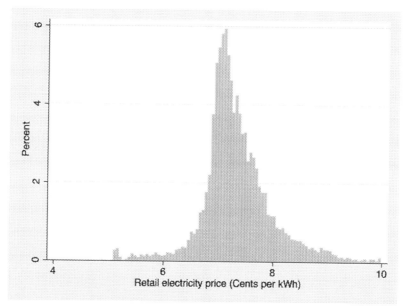

Fig. 14.4 The distribution of industrial retail electricity prices in the ERCOT region. *ERCOT*, Electric Reliability Council of Texas.

ERCOT Houston node region is depicted in Fig. 14.4. The mean calculated industrial price is $76.5 per MWh (Fig. 14.4), with a standard deviation of 2.95. With high overpotential values and low Faradaic efficiency, production of NH_3 using ENR would be challenging when the plant connects to the grid. However, with low overpotential and high Faradaic efficiency, the ammonia plant can become competitive with the existing technologies over large portions of the 24-h cycle.

Although high fixed costs describe the Haber–Bosch process, if the ENR plants are relatively efficient then these plants will probably connect to the grid and compete with the existing incumbent Haber–Bosch plants. However, if the electrochemical direct nitrogen reduction plant is not sufficiently efficient, then the plant may elect to make a larger upfront investment and connect to an off-grid solar/wind farm, paying larger upfront costs but lower marginal costs. Recall that the Haber–Bosch process does require very large upfront investments, but its operations and maintenance costs are relatively cheap because of the low natural gas prices.

14.5 Conclusions and perspectives

The goal of this chapter is to break down the ENR cost structure, separate out the different system components, and shed new light on the futuristic supply structures of ammonia. The analysis evaluated how improvements resulted in lower electricity feedstock costs and assessed the sensitivity of the ENR process to the assumptions on the Faradaic efficiency, overpotential, and electricity prices.

The chapter compared the outcome of the ENR process through its cost, energy use, and emissions to those generated by the Haber–Bosch and the E\H–B processes. It showed that the cost of electricity is determined via overall energy efficiency (i.e., Faradaic efficiency and overpotential) and the cost of MWh. Our results show that key to the success of this technology are improvements in efficiency and energy use. The comparison with the other processes highlights the importance of electricity costs. The analysis implies that the ENR process is much more flexible than the Haber–Bosch, and that its resource adequacy and usage may result in cleaner and more sustainable production processes, compared to other existing options.

This work depicts a futuristic production structures of nitrogen to ammonia, suggesting a supply chain alternative to the existing ammonia industry that may use solar or wind energy to supply electricity off the grid and work well with intermittency and renewable energy systems introduced. The ENR technology can be utilized to shift electric load, which increases the reliability of the supply of electricity, reduces the generation, and transmission capacity needed when electricity load peaks, and increases the electric power system's ability to handle intermittent resources such as wind and solar. Although the current state of ENR technology is far from economically viable, this work suggests that the development of an efficient ENR technology is an objective with tremendous potential.

References

[1] http://www.fao.org/faostat/en/#data/RFN. [Accessed May 8, 2020].
[2] V. Smil, Enriching the Earth: Fritz Haber, Carl Bosch, and the Transformation of World Food Production, MIT Press, Cambridge, MA, 2004.
[3] I. Rafiqul, C. Weber, B. Lehmann, A. Voss, Energy efficiency improvements in ammonia production—perspectives and uncertainties, Energy 30 (2005) 2487–2504.
[4] J.R. Bartels "A feasibility study of implementing an Ammonia Economy" Iowa State University—digital repository, 2008. http://lib.dr.iastate.edu/etd/11132/. [Accessed January 2019].
[5] S. Giddey, S.P.S. Badwal, C. Munnings, M. Dolan, Ammonia as a renewable energy transportation media, ACS Sustain. Chem. Eng. 5 (2017) 10231–10239.
[6] G. Hochman, A. Goldman, F.A. Felder, J. Mayer, A. Miller, P. Holland, L. Goldman, P. Manocha, Z. Song, S. Aleti, "The potential economic feasibility of direct electrochemical nitrogen reduction as a route to ammonia." ACS Sustainable Chemistry & Engineering (2020).
[7] J. Gosnell, Efficient ammonia production, 13 October 2005 Hydrogen Conference Argonne National Laboratory, 2006.
[8] Data available at the U.S. Energy Information Administration site: https://www.eia.gov/dnav/ng/hist/rngwhhdm.htm. [Accessed January 29, 2020].
[9] https://www.nrel.gov/hydrogen/h2a-production-models.html. [Accessed February 28, 2020].
[10] https://ammoniaindustry.com/freeport-tx-yara-basf/. [Accessed May 9, 2020].
[11] G. Saur, T. Ramsden, B. James, W. Colella Future distributed hydrogen production from grid PEM electrolysis NREL, 2018. https://www.nrel.gov/hydrogen/h2a-production-case-studies.html. [Accessed Febraury 2019].
[12] E.R. Morgan, Techno-economic feasibility study of ammonia plants powered by offshore wind, University of Massachusetts Amherst, 2013.

[13] K. Noelker, J. Ruether, Low energy consumption ammonia production: baseline energy consumption, options for energy optimization, Nitrogen + Syngas Conference, Duesseldorf, 2011.
[14] A. Makhlouf, T. Serradj, H. Cheniti, Life cycle impact assessment of ammonia production in Algeria: a comparison with previous studies, Environ. Impact Assess. Rev. 50 (2015) 35–41.
[15] B.H.R. Suryanto, H.-L. Du, D. Wang, J. Chen, A.N. Simonov, D.R MacFarlane, Challenges and prospects in the catalysis of electroreduction of nitrogen to ammonia, Nat. Catal. 2 (2019) 290–296.
[16] http://www.ercot.com/services/rq/credit. (Accessed January 28, 2020).
[17] https://www.eia.gov/electricity/data.php. (Accessed January 28, 2020).
[18] Capacity costs are not introduced, since the ERCOT region does not have capacity costs.
[19] Note that a 10 MW solar farm costs around 25 million USD and uses about 50 acres of land (recall that under our aspirational scenario −0.95 V overpotential and 95% Faradaic efficiency—energy input under the ENR process is 8.70 MWh per metric ton of NH3). http://innovativesolarfarms.com/solar-farm-cost-per-acre/. [Accessed October 2018].

CHAPTER FIFTEEN

Regional strategy of advanced biofuels for transportation in West Africa

Edgard Gnansounou, Bénédicte Nsalambi

Ecole Polytechnique Fédérale de Lausanne (EPFL), School of Environment, Civil Engineering, and Architecture, Institute of Civil Engineering, Bioenergy and Energy Planning Research Group, Lausanne, Switzerland

15.1 Introduction

West Africa is composed of 16 states: Benin, Burkina Faso, Cape Verde, Gambia, Ghana, Guinea, Guinea-Bissau, Ivory Coast, Liberia, Mali, Mauritania, Niger, Nigeria, Senegal, Sierra Leone, and Togo. Only countries of the Economic Community of West Africa were considered in this study so Mauritania was not. Furthermore, Cape Verde and Guinea-Bissau were also excluded due to missing data on biomass resources in both countries. This region mostly includes low living standard countries [1]. As most of Sub-Saharan Africa regions, West Africa shows one of the highest demographic growths worldwide and, due to the mineral resources endowment and the high level of unsatisfied needs of the people, the potential of economic growth of the region is strong. The most important economic activity is still agriculture. However, because of insufficient infrastructure, rainy season agriculture dominates, which means a high sensitivity to climate variability. The climate is highly diverse, showing three distinctive climatic zones. Besides the semi-arid Sahel in the North, the Sudanian Savanna with its high precipitation variability is neighbored at the South-West by the sub-humid Guinean zone along the Gulf of Guinea. The drought increases from the Atlantic Ocean to the Sahel.

The energy supply of the region is dominated by traditional fuels comprising fuelwood and charcoal, which fosters overexploitation of woody resources. The other important issue is the share of petroleum fuels in the transport sector. Presently, road transport is fueled exclusively by gasoline and diesel. The question about the potential alternative fuels that could be used to diversify the energy matrix of this sector is relevant. Indeed, with the demographic and economic growth, and the reduction of the woody resources, the current pressure on these resources will worsen, especially in the perspective of climate change.

The question concerning how much biofuels can contribute to the energy consumption of the transport sector at country, regional, or world levels has been addressed in many papers. The market of biofuels is presently dominated by edible-crops-based fuels, the so-called first-generation biofuels. The potential competition between mobility, food, and animal feed for biomass or land leads to the necessity to develop and use advanced biofuels. They comprise biofuels produced with wastes or feedstock that do not compete with food and feed. The International Renewable Energy Agency has fixed a world target of 18% lignocellulosic-based biofuels in the final energy consumption of the transport sector in 2050 [2]. This kind of target can be found by using partial equilibrium models that may address the issue of adequacy between sectors' energy demand and supply, linking bottom-up structure to global issues such as reduction of greenhouse gas emissions. As an example, Börjesson et al. [3] used MARKAL for modeling biofuel futures in road transport in the case of Sweden. They found that biofuels would account for 78% of the final energy use of the road transport of Sweden in 2050, contributing significantly to the reduction of the greenhouse gas emissions of the energy sector. General equilibrium models (GEM) are used as well. Karkatsoulis et al. [4] included in GEM specifications for European Union biofuels whose consumption was limited to 10% of the final energy of the transport sector. One important constraint considered in all these models concerns the availability and the nature of biomass feedstock.

Based on studies of the FAO regarding scenarios of agricultural lands potentials and productivities (website of FAO) [5], Gnansounou et al. [6] used a multiobjective optimization to assess the potential shares of bioelectricity, synthetic natural gas (SNG), and second-generation bioethanol for road passenger transport in 2050 for eight scenarios. The feedstock was agricultural residues. The model used was simple compared to MARKAL and GEM3E. The purpose of the work of Gnansounou et al. [6] was a prospective for a world region where only fewer data are available. The present study is the continuation of that work but in a perspective context of a federalization of this region, which implies the same biofuels strategy for the 13 countries by 2050. This strategy consists of the same percentage of mobility met by biomass and for each biofuel considered in the study, the same share in the total final biofuels from one country to the other. However, the implications of this biofuel strategy change from one country to the other depending, for example, on the availability of agricultural residues. Despite the low average income per capita across the 13 countries, there is, to some extent, an economic diversity with Nigeria representing above 70% of the regional GDP and 52% of the population in 2019. Ghana, Nigeria, and Ivory Coast are the top three in terms of GDP per capita.

The objective of this chapter was to find out and recommend a common strategy for advanced biofuels (bioelectricity, SNG, and Bioethanol) for the 13 countries and assess the implications for each country. In the next chapter of this book, another perspective was chosen where, each country was supposed to follow its own strategy while sharing with the other countries the same referential. Compared to Gnansounou et al. [6], a new methodology was developed in order to find only one strategy for all countries. However, the basic data and most of the assumptions remain the same. In particular, it was assumed that despite the present economic diversity within the region, there will be a convergence of the living standard by 2050, and thus, the same per capita average road passenger mobility will be demanded irrespective of the country, which also implies the same structure of transport infrastructure. This infrastructure was assumed dominated by road transport. The fuel chosen to estimate the reference final energy consumption for the passenger road transportation by 2050 was gasoline, and the fuel economy and its evolution were considered similar in all countries. Due to these assumptions, the reference fuel consumption from one country to the other was proportional to the population. Finally, technology progress and economic considerations were supposed to be one of the most important drivers of the introduction of advanced biofuels in this region. However, in order to take into consideration the issue of security of supply, a constraint of diversity was included, as well as in the problem setting.

15.2 Case of West Africa

15.2.1 Optimal biofuel strategies

For a given share $R_{st,j}$ of biomass to meet the demand of passenger road mobility, the biofuel strategy consisted in choosing the share of each selected biofuel (V). Once this share was found, the corresponding biomass resource Ebiom was estimated by considering the specific consumption of each biofuel, the energy transport efficiency (T), and the yield of the conversion of biomass to biofuel (Y). The annual mobility of vehicles is estimated taking into account the population (Pop), the proportion of people using road vehicles for their mobility (PUV), the annual average passenger mobility per user of road vehicle (mobility), and the average number of passengers per vehicle (filling). For a given strategy candidate k, the biomass resource adequacy was estimated by the ratio of the available biomass (TEbiom) to the used biomass (Ebiom). Eqs 15.1 and 15.2 describe these relations. The indexes i, j, k in these equations denote, respectively, the biofuel value chain, the biomass strategy, and the biofuel strategy candidate also referred to as portfolio. The problem consists of finding the optimal biofuel strategy ($V_{k,i,j}$). Perfbiom$_{k,j}$ is one of the objective functions.

$$\text{Ebiom}_{k,j} = \frac{R_{st,j}}{100} \times \text{Pop} \times \text{PUV} \times \frac{\text{mobility}}{\text{filling}} \times \sum_i \frac{V_{k,i,j}}{100} \times CS_i \times (T_i \times Y_i)^{-1} \quad (15.1)$$

$$\text{Perfbiom}_{k,j} = \frac{\text{TEbiom}}{\text{Ebiom}_{k,j}} \quad (15.2)$$

The second objective function is a relative cost function of the biofuels supply chains where the most important part is the contribution to new infrastructure. The objective is to minimize this function. In Eq. 15.3, c_i is the relative cost of the biofuel i. The reference biofuel was bioethanol and c_b was set to 1. Finally, A represents the area of the country.

$$\text{Perf inf } r_{k,j} = \text{Log}(A) \times \sum_i \frac{V_{k,i,j}}{100} \times CS_i \times c_i \times (CS_b \times c_b)^{-1} \quad (15.3)$$

The third objective function aims at maximizing the diversity of the final energy share ($SW_{k,i,j}$). These shares were estimated using Eqs 15.4 and 15.5.

$$\text{Perfdiversity}_{k,j} = S \times (1 - (1.5 \times \sum_{i=1}^{3}(SW_{k,i,j}^2) - 0.5)) \quad (15.4)$$

In Eq. 15.4, S represents a scaling factor of the diversity indicator. The implications in terms of energy consumption were assessed using Eq. 15.5.

$$W_{k,j} = \frac{R_{st,j}}{100} \times \text{Pop} \times \text{PUV} \times \frac{\text{mobility}}{\text{filling}} \times \sum_{i} \frac{V_{k,i,j}}{100} \times CS_i \quad (15.5)$$

where $W_{k,j}$ is the final bioenergy for portfolio k, according to strategy j. Their shares were estimated using Eq. 15.6.

$$SW_{k,i,j} = \frac{V_{k,i,j}}{100} \times CS_i \times (\sum_{m} \frac{V_{k,m,j}}{100} \times CS_m)^{-1} \quad (15.6)$$

where $SW_{k,i,j}$ is the share of the fuel pathway i in the total final bioenergy. Finally, the share of Ebiom that is converted in each biofuel was estimated using Eq. 15.7.

$$R_{k,i,j} = V_{k,i,j} \times CS_i \times (T_i \times Y_i)^{-1} \times (\sum_{m} V_{k,m,j} \times CS_m \times (T_m \times Y_m)^{-1})^{-1} \quad (15.7)$$

Similar to Gnansounou et al. [6], the performance of each portfolio was scored using an affine scaling function. The aggregate score of each portfolio is a weighted average of the three scores. For each weighting, the optimal portfolio was found and the selected optimal portfolio was the one the weighting of which maximized the Shannon entropy function described by Eq. 15.8.

$$E_{k,j} = -(Log(3))^{-1} \times \sum_{p=1}^{3} x_p \times Log(x_p) \quad (15.8)$$

where x_p is the relative weighted score of portfolio k for the objective p according to strategy j.

In the work of this chapter, the minimum and maximum values of each objective function were given by the same portfolios regardless of the country under study. Furthermore, these extremums scale the normalization process that converts values of the objective functions into scores. This leads to the same optimal portfolio for all countries.

Sixteen scenarios were assessed taking into account different values of the biofuel yield, specific consumption of biofuels, and relative cost. For each scenario, the optimal portfolio was found. Then an analysis of robustness was undertaken by evaluating the weighted score of each optimal portfolio found for a given scenario but considering the fifteen other scenarios. For each optimal biofuel strategy (V_e, V_g, V_b), two performances were estimated: the average performance over the 16 scenarios and the maximum regret. These two performances were normalized using, again, affine scaling functions, and the recommended optimal portfolio is the one that shows the maximum average value of scores, providing that each biofuel of the portfolio has a nonnull share.

15.2.2 The matrix of biofuels

The modeling of scores by using the country-specific minimum and maximum of the objective functions was effective. All countries have the same normalized scores by scenario, which leads to fewer common results and other country-specific results. They were analyzed separately and then, a general discussion was carried on. First, the 16 technological scenarios were outlined. Scenario 1 is the reference for the low technology progress. The group of four scenarios of this category shares the same assumption of specific consumption of fuel in 2050: 0.54, 2.16, and 2.72 MJ/km, respectively, for electric vehicles, SNG, and E85 vehicles. For the latter, only the bioethanol part was considered. Scenario 1 considers the low conversion efficiency of biomass into biofuels, $Y_e = 0.4$, $Y_g = 0.5$, and $Y_b = 0.44$, that are the energy efficiency of electricity, SNG and bioethanol respectively. In Scenario 2, the efficiency of electricity was assumed to improve ceteris paribus, $Y_e = 0.5$. In Scenario 3, it was assumed an improved yield of bioethanol, $Y_b = 0.5$, and it was assumed, in Scenario 4, an improvement of the yield of SNG, $Y_g = 0.74$. The second category of scenarios is composed of Scenarios 5–8. Besides the lower fuel economy 0.54, 1.85, and 2.35 MJ/km for the specific fuel consumption of electricity, SNG, and bioethanol, the improvement of the yield in 2050 was also assumed higher, Scenario 5 being the reference of this category. In Scenario 5, all yields improved, $Y_e = 0.5$, $Y_g = 0.74$, and $Y_b = 0.5$. In scenario 6, $Y_e = 0.4$; in Scenario 7, $Y_b = 0.44$; and $Y_g = 0.5$ in Scenario 8. Finally, the category of Scenarios 9–16 shares the same technology progress assumptions with Scenarios 1–8, the only difference being a relative cost $c_e = 15$ instead of $c_e = 10$. For all assessments, it was considered a biomass strategy where the contribution of biomass to meet passenger road mobility was set to 20%. This contribution was referred to as biomobility. Then, for each scenario, the optimal strategy was found. For simplicity, it is denoted by V_e, V_g, V_b that are, respectively, the shares of electricity, SNG and bioethanol in the biomobility. Similarly, SW_e, SW_g, and SW_b are the shares of electricity, SNG, and bioethanol in the final bioenergy. Finally, R_e, R_g, and R_b denote the shares of used feedstock for conversion into these biofuels.

15.2.2.1 Common results for the countries

Table 15.1 shows the results that are common to all countries. Scenarios 1 and 2 result in the same biofuel strategy, which means that the improvement of the Y_e from 0.4 to 0.5 has no effect on the strategy. The increase of Y_b in scenario 3 has a significant effect on the share of bioethanol, which increases from 8% to 11%; a slight increase of V_g also occurs. With the increase of Y_g the effect on the biofuel strategy is more noticeable, due to the magnitude of assumed yield progress, from 0.5 to 0.74.

The share of SNG increases from 10% to 15% from Scenario 1 to Scenario 4 and in consequence, Ve decreases from 82% to 77%. In Scenarios 3 and 4, the share of electricity in the final bioenergy decreases from 51% to 43%–44%. The remaining is shared between SNG and bioethanol with more bioethanol for Scenario 3, and the contrary in the case of Scenario 4. The distribution of the used feedstock between

Table 15.1 Details on common results to the countries.

	V_e[%]	V_g[%]	V_b[%]	SW_e[%]	SW_g[%]	SW_b[%]	P3	W/WRst	R_e[%]	R_g[%]	R_b[%]	w1	w2	w3	Max. score	Max.E
Scenario 1	82	10	8	51	25	25	46.67	4.6%	54	21	24	3	3	2	4.92	0.85
Scenario 2	85	10	8	51	25	25	46.67	4.6%	49	24	27	7	7	5	4.78	0.85
Scenario 3	78	11	11	44	25	31	48.57	5.1%	50	22	28	6	7	5	5.01	0.87
Scenario 4	77	15	8	43	34	23	48.38	5.1%	53	22	25	5	6	4	5.13	0.87
Scenario 5	74	17	9	43	34	23	48.45	4.9%	49	26	26	4	4	3	5.33	0.89
Scenario 6	69	20	11	37	37	26	49.36	5.3%	48	26	27	6	7	5	5.50	0.91
Scenario 7	76	17	7	46	35	18	47.09	4.7%	51	26	23	7	7	5	5.23	0.88
Scenario 8	76	13	11	45	26	28	48.42	4.8%	45	26	28	7	7	5	4.96	0.89
Scenario 9	81	12	7	49	29	21	46.89	4.7%	53	25	21	7	6	5	5.32	0.86
Scenario 10	82	11	7	51	27	22	46.43	4.6%	49	27	24	6	5	4	5.04	0.86
Scenario 11	77	13	10	43	29	28	48.96	5.1%	48	26	25	5	5	4	5.29	0.88
Scenario 12	76	17	7	42	38	20	47.82	5.1%	53	25	22	5	5	5	5.57	0.89
Scenario 13	75	18	7	45	37	18	47.19	4.8%	51	28	21	7	6	4	5.59	0.89
Scenario 14	69	21	10	37	39	24	48.92	5.3%	48	27	24	5	5	4	5.80	0.92
Scenario 15	76	18	6	46	38	16	46.31	4.7%	52	28	20	7	6	5	5.55	0.89
Scenario 16	78	13	9	48	28	24	47.46	4.6%	48	28	20	6	5	4	5.18	0.89

biofuels technologies is also consistent with the assumed technology progress in the different scenarios. The results of the first category of the scenario fully meet the expectations. The biofuel strategy that results from scenario 5 is characterized by a significant increase of V_g and a slight increase of V_b detrimental to V_e. Indeed, all energy yields improve in this scenario, and the fuel economy for SNG and bioethanol increases. These improvements are favorable to SNG and bioethanol whose share clearly improves, 8% change from Scenario 1. However, the improvement is in favor of SNG because of the comparative advantage assumed for that biofuel in the assumptions of this scenario. In Scenario 6, the efficiency of electrical generation decreases, which results in a higher increase of the technology advantage of SNG and bioethanol with, again, relative advantage of SNG. Biofuel strategies for Scenarios 7 and 5 are close with a slight decrease of the share of bioethanol due to the assumption of low yield of this biofuel. In Scenario 8, the yield of SNG decreases, enhancing the advantage of bioethanol whose share increases to 11% of the biomobility compared to 7% in Scenario 7, this is detrimental to electricity.

The shares of electricity in Scenarios 9–16 are expected to be lower compared to related Scenarios 1–8 due to the increase of c_e. The comparison of the results of Scenarios 9 and 10 with Scenarios 1 and 2 do not show any significant change of V_e. The analyses of the results of Scenarios 11 and 3 show a slight decrease of V_e and an increased share of SNG due to an improved advantage compared to electricity and bioethanol. The results of Scenarios 12–16 are close to the ones of their related scenarios from 4 to 8. Table 15.1 gives the detailed common results of all countries. V_e varies from 69% to 82%, the lowest value being for the optimal strategy in the case of Scenarios 6 and 14, and the highest for Scenarios 1, 2, and 10. Scenarios 6 and 14 are different only by c_e, the same for Scenarios 2 and 10, which evidences that increasing c_e by 50% does not significantly impact the results. The share of SNG ranges between 10% and 21%. The lowest value is explained by advantage of electricity due to a better relative energy efficiency; the same factor explains the higher value, the strong reduction of the efficiency advantage of electricity coupled with a minor effect of the increase of the relative cost of electricity. The biomass strategy that consists in a contribution of 20% to meet the mobility would result in a share of about 5% of the final energy. The level of this share is negatively correlated with the contribution of electromobility due to the better energy efficiency of this value chain. About 50% of used biomass was converted into electricity, the remaining being almost equally distributed between SNG and bioethanol technologies.

15.2.2.2 Country specific results for each scenario
Common results presented in the previous section have different implications from one country to the other, which are discussed with respect to the values of the first and second objective functions and the total final biofuel energy. For this discussion, countries

are split into two groups. Group 1 is composed of Burkina Faso, Gambia, Liberia, Mali, Niger, Senegal, and Sierra-Leone. This group is dominated by Sahelian countries. The remaining countries composed the Guinean Gulf Group; they are referred to as Group 2.

15.2.2.2.1 Pressure on available feedstock

As it is expected, the ratio between available and used feedstock is higher for Group 2 compared to Group 1. A low value of P_1 meaning a high pressure on the available feedstock, Table 15.2 shows that in Group 1, the pressure is higher for Gambia followed by Senegal, Burkina, Mali, and Niger. It is less severe for Liberia and Sierra-Leone. In the scenarios where the overall energy efficiency is low, the used biomass is high and values of P_1 are lower; these are observed for Scenarios 1, 3, 9, 11. The lowest value of P_1 in group 1 occurs in Scenario 11 in Gambia followed by Senegal. The highest values occur in Sierra Leone in Scenario 15 followed by Scenario 13. Both scenarios are characterized by a high fuel economy for the three biofuels and high energy yields for at least electricity and SNG.

To attain 20% of passenger road mobility, countries of group 1 use between 13% (Sierra Leone) and 29% (Gambia) of their available agricultural residues. The available feedstock is more abundant in countries of Group 2 where used biomass represents between 4.6% and 10.3%. The lowest value of P_1 for this group occurs in Guinea for Scenario 11 and the highest value in Benin for Scenario 15. For all 13 countries, values of P_1 are proportional from one scenario to the other, with the same proportionality for a given couple of scenarios regardless of the country. For example, the ratio between the maximum and minimum values of P_1 in each country is 1.35.

15.2.2.2.2 Economic considerations

The economic considerations in this study are rough. In the second objective function, relative costs including cost of infrastructure and operation cost were summed up and normalized with respect to the cost for bioethanol. Table 15.3 shows the results related to the value of P_2. Since area of countries and energy consumption were also considered in the objective function, larger countries show higher value of P_2. In group 1, values of P_2 range between 7.40 (Gambia) and 18.21 (Niger). Lowest values occur in scenario 3 while scenario 16 involves the highest values. Values of P_2 in group 2 vary between 8.73 in Togo for Scenario 3 and 17.80 in Scenario 16 for Nigeria.

Similarly to P_1, values of P_2 are proportional between scenarios from one country to the other. In particular, the ratio between the highest and the lowest value for each country equals 1.62. These proportionalities are due to the specifications of the model that is linear.

15.2.2.2.3 Final energy consumption

Due to the basic assumptions in the methodology, the final energy consumption depends on the population, the optimal portfolio, and the specific energy consumption. For countries of the group 1, values of energy consumption range from 0.42 PJ in

Table 15.2 Detailed results on the first objective function.

P1	Benin	Burkina Faso	Ivory coast	Gambia	Ghana	Guinea	Liberia	Mali	Niger	Nigeria	Senegal	Sierra Leone	Togo
Scenario 1	17.14	3.88	15.68	3.63	16.22	10.27	4.83	3.88	4.08	10.82	3.68	6.55	13.68
Scenario 2	19.23	4.36	17.59	4.07	18.20	11.52	5.42	4.36	4.58	12.15	4.13	7.35	15.35
Scenario 3	16.39	3.71	14.99	3.47	15.51	9.82	4.62	3.71	3.90	10.35	3.52	6.26	13.08
Scenario 4	17.67	4.00	16.17	3.74	16.72	10.59	4.98	4.00	4.21	11.16	3.80	6.75	14.11
Scenario 5	21.15	4.79	19.35	4.48	20.01	12.67	5.96	4.79	5.04	13.36	4.54	8.08	16.88
Scenario 6	17.88	4.05	16.36	3.79	16.92	10.71	5.04	4.05	4.26	11.29	3.84	6.83	14.27
Scenario 7	21.51	4.87	19.68	4.55	20.35	12.89	6.07	4.87	5.12	13.58	4.62	8.22	17.17
Scenario 8	19.15	4.34	17.52	4.05	18.12	11.48	5.40	4.34	4.56	12.10	4.11	7.32	15.29
Scenario 9	17.04	3.86	15.59	3.61	16.13	10.21	4.81	3.86	4.06	10.76	3.66	6.51	13.61
Scenario 10	19.43	4.40	17.78	4.11	18.39	11.64	5.48	4.40	4.63	12.27	4.17	7.42	15.51
Scenario 11	16.25	3.68	14.86	3.44	15.37	9.73	4.58	3.68	3.87	10.26	3.49	6.21	12.97
Scenario 12	17.83	4.04	16.31	3.77	16.87	10.68	5.03	4.04	4.25	11.26	3.83	6.81	14.23
Scenario 13	21.93	4.96	20.06	4.64	20.74	13.14	6.18	4.96	5.22	13.85	4.71	8.37	17.50
Scenario 14	18.08	4.10	16.54	3.83	17.11	10.83	5.10	4.10	4.31	11.42	3.88	6.91	14.44
Scenario 15	21.89	4.96	20.03	4.63	20.72	13.12	6.17	4.96	5.21	13.83	4.70	8.36	17.48
Scenario 16	19.95	4.52	18.25	4.22	18.88	11.95	5.63	4.52	4.75	12.60	4.28	7.62	15.92

Table 15.3 Detailed results for the second objective function.

P2	Benin	Burkina Faso	Ivory Coast	Gambia	Ghana	Guinea	Liberia	Mali	Niger	Nigeria	Senegal	Sierra Leone	Togo
Scenario 1	9.46	10.16	10.30	7.53	10.05	10.08	9.44	11.39	11.41	11.15	9.90	9.08	8.89
Scenario 2	9.46	10.16	10.30	7.53	10.05	10.08	9.44	11.39	11.41	11.15	9.90	9.08	8.89
Scenario 3	9.29	9.97	10.11	7.40	9.87	9.90	9.27	11.19	11.20	10.95	9.72	8.91	8.73
Scenario 4	9.36	10.05	10.19	7.45	9.95	9.97	9.33	11.27	11.29	11.03	9.79	8.98	8.79
Scenario 5	10.41	11.18	11.34	8.29	11.07	11.09	10.39	12.54	12.56	12.28	10.90	9.99	9.78
Scenario 6	10.17	10.92	11.07	8.10	10.81	10.84	10.15	12.25	12.27	11.99	10.64	9.76	9.56
Scenario 7	10.54	11.32	11.48	8.40	11.21	11.23	10.52	12.70	12.72	12.43	11.03	10.12	9.91
Scenario 8	10.43	11.20	11.35	8.30	11.08	11.11	10.40	12.56	12.58	12.30	10.91	10.01	9.80
Scenario 9	13.54	14.54	14.75	10.78	14.39	14.43	13.51	16.31	16.34	15.97	14.17	13.00	12.73
Scenario 10	13.61	14.62	14.82	10.84	14.47	14.50	13.58	16.40	16.42	16.05	14.24	13.07	12.79
Scenario 11	13.17	14.14	14.34	10.49	14.00	14.03	13.14	15.86	15.89	15.53	13.78	12.64	12.38
Scenario 12	13.19	14.16	14.36	10.50	14.02	14.05	13.16	15.89	15.91	15.55	13.80	12.66	12.40
Scenario 13	14.87	15.96	16.19	11.84	15.80	15.84	14.83	17.91	17.93	17.53	15.56	14.27	13.97
Scenario 14	14.21	15.26	15.47	11.32	15.11	15.14	14.18	17.12	17.14	16.76	14.87	13.64	13.35
Scenario 15	14.99	16.10	16.32	11.94	15.93	15.97	14.95	18.06	18.08	17.68	15.69	14.39	14.09
Scenario 16	15.09	16.21	16.43	12.02	16.04	16.08	15.06	18.18	18.21	17.80	15.79	14.49	14.18

Gambia for scenario 10 and 6.57 PJ in Niger for Scenario 6. Scenario 10 is characterized by low ä fuel economy for SNG and bioethanol, and an increased efficiency of electricity generation that would favor a higher contribution of electricity but a lower energy need. Adversely, Scenario 6 involved higher yields of SNG and Bioethanol and lower generation efficiency for electricity and thus an increase of the contributions of both fuels. Since, in spite of the reduction of electricity generation, electrical value chain remains more efficient compared to SNG and bioethanol, the energy need increased under scenario 6. For group 2, the final energy ranges from 1.34 PJ in Togo for scenario 10–40.18 PJ in Nigeria for scenario 6. Also, there is a perfect correlation between results of scenarios across countries for the final energy due to the same reasons of model specifications. For each country, the ratio between the highest value given by Scenario 6 and the lowest value in Scenario 10 equals 1.14. Table 15.4 shows detailed results for each country.

The total bioenergy consumption in the region would range from 68.18 PJ (Scenario 10) to 78.41 PJ (Scenario 6). It is worth noting that Scenario 10 involved the highest share of electricity and Scenario 6 the lowest. The shares of bioenergy in the total final energy in Scenarios 10 and 6 are 4.6% and 5.3%, respectively, corresponding both to a share of 20% of the biomobility. Common sense would suggest that it is better to get 5.3% contribution of renewable fuels in the final energy consumption instead of 4.6% and thus, that the optimal strategy from Scenario 6 is better. This common-sense inspired the general expression of energy and environmental policies where the targeted contribution of renewable energy is stated in terms of percentage of final energy. This kind of statement assumes equals energy efficiency of all fuels irrespective of their value chains. Since in reality, this is not the case, the energy content of electricity is often overestimated while calculating the compliance to this kind of requirement. For example in the Renewable Energy Directive of the European Union of 2018 (REDII) [7], the European Commission set a minimum requirement of 14% biofuels for 2030. However, in the calculation of this minimum requirement, the directive provides that *"the share of renewable electricity shall be considered to be four times its energy content when supplied to road vehicles."* Since this overestimation is not scientifically sound as far as final energy is concerned, it would be worthwhile to express policy requirements in terms of services, here mobility, which implies including service level or useful energy into the energy accounting. In the case of Europe, statistics on mobility exist based on assumptions, not in West Africa. Thus it is advised to improve the statistical bases in all countries of the region and enlarge the boundary of the system to the services taking into account the millage of the vehicles.

15.2.3 Recommendations

15.2.3.1 Recommended biofuels strategy

For each scenario one optimal strategy was found. However, since the scenarios were used to model uncertainties, it would be indicated to seek a robust strategy. This was done by estimating the score of each strategy for each of the 16 scenarios and then

Table 15.4 Detailed results of final energy consumption by country.

W [PJ]	Benin	Burkina Faso	Ivory Coast	Gambia	Ghana	Guinea	Liberia	Mali	Niger	Nigeria	Senegal	Sierra Leone	Togo
Scenario 1	2.13	3.80	4.49	0.43	4.56	2.28	0.82	3.82	5.75	35.16	2.91	1.13	1.35
Scenario 2	2.13	3.80	4.49	0.43	4.56	2.28	0.82	3.82	5.75	35.16	2.91	1.13	1.35
Scenario3	2.32	4.16	4.91	0.47	4.98	2.49	0.89	4.17	6.28	38.43	3.18	1.24	1.48
Scenario 4	2.32	4.16	4.91	0.47	4.98	2.49	0.89	4.17	6.28	38.41	3.18	1.24	1.48
Scenario 5	2.25	4.02	4.74	0.45	4.81	2.40	0.86	4.03	6.07	37.15	3.07	1.20	1.43
Scenario 6	2.43	4.35	5.13	0.49	5.21	2.60	0.94	4.36	6.57	40.18	3.32	1.30	1.54
Scenario 7	2.16	3.86	4.56	0.43	4.63	2.31	0.83	3.88	5.83	35.69	2.95	1.15	1.37
Scenario 8	2.21	3.95	4.66	0.44	4.73	2.36	0.85	3.96	5.97	36.50	3.02	1.18	1.40
Scenario 9	2.15	3.85	4.55	0.43	4.61	2.30	0.83	3.86	5.82	35.58	2.94	1.15	1.37
Scenario 10	2.11	3.78	4.46	0.42	4.53	2.26	0.81	3.79	5.71	34.93	2.89	1.13	1.34
Scenario 11	2.35	4.20	4.96	0.47	5.04	2.51	0.90	4.22	6.35	38.85	3.21	1.25	1.49
Scenario 12	2.35	4.20	4.96	0.47	5.03	2.51	0.90	4.22	6.35	38.84	3.21	1.25	1.49
Scenario 13	2.19	3.92	4.63	0.44	4.69	2.34	0.84	3.93	5.92	36.22	3.00	1.17	1.39
Scenario 14	2.42	4.33	5.11	0.49	5.18	2.59	0.93	4.34	6.53	39.97	3.31	1.29	1.54
Scenario 15	2.15	3.84	4.53	0.43	4.60	2.30	0.83	3.85	5.80	35.49	2.94	1.14	1.36
Scenario 16	2.12	3.79	4.48	0.43	4.54	2.27	0.82	3.81	5.73	35.04	2.90	1.13	1.35

characterizing each strategy by its average score and its maximum regret. The objective is to recommend the strategy that provides the best trade-off between both objectives. Table 15.5 shows the 16 strategies with their performances in case of each scenario. It comes out that Strategy 14 gets the maximum average score with a value of 5.74. However, its maximum regret is 0.213. Concurrently, Strategy 4 scores 5.33 in average and gets the minimum of the maximum regret, which is a value of −0.052.

After an affine rescaling, all strategies were scored based on the two criteria (mean multiscenarios score, max regret). The average score based on these two criteria elected Strategy 4 as the most robust strategy with a score of 7.73 followed by Strategy 6 with 7.24. Strategy 4 proposes the following portfolio: V_e = 77%, V_g = 15%, and V_b = 8%.

The regrets of strategy 4 are negative for each scenario different from scenario 4, which means its score is better under each of these scenarios.

Table 15.6 shows the mix of the bioenergy and the share of bioenergy in the total final energy for two groups of scenarios.

The share of electricity in the final bioenergy for scenarios 1–4 and 9–12 is 43% along with 34% and 23% for SNG and bioethanol, respectively. For other scenarios 5–8 and 13–16 these shares are 47%, 31%, and 22%, respectively, for electricity, SNG and bioethanol. For the first group, bioenergy represents 5.06% of the total final energy; the value is 4.66% for the second group of scenarios. These results are consistent with the definition of the different scenarios. The first group of scenarios is those with low

Table 15.5 Determination of the robust strategy.

		Average score	Max. regret	N1	N2	N
Strategies	Strategy 1	5.07	0.05	7.17	2.47	4.82
	Strategy 2	4.85	0.11	5.63	0.00	2.82
	Strategy 3	5.09	0.09	6.11	2.72	4.41
	Strategy 4	5.33	−0.05	10.00	5.46	7.73
	Strategy 5	5.34	0.16	4.25	5.48	4.87
	Strategy 6	5.58	0.09	6.27	8.21	7.24
	Strategy 7	5.30	0.12	5.34	5.05	5.19
	Strategy 8	5.03	0.08	6.29	2.03	4.16
	Strategy 9	5.19	0.32	0.00	3.80	1.90
	Strategy 10	4.96	0.25	1.86	1.26	1.56
	Strategy 11	5.22	0.23	2.45	4.16	3.31
	Strategy 12	5.51	0.24	1.92	7.42	4.67
	Strategy 13	5.47	0.29	0.67	6.94	3.81
	Strategy 14	5.74	0.21	2.79	10.00	6.40
	Strategy 15	5.43	0.30	0.34	6.54	3.44
	Strategy 16	5.12	0.21	2.87	3.06	2.96
Key results	Max.	5.74	0.32	10.00	10.00	7.73
	Min.	4.85	−0.05			
	Strategy	Strategy 14	Strategy 4	Strategy 4	Strategy 14	Strategy 4

Table 15.6 Energy characteristics of the strategy 4 for different scenarios.

Scenarios	SW_e [%]	SW_g [%]	SW_b [%]	W/WRst
1–4 and 9–12	43	34	23	5.06
5–8 and 13–16	47	32	21	4.66

fuel economies while the second group is composed of scenarios with higher fuel economies.

For the shares of the feedstock converted into electricity, SNG, and bioethanol, couples of scenarios shown in Table 15.7 are each associated with the same value of distribution between biofuel conversion technologies.

15.2.3.2 Specific characteristics per country of the recommended strategy

15.2.3.2.1 Pressure on the feedstock in the case of the recommended strategy

As it was discussed for the optimal strategy for each scenario, the pressure on the available feedstock is higher in countries of group 1 (Burkina Faso, Gambia, Liberia, Mali, Niger, Senegal, and Sierra Leone). For all countries, values are perfectly correlated between scenarios. The highest value of P_1—lowest pressure—occurs in the case of Scenario 5 due to the assumed higher technology progress. Most critical cases are with Scenarios 1 and 9. These two scenarios are similar except for the value of c_e.

Compared to the optimal strategy calculated for each scenario, the recommended strategy leads to a higher cost in the cases of Scenarios 1, 2, 7, 9, 10, 12, 15, 16. The maximum additional cost is 9% and occurs in the case of Scenario 10. Conversely, there is a reduced cost in seven scenarios. The maximum reduction occurs in the case of Scenario 14 with 7.6% of the value of P_1 estimated for the optimal strategy under Scenario 14. These results are common to all countries. In group 1, the minimum and maximum values of P1 occur in Gambia and Sierra Leone respectively. In group 2, the lowest is Guinea and the highest Benin as in the case of optimal strategies. Table 15.8 gives the detailed results of each country.

Table 15.7 Distribution of the used feedstock for different scenarios.

Scenarios	R_e[%]	R_g[%]	R_b[%]
6	14	58	21	21
4	12	53	22	25
5	13	52	24	24
7	15	51	23	26
3	11	49	31	20
1	9	48	30	22
8	16	47	32	21
2	10	42	33	25

Table 15.8 Pressure on the available feedstock in the case of the recommended strategy.

P1	Benin	Burkins Faso	Ivory Coast	Gambia	Ghana	Guinea	Liberia	Mali	Nigar	Nigeria	Senegal	Sierra Leone	Togo
Scenario 1	15.97	3.62	14.61	3.38	15.11	9.57	4.50	3.62	3.80	10.09	3.43	6.10	12.75
Scenario 2	17.65	4.00	16.15	3.74	16.70	10.58	4.98	4.00	4.20	11.15	3.79	6.74	14.09
Scenario 3	16.42	3.72	15.02	3.48	15.53	9.84	4.63	3.72	3.91	10.37	3.53	6.27	13.11
Scenario 4	17.67	4.00	16.17	3.74	16.72	10.59	4.98	4.00	4.21	11.16	3.80	6.75	14.11
Scenario 5	22.01	4.98	20.14	4.66	20.83	13.19	6.21	4.98	5.24	13.90	4.73	8.41	17.57
Scenario 6	19.46	4.41	17.80	4.12	18.41	11.66	5.49	4.41	4.63	12.29	4.18	7.43	15.53
Scenario 7	21.32	4.83	19.51	4.51	20.18	12.78	6.01	4.83	5.08	13.47	4.58	8.14	17.02
Scenario 8	19.77	4.48	18.08	4.18	18.70	11.84	5.57	4.48	4.71	12.48	4.24	7.55	15.78
Scenario 9	15.97	3.62	14.61	3.38	15.11	9.57	4.50	3.62	3.80	10.09	3.43	6.10	12.75
Scenario 10	17.65	4.00	16.15	3.74	16.70	10.58	4.98	4.00	4.20	11.15	3.79	6.74	14.09
Scenario 11	16.42	3.72	15.02	3.48	15.53	9.84	4.63	3.72	3.91	10.37	3.53	6.27	13.11
Scenario 12	17.67	4.00	16.17	3.74	16.72	10.59	4.98	4.00	4.21	11.16	3.80	6.75	14.11
Scenario 13	22.01	4.98	20.14	4.66	20.83	13.19	6.21	4.98	5.24	13.90	4.73	8.41	17.57
Scenario 14	19.46	4.41	17.80	4.12	18.41	11.66	5.49	4.41	4.63	12.29	4.18	7.43	15.53
Scenario 15	21.32	4.83	19.51	4.51	20.18	12.78	6.01	4.83	5.08	13.47	4.58	8.14	17.02
Scenario 16	19.77	4.48	18.08	4.18	18.70	11.84	5.57	4.48	4.71	12.48	4.24	7.55	15.78

15.2.3.2.2 Economic considerations in the case of the recommended strategy

The recommended strategy incurred a supplementary cost that is the cost of robustness. This additional cost occurs under the following scenarios: 3, 5, 6, 7, 8, 11, 12, 13, 14,15. These scenarios result in less electricity in the associated optimal strategy. The maximum additional cost is 5.74% and occurs in the case of Scenario 14 where the share of electricity is the lowest. Adversely there is a cost reduction in the cases of Scenarios 1, 2, 9, 10, and 16. The maximum reduction amounts to 2.8% and occurs in Scenario 10. As Table 15.9 shows, values of P_2 are common within the following categories of strategies: category 1 (Scenarios 1–4), category 2 (Scenarios 5–8), category 3 (Scenarios 9–12), and category 4 (Scenarios 13–16). Scenarios in each category share the same fuel economy and the same relative costs.

15.2.3.2.3 Final bioenergy consumption in the case of the recommended strategy

Compared to the optimal strategies found for the different scenarios, the recommended scenario induces an additional final energy consumption for Scenarios 1, 2, 9, 10, 16. The maximum additional energy consumption is 9.94% and occurs in the case of Scenario 10. The consumption of bioenergy reduces in the cases of Scenarios 3, 5, 6, 7, 8, 11, 12, 13, 14, 15. The maximum reduction is 12% in the case of Scenario 6, followed by Scenario 14. Both scenarios have the lowest share of electro-mobility. Table 15.10 presents values of biofuel consumption for different scenarios in the case of Strategy 4.

Once again, results are the same within each of the four categories of scenarios.

15.3 Conclusions and perspectives

West Africa will become a region of strong development in the long-term due to the growth of population, the progress of its human resources, and the endowment of natural resources. However, the region is threatened by climate variability and change. Although rich in petroleum and natural gas, West Africa must take a step in developing advanced biofuels based on valorization of residual biomass. In this chapter, strategies were compared under 16 scenarios for 2050 for a modest share of biomobility. Results can be adapted to a higher share. However, they are based on simple assumptions considering the lack of data and difficulty of anticipating changes in the long term. Thus, this study is prospective and provides fewer orders of magnitude. Finally, the study calls for efforts to collect data and to develop knowledge and technology transfers in biomass valorization. That is a prerequisite for anticipating the growing need of petroleum to fuel the unavoidable development of transportation in West Africa where urbanization is being galloped.

Table 15.9 Values of the second objective function in the case of the recommended strategy.

P2	Benin	Burkins Faso	Ivory Coast	Gambia	Ghana	Guinea	Liberia	Mali	Nigar	Nigeria	Senegal	Sierra Leone	Togo
Scenario 1	9.36	10.05	10.19	7.45	9.95	9.97	9.33	11.27	11.29	11.03	9.79	8.98	8.79
Scenario 2	9.36	10.05	10.19	7.45	9.95	9.97	9.33	11.27	11.29	11.03	9.79	8.98	8.79
Scenario 3	9.36	10.05	10.19	7.45	9.95	9.97	9.33	11.27	11.29	11.03	9.79	8.98	8.79
Scenario 4	9.36	10.05	10.19	7.45	9.95	9.97	9.33	11.27	11.29	11.03	9.79	8.98	8.79
Scenario 5	10.55	11.33	11.49	8.40	11.22	11.24	10.53	12.71	12.73	12.44	11.04	10.13	9.92
Scenario 6	10.55	11.33	11.49	8.40	11.22	11.24	10.53	12.71	12.73	12.44	11.04	10.13	9.92
Scenario 7	10.55	11.33	11.49	8.40	11.22	11.24	10.53	12.71	12.73	12.44	11.04	10.13	9.92
Scenario 8	10.55	11.33	11.49	8.40	11.22	11.24	10.53	12.71	12.73	12.44	11.04	10.13	9.92
Scenario 9	13.23	14.21	14.41	10.54	14.06	14.10	13.20	15.94	15.96	15.60	13.84	12.70	12.43
Scenario 10	13.23	14.21	14.41	10.54	14.06	14.10	13.20	15.94	15.96	15.60	13.84	12.70	12.43
Scenario 11	13.23	14.21	14.41	10.54	14.06	14.10	13.20	15.94	15.96	15.60	13.84	12.70	12.43
Scenario 12	13.23	14.21	14.41	10.54	14.06	14.10	13.20	15.94	15.96	15.60	13.84	12.70	12.43
Scenario 13	15.03	16.14	16.36	11.97	15.97	16.01	14.99	18.13	18.13	17.72	15.72	14.42	14.12
Scenario 14	15.03	16.14	16.36	11.97	15.97	16.01	14.99	18.13	18.13	17.72	15.72	14.42	14.12
Scenario 15	15.03	16.14	16.36	11.97	15.97	16.01	14.99	18.13	18.13	17.72	15.72	14.42	14.12
Scenario 16	15.03	16.14	16.36	11.97	15.97	16.01	14.99	18.13	18.13	17.72	15.72	14.42	14.12

Table 15.10 Bioenergy consumption in the case of the recommended strategy.

W [PJ]	Benin	Burkins Faso	Ivory Coast	Gambla	Ghana	Guinea	Liberia	Mali	Nigar	Nigeria	Senegal	Sierra Leone	Togo
Scenario 1	2.32	4.16	4.91	0.47	4.98	2.49	0.89	4.17	6.28	38.41	3.18	1.24	1.48
Scenario 2	2.32	4.16	4.91	0.47	4.98	2.49	0.89	4.17	6.28	38.41	3.18	1.24	1.48
Scenario 3	2.32	4.16	4.91	0.47	4.98	2.49	0.89	4.17	6.28	38.41	3.18	1.24	1.48
Scenario 4	2.32	4.16	4.91	0.47	4.98	2.49	0.89	4.17	6.28	38.41	3.18	1.24	1.48
Scenario 5	2.14	3.83	4.52	0.43	4.58	2.29	0.82	3.84	5.78	35.37	2.92	1.14	1.36
Scenario 6	2.14	3.83	4.52	0.43	4.58	2.29	0.82	3.84	5.78	35.37	2.92	1.14	1.36
Scenario 7	2.14	3.83	4.52	0.43	4.58	2.29	0.82	3.84	5.78	35.37	2.92	1.14	1.36
Scenario 8	2.14	3.83	4.52	0.43	4.58	2.29	0.82	3.84	5.78	35.37	2.92	1.14	1.36
Scenario 9	2.32	4.16	4.91	0.47	4.98	2.49	0.89	4.17	6.28	38.41	3.18	1.24	1.48
Scenario 10	2.32	4.16	4.91	0.47	4.98	2.49	0.89	4.17	6.28	38.41	3.18	1.24	1.48
Scenario 11	2.32	4.16	4.91	0.47	4.98	2.49	0.89	4.17	6.28	38.41	3.18	1.24	1.48
Scenario 12	2.32	4.16	4.91	0.47	4.98	2.49	0.89	4.17	6.28	38.41	3.18	1.24	1.48
Scenario 13	2.14	3.83	4.52	0.43	4.58	2.29	0.82	3.84	5.78	35.37	2.92	1.14	1.36
Scenario 14	2.14	3.83	4.52	0.43	4.58	2.29	0.82	3.84	5.78	35.37	2.92	1.14	1.36
Scenario 15	2.14	3.83	4.52	0.43	4.58	2.29	0.82	3.84	5.78	35.37	2.92	1.14	1.36
Scenario 16	2.14	3.83	4.52	0.43	4.58	2.29	0.82	3.84	5.78	35.37	2.92	1.14	1.36

References

[1] UNDP/United Nations Development Programme, Human development report. Human development for everyone, UNDP, New-York, NY, 2019.
[2] IRENA, Advanced biofuels. What holds them back? International Renewable Energy Agency, Abu Dhabi, 2019.
[3] M. Börjesson, E.O. Ahlgren, R. Lundmark, D. Athanassiadis, Biofuel futures in road transport—a modeling analysis for Sweden, Transp. Res. Part D 32 (2014) 239–252.
[4] P. Karkatsoulis, P. Siskos, L. Paroussos, P. Capros, Simulating deep CO2 emission reduction in transport in a general equilibrium framework: the GEM-E3T model, Transp. Res. Part D 55 (2017) 343–358.
[5] FAO, Global perspective studies. www.fao.org/global-perspectives-studies/food-agriculture-projections-to-2050 (accessed November, 2019).
[6] E. Gnansounou, E. Ruiz Pachón, B. Sinsin, O. Teka, E. Togbé, A. Mahamane, Using agricultural residues for sustainable transportation biofuels in 2050: case of West, Africa Bioresour. Technol. 305 (2020) 123080.
[7] REDII - L 328/82, Directive (EU) 2018/2001 of the European Parliament and of the Council of 11 December 2018 on the promotion of the use of energy from renewable sources, Official J. Eur. Union (2018).

CHAPTER SIXTEEN

Advanced biofuels for transportation in West Africa: Common referential state-based strategies

Edgard Gnansounou, Elia Ruiz Pachón
Ecole Polytechnique Fédérale de Lausanne (EPFL), School of Environment, Civil Engineering, and Architecture, Institute of Civil Engineering, Bioenergy and Energy Planning Research Group, Lausanne, Switzerland

16.1 Introduction

Biofuels strategies for contributing to meet the energy needs of passenger road transportation in West Africa were analyzed by Gnansounou et al. [1] and in the precedent chapter of the book. The present chapter shares all the assumptions and the methodology used for the case study in the precedent chapter except one. In the case study of this chapter, the assumption regarding the institutional organization of the region is that the present situation that seeks for a regional convergence through an Economic Community will remain in 2050. This assumption implies that each country will have its biofuel strategy even though the total percentage of biomass contribution to meet the mobility demand for passenger road transportation will be the same regardless of the country. There is a diversity in the population, the economy, and biomass resources between the countries that could justify country-specific biofuel strategy. Fig. 16.1 presents the population of each country in 2019.

Nigeria has the highest share of the population (52%) followed by Ghana (8%). The demographic growth is one of the highest worldwide. It is envisioned that the population will double in the next thirty years. Nigeria is also the regional economic leader with about 70% of the regional GDP. West Africa is rich in petroleum and natural gas resources, which can contribute to consolidate the market share of gasoline for road transport in the long-term. The availability of biomass resources varies from one climatic zone to the other. From a social point of view, the use of biomass for energy must avoid competition with the food and animal feed needs. In this work, only agricultural residues are targeted as feedstock for biofuels production. From the subhumid Guinean zone along the Gulf of Guinea to the Sahelian zone, the availability of agricultural lands and their productivity per hectare decrease. Another factor is the competition for lands between human establishments and agroforestry. Urbanization is growing fast in the region. These factors lead to different pressures on potential agricultural resources from one country to the other. The potentials of agricultural residues are estimated based on

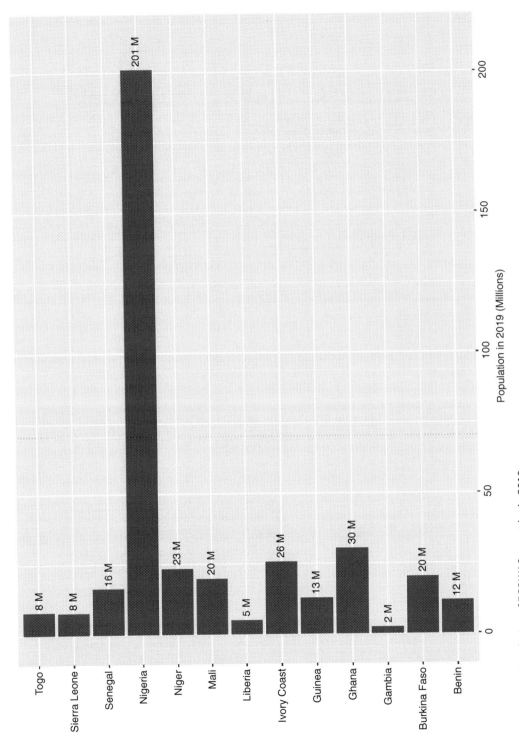

Fig. 16.1 Population of ECOWAS countries in 2019.

the data from the UN Food and Agriculture Organization [2]. Only a few crops were considered: banana, cassava, maize, oil palm fruit, rice, cotton, sorghum, soybeans, and sugarcane resulting in 196 million tons per year of residues in 2050 from which it was assumed that half will be left on the soil for fertility. The distribution by country of this resource is shown in Fig. 16.2.

The objective of this work is to propose biofuels strategies for contributing to meet the mobility demand of passenger road transportation in 2050 for each of the 13 countries considered in this study and enumerated in the precedent chapter. The strategies may be different from one country to the other depending on the availability of biomass, area, and population, which makes this chapter different from the precedent. Several scenarios were analyzed and one strategy was proposed by country as a result of robustness analyses. Biofuel strategy is defined as the shares V_e, V_g, and V_b of electricity, SNG, and bioethanol respectively into the biomobility. Before dealing with the case study of the Economic West African States, the next two sections are devoted to general aspects of biomass feedstock and biomass conversion into transportation biofuels.

16.2 Types of feedstock for advanced biofuels

Low availability of biomass feedstock may be a limiting factor for the development of advanced biofuels for use in transportation. Lignocellulosic biomass can be differentiated according to different criteria like sources, physicochemical properties, or levels of production, i.e., primary, secondary, and tertiary residues. Regarding the sources, the biomass feedstock can result from natural ecosystems (terrestrial or aquatic), energy dedicated crops, i.e., short rotation or perennial crops; primary residual biomass from agroforestry, secondary residues from processed biomass, and tertiary residues from postfinal consumption. The amount of lignocellulosic biomass produced annually seems to be enough for meeting the world's yearly energy demand. According to Balat and Ayar [3], the annual production of biomass is eight times higher than the world's annual energy consumption. However, only a relatively low share of this biomass can be mobilized due to physical, logistical, economic, and ecological constraints. Furthermore, there are direct and indirect competitions for biomass feedstock between different uses so that energy uses can expect to capture only a small share of the theoretical potential of lignocellulosic biomass. This section discusses the typology of lignocellulosic biomass for use as feedstock for biofuels and potential assessment.

Lignocellulosic biomass that can be used as feedstocks for advanced biofuels for transportation is classified into different categories. Based on the sources of the lignocellulosic biomass, the different categories are (1) woody biomass from forests, (2) energy dedicated crops, (3) primary residual biomass, (4) secondary residual biomass, (5) tertiary residual biomass. Due to availability, and the need for sustainability, only agricultural residues were assumed as feedstock and their harvest was supposed to comply with ecological requirements.

282 Biomass, Biofuels, Biochemicals

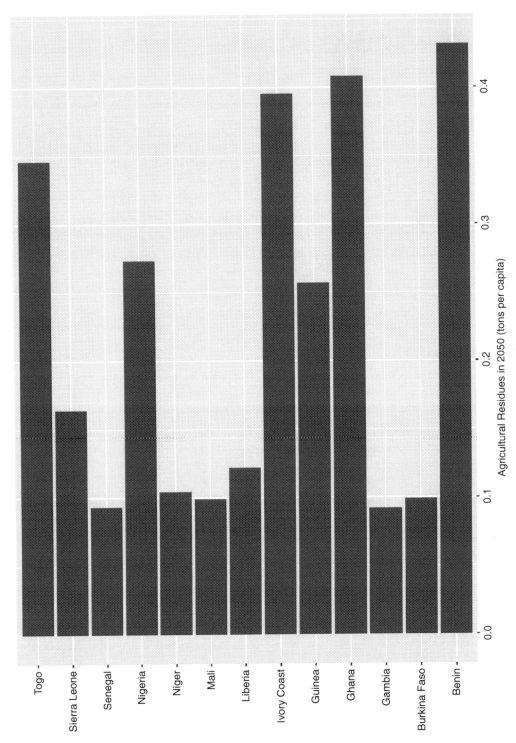

Fig. 16.2 Total agricultural residues selected in this work estimated by 2050.

16.3 Biofuels for transportation
16.3.1 Biofuels

Different renewable transportation fuels such as bioethanol, biobutanol, biomethanol, hydrogen, biomethane, natural gas, or electricity can be produced from biomass. Using agricultural residues for conversion into biofuels can be a promising choice for West Africa due to the high availability per capita in the following years.

16.3.1.1 Bioethanol

Bioethanol is a form of renewable energy that can be produced from agricultural feedstocks. According to the International Energy Agency, cellulosic ethanol could allow ethanol fuels to play a big role in the future [4]. After several decades of bench-scale, pilot and demonstration operations, biochemical conversion of lignocellulosic materials into ethanol started its "first of a kind" commercial stage in the earliest 2010s. Despite several troubles due to economic reasons such as low prices of petroleum oil and the high price of feedstock, technical issues like process inefficiencies or failures, lignocellulosic ethanol remains the most commercially mature and promising pathway for producing transportation biofuels. As examples of existing projects, a company in Slovakia is planning to set up a cellulosic ethanol plant that will use agricultural residues as feedstock [5]. The French pre-industrial plant Futurol is designed as a multi-feedstocks plant capable to use inter alia agricultural and forest residues [6].

16.3.1.2 Biobutanol

Biobutanol is a biofuel produced from biomass that can be used in an internal combustion engine. It is more similar to gasoline than it is to ethanol. Butanol is a drop-in fuel and thus works in vehicles designed for use with gasoline without modification [7]. The fuel is produced through the fermentation of biomass feedstock and the process is nearly identical to fuel ethanol production from biomass. Biobutanol companies produce a range of high-value products, including transportation fuel. Primary coproducts of biobutanol plants may include solvents/coatings, plastics, and fibers. Production of these coproducts helps biobutanol companies improve economic performance through diversification of product offerings. The first biobutanol plants were retrofits of existing corn ethanol plants. A challenge for biobutanol is that more ethanol than biobutanol can be produced from a bushel of corn. Thus, the near-term outlook for biobutanol production is limited, as production has been small and intermittent since 2012 [8].

16.3.1.3 Biomethanol

Biomethanol is identical to methanol, which is the simplest (and cheapest) of the alcohol. Longer-term, biomethanol is a good alternative to bioethanol for replacing petrol in automotive engines [9]. Like the other biofuels, biomethanol should be produced without taking up the agricultural land that is increasingly needed to feed the world's population. It has been successfully produced from feedstocks like wood waste, grass, algae, black liquor from pulping processes, and methane gas from landfills and animal waste [9]. Biomethanol is produced by gasifying organic materials and can produce a variety of fuels such as biohydrogen, bio-dimethyl ether, methanol-to-gasoline, petrol blends, and biodiesel. That versatility is attractive to countries looking for the security of fuel supplies and makes biomethanol a good contender for a low-cost sustainable fuel of the future [9].

16.3.1.4 Hydrogen

Hydrogen is one of the most abundant elements in the universe and could play a significant role in the transition to a clean and low-carbon energy system. Hydrogen can be produced from renewable sources, using biogas, or through electrolysis using electricity generated by renewable sources.

As previously seen, the biogas mixture obtained through anaerobic digestion is a product composed mainly of CH_4 and CO_2, associated with traces of other gases such as H_2S, NH_3, H_2, N_2, O_2, and water vapor. The reforming process of biogas can produce H_2 with high purity. The production of H_2 from biogas has gained much attention lately.

Fuel cells are devices that can convert the chemical energy of a fuel directly into electricity, without combustion, with high efficiency, and with lower polluting emissions than conventional equipment/techniques. The integration of biogas reforming processes and the activation of a fuel cell using H_2 represent an important route for generating clean energy, with added high-energy efficiency [10]. Fuel cells in vehicles generate electricity to power the motor using O_2 from the air and compressed H_2 emitting only water and heat. Fuel cell hydrogen vehicles are classified as zero-emissions vehicles.

16.3.1.5 Biomethane

In order to achieve the target of CO_2 emission reduction, there is a rapidly growing interest in using biomethane as fuel for transport applications. Biomethane can be produced through anaerobic digestion or biomass gasification.

16.3.1.5.1 Biomethane production from anaerobic digestion

Anaerobic digestion is a biochemical process, where complex organic matter is decomposed in absence of oxygen by various types of anaerobic microorganisms, with a low energy demand of heat and electricity; hence, it normally has high energy efficiency. Biogas is produced from anaerobic digestion of the organic material containing a big fraction of CO_2 (about 35%–45vol%). In order to use biogas as vehicle fuel, it has to be upgraded to remove CO_2 and achieve more than 95vol% purity of biomethane [11]. However, since the costs of common technologies of biogas upgrading including water scrubbing, pressure swing adsorption, chemical absorption, etc., are relatively high due to the use of either energy, chemicals, or both [11], there is a big challenge to increase the production of biomethane in energy and cost-efficient way. Biomethane is similar to natural gas and can have the same applications as transportation biofuel while reducing GHG.

16.3.1.5.2 Biomethane production from biomass gasification

Biomass gasification is another technology for efficient biomass use. The gasification of organic matter produces syngas. The produced syngas can be further used to generate power, heat, and fuels, such as hydrogen and ethanol. It is also possible to produce biomethane. Due to the low CH_4 content in the syngas, a methanation process is further required. In this process, CO and H_2 are converted into CH_4. Before methanation, the water-gas-shift process is also needed, which can control the C/H ratio of syngas. The additional WGS and methanation processes result in lower overall efficiency of gasification, a more complicated system configuration, and higher investment costs and operation costs compared to the previous method. However, biomethane can be produced on a larger scale through biomass gasification than through an anaerobic digestion process.

16.3.1.6 Electricity

Electricity for transport is being considered as an alternative to transportation fuels as an energy source. A pure battery electric vehicle is considered a more efficient alternative to a hydrogen fuel propelled vehicle as there is no need to convert energy into electricity since the electricity stored in the battery can power the electric motor. Besides an all-electric car is easier and cheaper to produce than a comparable fuel-cell vehicle.

Electricity can be generated form all renewable sources. In this section, the electricity generated from agricultural residues in a biomass integrated gasification combined cycle (BIGCC) plant will be discussed. In a BIGCC, the biomass gasification process is integrated into a combined cycle to generate the electricity that would be incorporated into the national grid [12]. This electricity mix is the source to fuel electric vehicles. The main barriers to the development of electric cars are the lack of storage systems capable of providing driving ranges and speeds comparable to those of conventional vehicles [13]. The low energy capacity of batteries makes the electric car less competitive than internal combustion engines using gasoline or other biofuels. As technology improves, the energy and cost-effectiveness of batteries are getting better. Electric vehicles are eminently suitable for urban transportation because of the lower ranges involved.

16.3.2 Multifeedstock plants

The conversion of the agricultural residues should be performed in multi-feedstock plants since the different mass composition for each residue plays a key role in the downstream processes. Depending on the biochemical or thermo-chemical processes taking place to produce a specific biofuel, a proper multifeedstock plant should be designed.

Considering the different biofuels discussed here, a lignocellulosic alcohol production plant, a biomethane biological production plant, a biomethane production plant from biomass gasification, a hydrogen production plant from biogas reforming or a BIGCC plant would be required.

The hydrogen vehicle technology is still not mature so rarely would be extended by 2050. At the same time, the high cost associated with its implementation in West Africa makes this option not attractive to be considered in the case study treated in this chapter.

The biomethane produced through anaerobic digestion biogas is less efficient than that produced through gasification.

The lignocellulosic ethanol production plant is the more mature and efficient among the other lignocellulosic alcohol biofuels.

For all this, three biofuels were well-thought-out for transport in West Africa by 2050 as a promising strategy. The potential selected biofuels are bioethanol, biomethane from biomass gasification, and electricity. In this chapter, the designs of three multi feedstock plants are presented to produce these auspicious biofuels.

16.3.2.1 Lignocellulosic bioethanol plant

The existing technologies for bioethanol production were extrapolated in the future based on the 2011 design by the National Renewable Energy Laboratory and documented in Humbird et al. [14]. This extrapolation keeps on the high variety of possible designs and the potential technology progress by 2050. However, whatever progress could occur by 2050, it was considered that both hexoses and pentoses would be converted into bioethanol. From the technology in Humbird et al. [14], a reference overall yield of 87 gallon/ton of biomass, dry basis was found that was assumed as the lowest value in 2050. Assuming a full conversion and stoichiometric reactions, a theoretical yield of 105 gallons/ton d.b. was calculated. The highest value of yield was chosen as 95% of the theoretical value. The yields are then expressed in terms of energy ratio (energy value of bioethanol/energy value of feedstock). It is worth noting that the terms "energy ratio" only consider the ratio between the energy content of the bioethanol and the one in the biomass feedstock. This ratio that only expresses the ethanol yield per ton of feedstock, does not account for the energy used in the conversion process. That is not energy efficiency. The yield in Humbird et al. [14] says 87 gallons/ton d.b or an energy ratio of 0.44 represents 83% of the theoretical yield. The latter corresponds to an energy ratio of 0.53 and the assumed the highest value in 2050 to an energy ratio of 0.50. As a comparison with similar assumptions, Sandia National Laboratories and General Motors' R&D Center [15] assumed an average yield of 95 Gallon /ton d.b. for the conversion of cellulosic material to ethanol by 2030 or an energy yield of 0.48.

16.3.2.2 Synthetic natural gas

The agricultural residues feedstocks are converted into biomethane-rich gas that is purified to release a substitute to fossil-based natural gas: the Bio-Synthetic Natural Gas (Bio-SNG). The thermochemical conversion process involves the following phases: pretreatment of the feedstock, gasification, syngas conditioning, methanation, and conditioning of the Bio-SNG. First, the solid biomass is grinded and dried. The small particles are then conveyed to a Milena-type gasifier. The indirect gasifier operates at a temperature in the range of 700–900°C and a pressure of 7 bar [16]. It generates a raw syngas that is a mixture of CO, CO_2, CH_4, H_2, ethylene, benzene, and contaminants like nitrogen, sulfur compounds, and heavy metals). The conditioning of the raw syngas consists of cooling, cleaning, and adjusting the composition (CO_2 content and H_2/CO ratio in order to comply with the requirements of the subsequent phase. At the methanation stage, the cleaned syngas is converted into Bio-SNG through an exothermic reaction. The raw Bio-SNG still contains water and CO_2 that need to remove. The gas is then cooled and its pressure is adjusted to the application requirements. Bio-SNG is at the stage of pilot and demonstration stages. The efficiency in the case of the Milena project was reported to be 70%. In the following calculations, a minimum efficiency of 54% and a maximum of 74% [17] were considered.

16.3.2.3 Biomass integrated gasification combined cycle

While the bio-SNG plant issues a substitute natural gas that can be converted into electricity in a Combined Cycle Power Plant, it is expected that direct integration of the gasification stage with a combined cycle power generation will improve the whole energy efficiency. Like the development of bio-SNG that learns from Coal-SNG, BIGCC inherits from Coal Integrated Gasification and Combined Cycle. In a BIGCC, the syngas produced after the gasification is directly used to fuel a gas turbine that is equipped with a Heat Recovery Steam Generator. Then the generated heat drives a steam turbine. Whereas a combination of bio-SNG and Combined Cycle Power Plant would give a 45% maximum efficiency at present, the range of 40%–50% BIGCC efficiency has been assumed by 2050 in further sections of this chapter.

16.4 Cases of West African states

16.4.1 Influences of the methodology

The methodology adopted in the precedent chapter was followed in the case study of this chapter with only one difference. When performing the affine scaling to get the score for each portfolio, all countries were assumed to share the same minimum and maximum values of the performances for each objective function. Let $P_{1,k,c}$, $P_{2,k,c}$ and $P_{3,k,c}$ be the performance of the portfolio k for a country c. Let $MinP_{v,c}$ and $MaxP_{v,c}$, Mi_{nv}, Max_v be such as:

$$MinP_{v,c} = \min_{k}(P_{v,k,c}), v = 1,2,3 \qquad (16.1)$$

$$MaxP_{v,c} = \max_{k}(P_{v,k,c}), v = 1,2,3 \qquad (16.2)$$

$$Min_{v} = \min_{c}(MinP_{v,c}), v = 1,2,3 \qquad (16.3)$$

$$Max_{v} = \max_{c}(MaxP_{v,c}), v = 1,2,3 \qquad (16.4)$$

Then the scores $N_{v,k,c}$ according to the performance $P_{v,k,c}$ are estimated using the following affine scaling functions:

$$N_{v,c} = 10 \times \frac{P_{v,k,c} - \text{Min}_v}{\text{Max}_v - \text{Min}_v}, v = 1;3 \qquad (16.5)$$

$$N_{2,c} = 10 \times \frac{\text{Max}_2 - P_{2,k,c}}{\text{Max}_2 - \text{Min}_2} \qquad (16.6)$$

The aggregate performance N_c was a weighted average of the $N_{v,c}$, $v = 1,2,3$, where the weights were estimated by maximizing the Shannon Entropy of the related weighted scores.

The difference with the method used in Chapter 13 is that the scores were estimated using the Min and Max of each country. The next steps of the methodology are similar to the one used in Chapter 13.

16.4.2 Evaluation of the available feedstock

The total primary energy demand for Africa is predominantly determined by biomass demand, with almost half of the energy demand being covered by biomass and waste. The different biomass resources include energy crops, forestry biomass and residues and waste.

The major share of the bioenergy potential in 2050 would come from surplus land. Thus, the evaluation of the availability of residues at long term in WA is appropriate to assess their potential contribution to the production of second-generation biofuels for transport. One of the main advantages of forestry and agricultural residues as 2G feedstock is their short-term availability. These types of residues are also attractive since they do not need additional land use to produce them, avoiding competition with food production. To assess the amount of forestry and agricultural residues, data on roundwood and crop production as well as on the ratio between main products and residues, are required.

For forestry residues, an approach presented by Smeets and Faaij [18] shows a residue-to-product ratio (RPR) of 0.6 for primary residues. This means that for every solid cubic meter of industrial roundwood, 0.6 solid cubic meters of logging residues are left behind as residue. For secondary residues, which are produced during wood processing, an RPR of 0.5 was indicated by Smeets and Faaij [18]. The moisture content of the wood residues is assumed to be 50% and the weight of 1 m³ solid roundwood is set at 0.5 t [19,20]. However, logging residues from fuelwood harvesting are already a traditional energy source especially in African countries, so they should not be considered as feedstock for transport.

Using agricultural residues for biofuel production would directly offer revenues to farmers, which could positively affect rural development. They are also deprived of new cultivation technique development. For the purpose of this Chapter, the potential of agricultural residues as feedstock for transportation biofuels in WA is assessed. The crop data in such regions were adapted from [21]. FAOSTAT is a site providing free access to food and agriculture data for over 245 countries (including those considered in this chapter), and covers all FAO regional groupings from 1961 to the most recent year available. Data on the harvesting area for a given crop in 2020 were extracted for the countries considered in Chapter 13. More than 70 crops were identified in 2020 in WA. Some of them were available in just one country of the studied region, some were available in more than 10 countries. A filtered criterion was established based on the availability of a given crop in at least 5 countries of the studied region. The selected crops available in more than 5 countries of WA in 2020 are banana, cassava, cocoa beans, grain maize, oil palm fruit, paddy rice, plantains, potatoes, raw cotton, sorghum, soybean, and sugarcane.

Agricultural data from 2012 to 2050 in 5-year intervals are also available for visualization and download at country level by scenario and by crop in the Global Perspectives Studies made by FAO [21]. The projected 2050 data available in the site were estimated by FAO for the preparation of the "The future of food and agriculture - Alternative pathways to 2050" report [2]. The FAO database includes projections considering three different cases: business as usual (BAU), toward sustainability (TS), and stratified societies. The BAU case shows a perspective where despite the efforts of many countries, several outstanding challenges facing food and agriculture are left unaddressed. The TS case ensures universal and sustainable access to food mostly produced with environmentally sustainable methods. The challenges for both access and utilization, as well as sustainable food stability and availability, are lower than under the BAU case. TS involves a more sustainable use of natural resources and climate change mitigation compared to BAU. Food systems generating low GHG emissions are favored. Adopting conservation agriculture systems, agro-ecological approaches, agroforestry, and other environmentally friendly techniques allows yields to increase against current levels—albeit more moderately than under BAU—and to converge across countries, while food systems drastically reduce GHG emissions compared with current levels.

In this chapter, the potential of the residues of the selected crops for the studied region in 2020 was evaluated by 2050. Since the case focuses on prospective scenarios in 2050 where the situation in WA is carried out by sustainable methods, the TS case was chosen and evaluated.

To assess the number of agricultural residues for each crop type, RPR was adopted from a free access tool provided by FAO where they are presented per crop and per country [21]. Data for processing residues were not available for all the selected crops, due to a wide variety of processing techniques which result in different shares of residues. Therefore, 9 crops with a process relatively standardized were considered in this chapter: banana, cassava, grain maize, oil palm fruit, paddy rice, raw cotton, sorghum, soybean, and sugarcane.

The ratio between main product and residue can vary depending on crop variety, water and nutrient supply, and the use of chemical growth regulators [20].

The RPR and projected crop data by 2050 were used to calculate the availability of the selected residues by 2050 in the studied region. As shown in Fig. 16.2, the availability of total residues depends on the selected country in 2050.

All the crop residues should not be considered as potential feedstock for conversion due to soil conservation and organic nutrients preservation issues. Half of the residues are recommended to be kept on the soil [22] because a lesser fraction would lead to extremely low soil organic carbon fluxes.

The mass composition of the agricultural residues is required to calculate its energy potential for conversion into transportation biofuels. Table 16.1 shows the mass composition values calculated from average data collected from literature [23].

The mass compositions of the selected residues were used to calculate the cellulose, hemicellulose, and lignin availability in 2050 per country. Table 16.2 shows these amounts projected by 2050 and their energy potential.

The high energy potential of the selected residues estimated for each country by 2050 makes possible a strategy for their use as potential feedstock to produce transportation biofuels.

16.4.3 Optimal biofuel strategies

For all the assessments, the share of biomass to meet the demand of passenger road mobility was 20%. That is also referred to as biomobility. For scenarios 1–4, the specific fuel consumptions were 0.54, 2.16, and 2.72 MJ/km, respectively, for electricity, SNG, and bioethanol. These values were 0.54, 1.85 and 2.35 MJ/km for scenarios 5–8.

The scenarios 9–16 shared the same assumption with respectively the scenarios 1–8 except the relative cost for electricity that was 10 for the 8 first scenarios and 15 for the scenarios 9–16. All these conditions for each scenario are summarized in Table 16.3.

Table 16.1 Residues mass composition.

Crops	Residues	Cellulose (%)	Hemicellulose (%)	Lignin (%)
Bananas	Top/leaves	33.8	20.4	18.3
Cassava	Stalk	32.9	18.2	20.2
Grain maize	Stover	37.9	23.5	19.1
Oil palm fruit	Empty bunches	34.4	22.8	22
Oil palm fruit	Fronds	38.2	23.3	21.9
Oil palm fruit	Shell	28.8	18.9	46.3
Paddy rice	Straw	34.9	22.1	17.6
Raw cotton	Stalk	35.5	15.6	27.4
Sorghum	Stalk	34.6	22.9	13.5
Soybeans	Straw	39	19.6	20.2
Sugarcane	Top/leaves	35.3	27.5	19
Sugarcane	Bagasse	39	26.4	21.8

Table 16.2 Amount of cellulose, hemicellulose, and lignin available in the selected WA area by 2050 considering that 50% residues were left on the field.

Country	Cellulose (10⁶t/y)	Hemicellulose (10⁶t/y)	Lignin (10⁶t/y)	Total Energy content (PJ/year)
Benin	2.4	1.5	1.3	94
Burkina Faso	4.6	2.9	2.2	38
Ivory Coast	3.3	2.0	1.9	182
Gambia	0.11	0.067	0.054	4
Ghana	4.9	3.0	2.7	191
Guinea	1.5	0.9	0.8	60
Liberia	0.26	0.16	0.15	1
Mali	3.6	2.2	1.7	38
Niger	1.7	1.1	0.66	60
Nigeria	25.6	16.0	12.8	981
Senegal	0.73	0.5	0.4	27
Sierra Leone	0.48	0.29	0.27	19
Togo	1.3	0.78	0.63	47

Energy content in the selected biomass.

Table 16.3 Yields, specific fuel consumptions, and relative costs assumed for each scenario.

Scenario	$Y_{e(-)}$	$Y_{g(-)}$	$Y_{b(-)}$	$CS_{e(MJ/km)}$	$CS_{g(MJ/km)}$	$CS_{b(MJ/km)}$	$C_{e(-)}$	$C_{g(-)}$	$C_{b(-)}$
1	0.40	0.50	0.44	0.54	2.2	2.7	10	2	1
2	0.50	0.50	0.44	0.54	2.2	2.7	10	2	1
3	0.40	0.50	0.50	0.54	2.2	2.7	10	2	1
4	0.40	0.74	0.44	0.54	2.2	2.7	10	2	1
5	0.50	0.74	0.50	0.54	1.85	2.35	10	2	1
6	0.40	0.74	0.50	0.54	1.85	2.35	10	2	1
7	0.50	0.74	0.44	0.54	1.85	2.35	10	2	1
8	0.50	0.50	0.50	0.54	1.85	2.35	10	2	1
9	0.40	0.50	0.44	0.54	2.2	2.7	15	2	1
10	0.50	0.50	0.44	0.54	2.2	2.7	15	2	1
11	0.40	0.50	0.50	0.54	2.2	2.7	15	2	1
12	0.40	0.74	0.44	0.54	2.2	2.7	15	2	1
13	0.50	0.74	0.50	0.54	1.85	2.35	15	2	1
14	0.40	0.74	0.50	0.54	1.85	2.35	15	2	1
15	0.50	0.74	0.44	0.54	1.85	2.35	15	2	1
16	0.50	0.50	0.50	0.54	1.85	2.35	15	2	1

16.4.3.1 Scenario 1

This scenario is the reference to those with high specific consumption of fuel. For this scenario, the conversion yields from biomass to biofuels were chosen with the following values respectively for biomass to electricity, gas, and bioethanol: 40%, 50%, and 44%. As Table 16.4 shows, the optimal biofuel strategy in terms of share of each fuel value chain in the biomobility is close from one country to the other. Electricity has the highest

Table 16.4 Detailed results for Scenario 1.

Country	w₁ (_)	w₂ (_)	w₃ (_)	Vₑ (%)	Vg (%)	Vb (%)	SWₑ (%)	SWg (%)	SWb (%)	P1 (_)	P2 (_)	P3 (_)	W (PJ)	Re (%)	Rg (%)	Rb (%)
Benin	5	7	3	80	11	9	47	26	27	16.5	9.4	47.8	2.2	51	23	26
Burkina F.	7	2	1	80	11	9	47	26	27	3.7	10.1	47.8	4.0	51	23	26
Ivory C.	5	6	3	80	11	9	47	26	27	15.1	10.2	47.8	4.7	51	23	26
Gambia	7	1	1	84	10	6	54	26	20	3.8	7.6	44.8	0.4	59	22	19
Ghana	3	4	2	78	12	10	44	27	29	15.1	9.9	48.7	5.0	48	24	28
Guinea	7	6	3	78	12	10	44	27	29	9.5	9.9	48.7	2.5	48	24	28
Liberia	6	2	1	82	10	8	51	25	25	4.8	9.4	46.7	0.8	54	21	24
Mali	6	1	1	81	11	8	49	27	24	3.8	11.4	47.2	3.9	53	23	24
Niger	7	2	1	80	11	9	47	26	27	3.9	11.3	47.8	6.0	51	23	26
Nigeria	5	4	2	80	11	9	47	26	27	10.4	11.1	47.8	36.7	51	23	26
Senegal	7	2	1	80	11	9	47	26	27	3.5	9.8	47.8	3.0	51	23	26
S. Leone	4	2	1	81	11	8	49	27	24	6.5	9.1	47.2	1.2	53	23	24
Togo	6	7	3	80	11	9	47	26	27	13.2	8.8	47.8	1.4	51	23	26

share with about 80% followed by SNG 11% and bioethanol 9%. The values of the objective functions vary from one country to the other. The value of the first objective function (P_1) is higher than 8 for the countries of the sub-humid Guinean zone such as Benin, Ghana, Nigeria, Ivory-Coast, and Togo, whereas it is less than 4 for Sahelian countries like Burkina Faso, Mali, Niger, and Senegal. The value of the second objective function that is a relative cost function to be minimized (P_2) is as expected higher for vast countries such as Niger, Mali, and Nigeria whereas it shows lower values for small areas countries such as Gambia, and Togo. The values of the third objective function (P_3) were close from one country to the other. Niger has the lowest overall score and Benin the highest followed by Togo, Ghana, and Ivory-Coast. About 50% of the mobilized biomass were converted into electricity and about 26% went to bioethanol refinery. In terms of energy, the 20% biomobility corresponded to only 5% share of final energy. This is due to the low fuel economy of electromobility that represents about 50% of the total final bioenergy.

16.4.3.2 Scenario 2
Compared to Scenario 1 the energy efficiency of biomass conversion to electricity increases from 40% to 50%. For all the countries, the share of electricity in the biomobility remains stable or increases, except Nigeria. In that case, the decrease of the share of electricity that would increase the required biomass (Ebiom) was compensated by the increase of the electricity generation efficiency and the improvement of the P_2 and P_3 due to the reduction of the cost and the increase of the diversity of the final bio-energy. Ghana shows the highest increase in the share of electricity from 78% in Scenario 1 to 82% in Scenario 2.

16.4.3.3 Scenario 3
The yield of bioethanol increases from 0.44 in Scenario 1 to 0.5 in Scenario 3. As expected, the share of bioethanol in the biomobility increases for all countries. However, the change is slight and detrimental to SNG.

16.4.3.4 Scenario 4
The share of SNG increases significantly detrimental to electricity and, to a lesser extent, to bioethanol. These results stem from the increase of the yield of the SNG from 0.5 to 0.74 in this scenario. The share of SNG is around 15% for all of the countries and for most of them, electricity's share decreases from 80% (Scenario 1) to around 78%.

16.4.3.5 Scenario 5
All the yields increase, from 0.4, 0.5, 0.44 to 0.5, 0.74, and 0.5 for electricity, SNG, and bioethanol, respectively. The share of electricity drops significantly, in favor of SNG. In the particular case of Togo, only SNG and ethanol are used henceforth. Two groups emerge among the countries. Group 1 is dominated by Sahelian countries (Burkina

Faso, Gambia, Liberia, Mali, Niger, Senegal, and Sierra Leone); their electricity shares in the biomobility ranges from 74% (Niger) to 77% (Burkina Faso). The shares of SNG and bioethanol are about 16% and 8%, respectively. Despite a relatively higher share of electricity, this group presents higher pressure on feedstock- lower values of the first objective function (P1). That is explained by their lower available feedstock. The higher values of the second objective function are also in this group, with Mali (12.66) and Niger (12.56). The shares of electricity, SNG, and bioethanol in the final bio-energy are about 45%, 33%, and 22%, respectively. About 50% of the used feedstock (Ebiom) are converted into Electricity, whereas the remaining is equally converted into SNG and bioethanol. This group presents the lower value of the overall score, ranging from 2.19 (Mali) to 3.75 (Sierra-Leone).

The second group is composed of the countries of the Sub-humid Guinean zone with Benin, Ghana, Ivory-Coast, Guinea, and Nigeria. Their shares of electricity in the biomobility and bio-energy range from 70% to 72%, and 36% to 41%. About 45% of the used feedstock are converted into electricity, the remaining being almost equally converted into SNG and bioethanol.

16.4.3.6 Scenario 6

With this scenario, the SNG and bioethanol remain at their higher yield with 74% and 50%, respectively, while the yield of electricity decreases to its reference value of 40%. Three groups can be identified. Group 1 is now composed of Burkina, Gambia, Liberia, Mali, and Senegal. Group 2 comprises Benin, Ghana, Ivory-Coast, Guinea, Sierra-Leone, and Togo. In fact, Sierra Leone is between Groups 1 and 2. Finally, Niger and Nigeria form Group 3.

For Group 1, the shares of electricity in the biomobility and the final bioenergy range from 73% (Gambia and Liberia) to 75% (Burkina Faso), and from 42% to 45%, respectively. The overall performances of this group are the low between 2.23 (Mali) to 2.92 (Liberia). From 49% to 56% of the used feedstock are converted into electricity, and the remaining almost equally into SNG and bioethanol.

In Group 2, electricity shares around 70% of the biomobility and 33%–37% of the final bioenergy. This group presents the highest overall performances ranging from 3.87 (Sierra-Leone) to 6.15 (Benin). In terms of share of final bioenergy, SNG and electricity play equally. The performance P_1 of the countries of this group is higher than the one of Group1 and ranges from 6.87 (Serra-Leone) to 17.88 (Benin). The share of used feedstock converted into electricity ranges from 44% in Togo to 48% in Benin and Ghana. The remaining feedstock is shared between bioethanol and SNG with more biomass sent to bioethanol refinery.

Finally, Group 3 is composed of Niger and Nigeria. The common point of these two countries is their high area that penalizes electromobility. However, these two countries differ by their value of P_1. By the value P_1, Niger would be in Group 1 and Nigeria in Group 2. However, due to the sensitivity of P_2 to the area, the share of electricity in the biomobility drops to zero; SNG and bioethanol share the biomobility with a slight advantage to SNG. However, in terms of share of the final bioenergy, bioethanol "takes over" with 55% due to its lower fuel economy. In total, 63% of the used feedstock is converted into bioethanol. The zero value of V_e for Niger results from a tradeoff between a strategy with, on one side, a value like the one of the Sahelian zone that would result in a higher value of P_1 and P_3 but a lower value of P_2, and on the other side, a zero value that decreases the values of P_1 and P_3, but increases P_2. Due to the data of this country, the second solution prevails. The case of Nigeria results from the same kind of tradeoff. Table 16.5 gives the details of the results for Scenario 6.

16.4.3.7 Scenario 7

With Scenario 7, the yield of bioethanol was set at its reference value and the efficiency of electricity and SNG was set at their high level. As expected, the shares of SNG in the biomobility and final bioenergy rise as in this scenario the technology progress of SNG is better than the one of the two competitors. Concerning the shares of electricity, the difference between the Sudano-Sahelian and Guinean Groups remains. For Group 1, the shares of electricity are higher, ranging from 74% (Mali and Sierra Leone) to 78% (Burkina Faso) and 43% to 49% for the biomobility and bioenergy, respectively, while le values for Group 2 are 63%–73% and 31%–42%. The shares of bioethanol are lower for countries of Group 1. Concerning the share of the used biomass converted into electricity, it amounts to 47%–54% and 33%–46% for Group 1 and Group 2, respectively.

16.4.3.8 Scenario 8

Compared to Scenario 5 from which this scenario differs by the lower yield of SNG and ethanol, what could be expected concerning the shares of electricity, SNG, and bioethanol in the biomobility and bioenergy? The results show that for Group 1, the shares of SNG and bioethanol in the biomobility are 10%–12% and 8%–10% against 15%–17% and 8%–9% for bioethanol. So there a slight improvement in the share of bioethanol and the expected effect for SNG. The value of P_1 ranges from 4.45 to 6.84 in this scenario against 4.62–8.08 in the case of Scenario 5. For Group 2, the results related to the shares are more erratic. These shares are in the ranges of 13%–47% and 11%–60% for SNG and bioethanol respectively in this Scenario against 18%–48% and 9%–31% for Scenario 5. The cases of 47% and 60% occur only in Togo and Guinea where the optimal portfolios are without electricity. Most of the other countries in Group 2 are in ranges of 13%–21% and 11%–18% for SNG and ethanol respectively, which denotes more attractiveness for SNG and ethanol for countries of Group 2.

16.4.3.9 Scenario 9

This scenario shares the same assumptions with Scenario 1 except for the relative cost c_e of electricity that is 15 instead of 10 for Scenario 1. Hence, the share of electricity would be expected lower. As Table 16.6 shows, for the countries of Group 1, the

Table 16.5 Detailed results for Scenario 6.

Country	w₁ (_)	w₂ (_)	w₃ (_)	Vₑ (%)	V_g (%)	V_b (%)	SWₑ (%)	SW_g (%)	SW_b (%)	P1 (_)	P2 (_)	P3 (_)	W (PJ)	Re (%)	Rg (%)	Rb (%)
Benin	3	4	2	69	20	11	37	37	26	17.9	10.2	49.4	2.4	48	26	27
Burkina F.	6	1	1	75	17	8	45	35	21	4.4	11.3	47.8	3.9	56	23	21
Ivory C.	6	7	4	68	20	12	36	36	28	16.1	11.0	49.6	5.2	46	25	28
Gambia	7	2	1	73	18	9	42	35	23	4.0	8.3	48.5	0.5	53	24	23
Ghana	5	6	3	69	20	11	37	37	26	16.9	10.8	49.4	5.2	48	26	27
Guinea	6	5	3	65	21	14	33	36	31	10.1	10.6	49.9	2.8	43	25	32
Liberia	6	2	1	73	18	9	42	35	23	5.3	10.4	48.5	0.9	53	24	23
Mali	6	1	1	74	17	9	43	34	23	4.3	12.5	48.4	4.0	54	23	23
Niger	7	2	1	0	51	49	0	45	55	2.3	7.9	37.1	13.7	0	36	64
Nigeria	6	5	3	0	52	48	0	46	54	6.2	7.7	37.3	83.9	0	37	63
Senegal	6	1	1	74	17	9	43	34	23	4.1	10.9	48.4	3.1	54	23	23
S. Leone	4	2	1	70	19	11	38	36	26	6.9	9.8	49.4	1.3	49	25	27
Togo	5	6	3	66	21	13	34	37	29	13.7	9.4	49.8	1.6	44	26	30

Table 16.6 Detailed results for Scenario 9.

Country	w_1 (_)	w_2 (_)	w_3 (_)	V_e (%)	V_g (%)	V_b (%)	SW_e (%)	SW_g (%)	SW_b (%)	P1 (_)	P2 (_)	P3 (_)	W (PJ)	Re (%)	Rg (%)	Rb (%)
Benin	5	5	3	80	12	8	48	29	24	16.7	13.4	47.6	2.2	52	25	24
Burkina F.	6	1	1	80	12	8	48	29	24	3.8	14.4	47.6	3.9	52	25	24
Ivory C.	4	4	3	74	15	11	39	32	29	13.7	14.1	49.6	5.2	43	28	29
Gambia	7	2	1	78	13	9	45	30	26	3.4	10.6	48.5	0.5	49	26	26
Ghana	3	6	3	0	50	50	0	44	56	6.3	7.0	37.0	12.7	0	41	59
Guinea	5	3	2	78	13	9	45	30	26	9.6	14.1	48.5	2.5	49	26	26
Liberia	7	2	1	83	11	6	53	28	19	5.0	13.7	45.4	0.8	57	24	19
Mali	6	1	1	80	12	8	48	29	24	3.8	16.2	47.6	4.0	52	25	24
Niger	5	1	1	76	14	10	42	31	28	3.7	15.8	49.2	6.5	46	27	27
Nigeria	5	3	2	0	52	48	0	46	54	4.2	7.8	37.3	97.4	0	43	57
Senegal	7	1	1	82	11	7	51	27	22	3.7	14.2	46.4	2.9	55	24	21
S. Leone	5	2	1	82	11	7	51	27	22	6.6	13.1	46.4	1.1	55	24	21
Togo	6	5	3	80	12	8	48	29	24	13.3	12.6	47.6	1.4	52	25	24

shares of electricity in the biomobility and bio-energy are in the ranges of 76%–83% and 42%–53%, respectively, compared to 80%–84% and 47%–54% for Scenario 1. Thus, the shares of electricity decrease slightly for the countries of this Group, even though the situations differ from one country to the other. The value of the second objective function P_2 worsens and needs to be compensated to some extent by the two other objective functions and by appropriate weightings. For Group 2, the trend is more visible. The share of electricity in the biomobility decreases from 78%–80% in Scenario 1 to 0%–80%. The higher decreases are with Ghana and Nigeria where it drops to zero.

16.4.3.10 Scenario 10
The comparison between scenarios 2 and 10 do not show any significant change. For Group 1, the share of electricity in the biomobility ranges from 80% to 85% in Scenario 2 compared to 78%–85% for Scenario 10. For Group 2, the change is from 80%–82% in Scenario 2 to 76%–82% for Scenario 10. The disincentive of electricity due to a 50% increase of c_e is not enough to reduce the attractiveness of the electromobility due to the large advantage of its efficiency. The share of electricity even increases in fewer countries such as Nigeria from 77% to 80% where the increase in P_2 is compensated by the higher value of P_1 and the new weighting that is slightly more favorable to the first objective.

16.4.3.11 Scenario 11
From the comparison with Scenario 3, fewer points can be highlighted. From countries in Group 1, V_e is stable from 79%–84% in Scenario 3 to 79%–83% in Scenario 11. The change is more remarkable in the Group 2 where for Ghana, it drops to zero. However, in average it is also relatively stable, except for Benin and Togo where it decreases from 80% to 77%.

16.4.3.12 Scenario 12
As it is noted for the comparison between scenarios 3 and 11, Scenario 12 results in lower changes compared to Scenario 4 in the case of Group 1. For Burkina and Liberia, the share slightly increases, whereas it slightly decreases for Gambia, Mali, Niger, Senegal, and Sierra Leone. In the case of Group 2, it drops to zero for Ivory Coast and Nigeria and remains stable for Benin and Ghana and slightly decreases in Togo (Table 16.7). In the case of Ivory Coast the reduction of P_1 is compensated by the improvement of P_2 enhanced by a change of the weighting in favor of the second objective. In the case of Nigeria, the improvement of P_2 is more influencing due to the area of the country.

16.4.3.13 Scenario 13
Compared to Scenario 5, V_e is relatively stable for the countries of the Group 1. It slightly decreases in Burkina, Mali, and Niger, whereas in the other countries it slightly increases. For countries of Group 2, V_e drops to zero in Benin and Nigeria, whereas for

Table 16.7 Detailed results for Scenario 12.

Country	w_1 (_)	w_2 (_)	w_3 (_)	V_e (%)	V_g (%)	V_b (%)	SW_e (%)	SW_g (%)	SW_b (%)	P1 (_)	P2 (_)	P3 (_)	W (PJ)	Re (%)	Rg (%)	Rb (%)
Benin	6	7	4	76	17	7	42	38	20	17.8	13.2	47.8	2.3	53	25	22
Burkina F.	6	1	1	79	15	6	47	35	18	4.2	14.4	46.8	4.0	57	23	20
Ivory C.	3	6	3	0	52	48	0	46	54	7.1	7.2	37.3	12.4	0	34	66
Gambia	7	2	1	77	16	7	44	36	20	3.8	10.6	47.8	0.5	54	24	22
Ghana	5	5	3	76	17	7	42	38	20	16.9	14.0	47.8	5.0	53	25	22
Guinea	5	3	2	77	16	7	44	36	20	10.8	14.1	47.8	2.5	54	24	22
Liberia	7	2	1	81	15	4	50	37	12	5.5	13.6	44.5	0.8	61	25	14
Mali	5	1	1	74	17	9	40	36	24	3.8	15.6	49.0	4.4	49	24	27
Niger	5	1	1	76	16	8	42	36	22	4.2	15.9	48.5	6.4	52	24	25
Nigeria	6	4	3	0	55	45	0	49	51	5.0	7.9	37.5	96.7	0	37	63
Senegal	6	1	1	78	15	7	45	35	20	3.9	13.9	47.7	3.1	55	23	22
S. Leone	7	3	2	75	17	8	41	37	22	6.6	12.6	48.5	1.3	51	25	25
Togo	7	7	4	75	17	8	41	37	22	13.9	12.3	48.5	1.5	51	25	25

Advanced biofuels for transportation in West Africa: Common referential state-based strategies 299

Togo, it climbs from 0% to 74%. The cases of Benin, Nigeria, and Togo can be explained as follows.

16.4.3.13.1 Cases of Benin and Nigeria

The share of electricity in the biomobility was 72% in Scenario 5 with, in the case of Benin, the values of the objective functions of $P_1 = 20.52$, $P_2 = 10.31$, $P_3 = 49.01$. The overall score was 5.89. In Scenario 13, with the increase of c_e from 10 to 15, the value of P_2 would increase. The fact to have V_e dropped to zero, improves the performance related to the second objective function, but deteriorates the values of P_1 and P_3: $P_1 = 9.92$, $P_2 = 6.63$, $P_3 = 37.44$. The total score was 6.79. Due to the affine scaling based on the Min. and Max. of the whole region, the improvement of P_2 dominates the deterioration of P_1 and P_3. The overall score improves.

16.4.3.13.2 Case of Togo

Contrary to Benin and Nigeria, the value of V_e in Togo increased from zero in Scenario 5%–74% in Scenario 13. The comparison of the values of P_1, P_2, P_3 in both scenarios reveals the following: Scenario 5 $P_1 = 7.68$, $P_2 = 6.09$, $P_3 = 36.78$; Scenario 13 $P_1 = 17.11$, $P_2 = 13.85$, $P_3 = 47.93$. The increase of c_e deteriorated P_2. However, the improvement of P_1 and P_3 overcame this deterioration.

16.4.3.14 Scenario 14

The comparison of scenarios 6 and 14 shows that for most of the countries such as Benin, Burkina, Ivory Coast, Ghana, Liberia, Mali, Senegal, and Togo, V_e reduces. In fewer countries such as Gambia, Guinea, Niger, Sierra Leone it increases.

16.4.3.15 Scenario 15

For Scenario 15, the comparison with Scenario 7 shows the same trend. The increase of c_e leads to a decrease of V_e for most of the countries. However, this is not a general rule due to the opposite trends between the values of the objective functions. In this scenario, the particular cases are Benin and Nigeria where V_e drops from 73% to 0%.

16.4.3.16 Scenario 16

In the case of Scenario 16, the comparison with Scenario 8 highlights fewer cases with noticeable changes. In Ivory coast, V_e increases from 61% in Scenario 8%–76% in Scenario 16. For Guinea and Togo, the change is from 0% to 76%. On the other side, the change for Nigeria is from 66% to 0%. These cases confirm that even though the influence of the increase of c_e is effective for most of the scenarios, for fewer, this influence is not noticeable or can be in different directions. This can be explained by the fact that objectives 1 and 2 work in opposite directions. A high value of V_e would improve P_1 while deteriorating P_2. Conversely, when $V_e = 0\%$, P_2 improves, but P_1 and P_3 deteriorate. In conclusion, the balance between these two effects would depend on the magnitude of improvements and deteriorations.

16.4.4 Robustness analyses

Based on the method described in the precedent chapter, one robust strategy was selected for each country. Table 16.8 shows the different recommended strategies. Compared to the results in the precedent chapter where the recommended strategy was: $V_e = 77\%$, $V_g = 15\%$, $V_b = 8\%$, there is a diversity of strategies from one country to the other. For Ghana, Ivory Coast, Mali, and Nigeria, the recommended biofuel strategy is the one found in the precedent chapter for all the countries. For other countries, it is significantly different. Few cases are discussed to explain the main reasons for the differences.

The recommended biofuel strategy for Benin is 76%, 13%, and 11% for V_e, V_g, and V_b, respectively. Compared to the strategy that is proposed in the precedent chapter, it plays equally for the average score of the 16 scenarios, but takes over in terms of maximum regret, 0.35 against 0.57. That is similar for other countries where the recommended strategy is different from the one of the precedent chapter. For example, in the case of Burkina Faso, the average score and the maximum regret are 2.29 and 0.01, respectively, against 2.29 and 0.32 for the recommended strategy in the precedent chapter. In the case of Sierra Leone, the comparison gives 3.60 and 0.20 against 3.61 and 0.56.

The detailed results found for the different countries are discussed concerning various indicators.

16.4.4.1 Matrix of biofuels and share of the biofuels in the final energy

For the recommended strategies, Table 16.9 shows the biofuels matrix in terms of shares to the total final bioenergy consumed. For all scenarios, the maximum value of the share of electricity is found in Liberia followed by Sierra Leone. The minimum value

Table 16.8 Recommended strategies for West Africa in 2050.

Country	Recommended Strategy Related scenario for which the strategy is optimum	V_e (%)	V_g (%)	V_b (%)
Benin	8	76	13	11
Burkina Faso	16	80	12	8
Ivory Coast	4	76	15	9
Gambia	16	81	11	8
Ghana	4	77	15	8
Guinea	16	72	16	12
Liberia	9	83	11	6
Mali	4	77	15	8
Niger	1	80	11	9
Nigeria	4	76	15	9
Senegal	16	79	12	9
Sierra Leone	9	82	11	7
Togo	16	76	14	10

Table 16.9 Shares of the three types of biofuels to the total final bioenergy consumed.

Country	Scenarios 1–4 and 9–12				Scenarios 5–8 and 13–16			
	SW_e (%)	SW_g (%)	SW_b (%)	W/Wref (%)	SW_e (%)	SW_g (%)	SW_b (%)	W/Wref (%)
Benin	41	28	30	5.2	45	26	28	4.8
Burkina Faso	48	29	24	4.8	51	26	22	4.5
Ivory Coast	42	33	25	5.2	46	31	24	4.8
Gambia	49	27	24	4.7	53	25	23	4.4
Ghana	43	34	23	5.1	47	31	21	4.7
Guinea	37	33	31	5.6	40	31	29	5.1
Liberia	53	28	19	4.5	57	26	18	4.2
Mali	43	34	23	5.1	47	31	21	4.7
Niger	47	26	27	4.8	51	24	25	4.5
Nigeria	42	33	25	5.2	46	31	24	4.8
Senegal	46	28	26	4.9	50	26	25	4.6
Sierra Leone	51	27	22	4.6	55	25	20	4.3
Togo	42	31	28	5.2	45	29	26	4.8

of this share is for Guinea followed by Benin. The maximum shares of SNG occur in Ghana and Mali for scenarios 1–4 and 9–12 and in Ivory Coast, Ghana, Guinea, Mali, and Nigeria for scenarios 5–8 and 13–16. The minimum share of SNG occurs in Niger for all the scenarios.

The share of biofuels in the total final energy is maximum in Guinea 5.6% for the first group of scenarios, and 5.1% for the second group of scenarios. The minimum share is found for Liberia. This share is perfectly negatively correlated with the share of electricity in the final bioenergy.

Table 16.10 presents the values of the shares of the feedstock converted into electricity (Re), SNG (Rg), and bioethanol (Rb) for each country. For all countries, the maximum value of Re occurs for scenarios 6 &14 and the minimum share for scenarios 2 &10. For Re, the maximum and minimum shares are found for scenarios 2 &10 and 7 &15, respectively. For R_b, the maximum and minimum are found for scenarios 7 &15 and 8 &16. The maximum values of R_e, R_g, and R_b are found in Liberia, Ghana/Mali, and Guinea, respectively. The minimum values occur in Guinea, Niger, and Liberia.

16.4.4.1.1 Values of P_1, P_2, and W

The values of the first objective function (P_1) are highest for scenarios 5 &13 and lowest for scenarios 1 & 9. The level of yields explains these results. Indeed, the yields are highest in scenarios 5 & 13 and lowest for scenarios 1 &9. Concerning P_2, the highest values occur for the scenarios 13–16 due to the improvement of the fuel economy of bioethanol and increase of the relative cost of electricity that decreases the denominator more than the numerator of P_2. The lowest values occur for scenarios 1–4 due to the higher value of the denominator due to the low fuel economy of bioethanol.

Table 16.10 Shares of the feedstock converted into the three types of biofuels for each country.

Country	Scenarios	Re (%)	Rg (%)	Rb (%)
Benin	1 & 9	45	25	30
	2 & 10	40	27	33
	3 & 11	47	26	27
	4 & 12	49	18	33
	5 & 13	49	20	31
	6 & 14	55	17	28
	7 & 15	47	19	34
	8 & 16	45	26	28
Burkina Faso	1 & 9	52	25	24
	2 & 10	46	28	26
	3 & 11	53	26	21
	4 & 12	56	18	26
	5 & 13	56	19	24
	6 & 14	62	17	21
	7 & 15	54	19	27
	8 & 16	51	26	22
Ivory Coast	1 & 9	46	29	25
	2 & 10	41	32	27
	3 & 11	47	30	23
	4 & 12	51	22	28
	5 & 13	51	23	26
	6 & 14	56	21	23
	7 & 15	49	22	29
	8 & 16	46	31	24
Gambia	1 & 9	53	23	24
	2 & 10	47	26	27
	3 & 11	55	24	22
	4 & 12	57	17	26
	5 & 13	57	18	25
	6 & 14	63	16	22
	7 & 15	55	17	27
	8 & 16	53	25	23
Ghana	1 & 9	48	30	23
	2 & 10	42	33	25
	3 & 11	49	31	20
	4 & 12	53	22	25
	5 & 13	53	24	24
	6 & 14	58	21	21
	7 & 15	51	23	26
	8 & 16	47	31	21
Guinea	1 & 9	40	29	31
	2 & 10	35	31	34
	3 & 11	42	30	28
	4 & 12	45	21	34

(continued)

Table 16.10 (Cont'd)

Country	Scenarios	Re (%)	Rg (%)	Rb (%)
	5 & 13	45	23	32
	6 & 14	50	21	29
	7 & 15	43	22	35
	8 & 16	40	31	29
Liberia	1 & 9	57	24	19
	2 & 10	51	27	21
	3 & 11	58	25	17
	4 & 12	62	18	20
	5 & 13	62	19	19
	6 & 14	67	16	17
	7 & 15	60	18	21
	8 & 16	57	26	18
Mali	1 & 9	48	30	23
	2 & 10	42	33	25
	3 & 11	49	31	20
	4 & 12	53	22	25
	5 & 13	53	24	24
	6 & 14	58	21	21
	7 & 15	51	23	26
	8 & 16	47	31	21
Niger	1 & 9	51	23	26
	2 & 10	46	25	29
	3 & 11	53	23	24
	4 & 12	55	16	28
	5 & 13	55	18	27
	6 & 14	61	15	24
	7 & 15	53	17	30
	8 & 16	51	24	25
Nigeria	1 & 9	46	29	25
	2 & 10	41	32	27
	3 & 11	47	30	23
	4 & 12	51	22	28
	5 & 13	51	23	26
	6 & 14	56	21	23
	7 & 15	49	22	29
	8 & 16	46	31	24
Senegal	1 & 9	50	24	26
	2 & 10	44	27	29
	3 & 11	51	25	24
	4 & 12	54	18	28
	5 & 13	54	19	27
	6 & 14	60	17	24
	7 & 15	52	18	29
	8 & 16	50	26	25

(continued)

Country	Scenarios	Re (%)	Rg (%)	Rb (%)
Sierra Leone	1 & 9	55	24	21
	2 & 10	49	27	24
	3 & 11	56	24	19
	4 & 12	60	17	23
	5 & 13	59	18	22
	6 & 14	65	16	19
	7 & 15	58	18	24
	8 & 16	55	25	20
Togo	1 & 9	46	27	27
	2 & 10	40	30	30
	3 & 11	47	28	25
	4 & 12	50	20	30
	5 & 13	50	21	29
	6 & 14	56	19	25
	7 & 15	48	21	31
	8 & 16	45	29	26

For W, the scenarios 5–8 and 13–16 give the lowest value due to better fuel economy compared to the scenarios 1–4 and 9–13. The values of P_1, P_2, and W are then compared between the recommended strategy and the optimal strategy for the various scenarios.

Recommended strategy versus optimal strategy in the case of Scenario 1

The value of P_1 deteriorates compared to the one that it would be in the case of the optimal strategy for most of the countries. The decrease ranges from 1.39% (Senegal) to 8.95% (Guinea). There is an increase in fewer countries that ranges from 0.34% (Ghana) to 3.39% (Liberia). In the case of Niger, the recommended strategy is similar to the optimal strategy of this scenario. The value of P_2 of the recommended strategy compared to what it would be for the optimal strategy decreases for eight scenarios in the range of 0.21% (Senegal) to 1.94% (Guinea). For Niger, the optimal strategy is similar to the recommended strategy. For five countries the value of P_2 deteriorates, increasing from 0.32% (Benin) to 0.84% (Liberia).

Finally, the value of W increases for 10 scenarios in the range of 0.53% (Ghana) to 11.38% (Guinea), which means a deterioration of the fuel economy of the recommended strategy compared to the optimal scenario. For Burkina Faso (0.61%), Liberia (3.12%), and Sierra Leone (2.44%), the fuel economy improves from the optimal strategy to the recommended strategy resulting in a decrease of the final biofuel compared to the optimal strategy.

Recommended strategy versus optimal strategy in the case of Scenario 2

The value of P_1 varies from 3.88 (Senegal) to 16.90 (Benin). There is a decrease of P_1 compared to the optimal strategy for 11 of the countries in the range from 0.60% (Liberia) to 14.25% (Guinea). For Nigeria and Sierra Leone, P_1 values increase slightly.

P_2 decreases for ten countries in the range of 0.21% (Senegal) to 2.67% (Guinea). The second objective function deteriorates, increasing for Liberia (0.10%), Nigeria (0.43%), and Sierra Leone (0.32%). The final biofuel W increases for 12 countries as a result of a deterioriation of the fuel economy for the recommended strategy compared to the optimal strategy related to Scenario 2. The range of increase is from 0.52% (Nigeria) to 16.03% (Guinea). W decreases only for Sierra Leone (0.63%).

Recommended strategy versus optimal strategy in the case of Scenario 3

P_1 ranges between 3.61 (Senegal) and 15.94 (Benin). Compared to the optimal strategies for this scenario, the value for the recommended strategy decreases for five countries and increases for eight countries. The decrease ranges from 1.72% (Nigeria) to 8.18% (Guinea), and the increase from 0.17% (Ghana) to 3.28% (Liberia). P_2 decreases for five countries in the range of 0.54% (Mali) to 1.62% (Guinea). The increase of P_2 for the other eight countries ranges from 0.32% (Senegal) to 1.16% (Liberia). W increases for six countries in the range of 2.22% (Nigeria) to 10.73% (Guinea). It decreases for seven countries in the range of 0.05% Ghana) to 3.73% (Liberia).

Recommended strategy versus optimal strategy in the case of Scenario 4

The value of P_1 decreases for six countries in the range of 2.59% (Niger) to 9.57% (Guinea). For Ivory Coast, Ghana, Mali, and Nigeria, the recommended strategy is similar to the optimal strategy under Scenario 4. Thus there is no change in the value of P_1. For Burkina Faso, Liberia and Sierra Leone, P_1 increases. P_2 decreases for six countries in the range of 0.42% (Niger/Gambie) to 2.35% (Guinea). It slightly increases in three countries. For four countries the value of W increases in the range of 1.13% (Benin) to 10.79% (Guinea). For five countries, it decreases in the range from 0.55% (Niger) to 11.32% (Liberia).

Recommended strategy versus optimal strategy in the case of Scenario 5

For all the countries except Guinea, P_1 increases, which means that the recommended strategy is more energy-efficient compared to the optimal strategy under Scenario 5. Indeed in this scenario, the improvement of the fuel economy of bioethanol and SNG and the yield of the three fuels result in the decrease in the share of electricity in the biomobility in the optimal strategy. The recommended strategy included more electricity for most of the countries enhancing the decrease of P_1. The maximum increase of P_1 is given by the case of Togo where V_e is null for the optimal strategy.

The same reason explains why for twelve countries P_2 increases for the recommended strategy compared to the optimal scenario. The range of increase is from 0.35% (Mali) to 61.28% (Togo) where the share of electricity in the optimal strategy is zero. For Guinea P_2 decreases by 0.56%. The final biofuels decrease for the same reason for the ten countries. The range of decrease is from 1.46% (Mali) to 57.04% (Togo). It increases for Guinea by 1.05%.

Recommended strategy versus optimal strategy in the case of Scenario 6

In this scenario, due to the lower value of the efficiency of the electricity generation, the share of electricity in the biomobility decreases for all countries in the optimal strategy and is lower compared to the recommended strategy. Thus, P_1 increases, P_2 increases, and W decreases for all countries. The ranges are 3.16% (Mali)–101.24%(Niger), 1.33% (Mali)–62.53% (Niger), and 4.79% (Mali)–59.57% (Niger) for the increase of P_1, P_2, and the decrease of W, respectively.

Recommended strategy versus optimal strategy in the case of Scenario 7

For four countries, P_1 decreases in the range of 0.82% (Benin) to 5.47% (Guinea). For the other countries, it increases in the range of 2.54% (Nigeria) to 20.83% (Togo). For P_2, the variation between the recommended strategy and the optimal strategy is an increase for 12 countries by 0.14% (Senegal) to 6.63% (Togo). The value of P_2 decreases for Guinea by 1.19%. Consistently to the values of P_2, the value of W increases for Guinea (2.99%) and decreases for all the other countries. The range of the decrease is from 1.85% (Senegal) and 18.13%(Togo).

Recommended strategy versus optimal strategy in the case of Scenario 8

The value of P_1 decreases for Burkina Faso, Mali, and Senegal. It is the same for Benin and Gambia. For the other countries, the increase is in the range of 1.55% (Niger)–133.8% (Togo). The value of P_2 decreases for Senegal and Burkina Faso by 0.96% and 0.68%, respectively. It increases for nine countries in the range of 0.20% (Mali)–64.78% (Guinea). The value of W increases for Burkina Faso (3.2%), Mali(0.35%), and Senegal (3.76%). For eight countries, it decreases in the range of 1.52% (Niger) and 57.24% (Togo).

Recommended strategy versus optimal strategy in the case of Scenario 9

P_1 decreases for five countries in the range from 4.08 (Mali) to 9.72 (Guinea). The recommended strategy being the optimal one under this scenario for Burkina Faso, Liberia, and Sierra Leone, the value of P_1 is the same for both strategies. For the other five countries, the range of increase is from 4.32% (Ivory Coast) to 140.51% (Ghana) for which V_e is null for the optimal strategy. For P_2, the increase is in the range from 1.55 (Ivory Coast) to 101.89 (Ghana) and five countries show this case. For the five remaining countries, the decrease ranges from 1.57% (Mali) to 3.86% (Guinea). For W, the decrease is consistent with the case of P_2 and ranges from 4.26% (Ivory Coast) to 60.72% (Ghana). For the other five countries, the increase of W ranges from 5.35 (Mali) to 12.03 (Guinea).

Recommended strategy versus optimal strategy in the case of Scenario 10

P_1 increases for Gambia, Ghana, and Niger. It decreases in the 10 remaining countries. The range of the decrease is 4.44 (Burkina Faso)–15.08 (Guinea). For the same 10 countries, P_2 decreases as a consequence of a lower share of electricity in the recommended strategy compared to the optimal strategies. The range of the decrease is from 1.23% (Liberia) to 5.08% (Guinea). For Gambia, Ghana, and Niger, the increase of P_2 is slight 0.52%, 0.99%, and 1.06%, respectively. For the final biofuel, the value of W increases for the ten countries and decreases for Gambia (1.78%), Ghana (2.77%), and Niger (3.42%). The range of the increase is 4.36% (Burkina Faso)–16.71% (Guinea).

Recommended strategy versus optimal strategy in the case of Scenario 11

For Niger, the recommended strategy is the optimal strategy under Scenario 11, which implies no variation for this country. For five counties, P_1 decreases in the range of 0.86 (Nigeria)–7.38% (Guinea). For the remaining seven countries, the increase ranges from 0.55 (Burkina Faso) to 130.29% (Ghana). For P_2, there is a decrease in the value of five countries ranging from 0.31% (Nigeria) to 3.13% (Guinea). For other seven countries, the value of P_2 increases with the highest increase for Ghana (102.82%). Within this group, the increase is low for the other countries. The final consumption of biofuels increases for five countries in the range from 1.1% (Nigeria) to 9.51% (Guinea). For the other seven countries, the decrease is low except for Ghana where it is 60.83%.

Recommended strategy versus optimal strategy in the case of Scenario 12

P_1 increases for six countries. Among them, Ivory Coast (121.88%) and Nigeria (117.05%) show the highest increase due to the V_e that is null for the optimal strategy. The increase for the other countries is low except for Sierra Leone (7.65%). For seven countries, P_1 slightly decreases. For P_2, the value decreases for Benin (0.91%) and Guinea (3.78%). For Ivory Coast and Nigeria, it highly increases, 98.55% and 95.89%, respectively. The increase is low for the remaining countries. The final consumption of biofuels variation is consistent with that of P_2. For Benin and Guinea it increases by 2.30% and 11.43%, respectively. For the other countries, it decreases. For Ivory Coast and Nigeria the decrease is the highest due to the same reason as for the variation of P_2 that is the increase of V_e.

Recommended strategy versus optimal strategy in the case of Scenario 13

P_1 increases for nine countries. The highest increases are for Nigeria (114.23%) and Benin (111.21%). For other countries of this group, the increase is in the range of 1.27% (Ivory Coast)– 11.35% (Ghana). For the remaining four countries P_1 decreases from 0.95% (Togo) to 3.79% (Guinea). P_2 decreases only for Guinea (0.60%). The increase in P_2 is the highest for Benin (123.95%) and Nigeria (122.88%). For the remaining countries, the increase is in the range of 0.86% (Gambia) to 6.12% (Niger). The final consumption of biofuels W increases only in the case of Guinea (1.58%) while for the 12 other countries it decreases. The highest decreases are for Nigeria (56.55%) and Benin (56.28%). For other countries in this group, the decrease is in the range from 1.76% (Togo) to 12.67% (Niger).

Recommended strategy versus optimal strategy in the case of Scenario 14

P_1 increases for all the countries. The highest increases are for Togo (90.25%) and Nigeria (87.72%). For the other countries, the range of increase is 1.83% (Guinea)–12.39% (Ivory Coast). P_2 increases and W decreases for all the countries. These results are comparable to the case of Scenario 6 despite the increase of c_e.

Recommended strategy versus optimal strategy in the case of Scenario 15

P_1 decreases for Ivory Coast (1.70%), Guinea (6.25%), and Togo (0.83%). For Senegal the recommended strategy is the optimal scenario under Scenario 15. P_1 increases for the other countries with the highest increase for Nigeria (122%) and Benin (119.63%). P_2 decreases only for Guinea (0.80%). For the remaining countries, Benin and Nigeria also show the highest increases. W increases for Guinea (2.11%) and decreases for all the other countries. The highest increase, after Guinea, are also in Benin and Nigeria because the value of V_e is null in the optimal strategies of both countries under Scenario 15.

Recommended strategy versus optimal strategy in the case of Scenario 16

The recommended strategy is the optimal strategy under Scenario 16 for five countries: Burkina Faso, Gambia, Guinea, Senegal, and Togo. P_1 decreases for Benin (0.55%) and Sierra Leone (3.85%). It increases for the other countries. For Nigeria, the increase is the highest because V_e is null in the optimal strategy under Scenario 16. The increase ranges from 0.56% (Ivory Coast) to 10.16% (Liberia) for the remaining countries. P_2 shows the same trend as P_1. Finally, w shows the opposite trend. All the trends regarding increase and decrease can be explained by the share of the electromobility in the recommended strategy compared to the optimal strategy under Scenario 16. In the cases of Benin and Sierra Leone, this share is lower in the recommended strategy, whereas it is higher for eight countries.

16.5 Conclusions and perspectives

In this chapter, optimal biofuels strategies were found for each of the 13 West African countries considered, using a multiobjective optimization in an energy prospective context. The models are simple because of the very uncertain context characterized by a lack of information and a long-term horizon (2050). Because of the simplicity of the models, the detailed results can be easily interpreted. However, these preliminary results must be refined by collecting reliable data on the agricultural residues and the environmental impacts of their uses, improving technoeconomic data on the various energy value chains considered and including explicitly the detailed cost and environmental impacts in the formalization of the mathematical problem. Despite all these limitations, the preliminary results highlight the potential roles of bioelectricity, SNG, and bioethanol in the passenger road mobility in the states of West Africa. They call for a long-term policy leadership in order to anticipate and harness the opportunities brought by the future bio-economy for loosening the energy and environmental burdens of the growing urbanization and transportation in this region.

References

[1] E. Gnansounou, E.R. Pachón, B. Sinsin, O. Teka, E. Togbé, A. Mahamane, Using agricultural residues for sustainable transportation biofuels in 2050: case of West Africa, Bioresour. Technol. 305 (2020).
[2] FAO, Global perspective studies. Food and agriculture projections to 2050. Food and agriculture 2050 data portal. https://www.fao.org/global-perspectives-studies/food-agriculture-projections-to-2050 (Accessed November 28, 2019).
[3] M. Balat, G. Ayar, Biomass energy in the world, use of biomass and potential trends, Energy Sources 27 (2005) 931–940.

[4] OECD/IEA, International Energy Agency, World energy outlook, 2006. (612006231P1) ISBN 92-64-10989-7 - 2006. Second Edition. https://iea.blob.core.windows.net/assets/390482d0-149a-48c0-959b-d5104ea308ca/weo2006.pdf (Accessed November 28, 2019).

[5] O. Rosales-Calderon, V. Arantes, A review on commercial-scale high-value products that can be produced alongside cellulosic ethanol, Biotechnol. Biofuels 12 (1) (2019) 240.

[6] T. Stadler, J.M. Chauvet, New innovative ecosystems in France to develop the Bioeconomy, New Biotechnol. 40 (2018) 113–118.

[7] S. Atsumi, T. Hanai, J.C. Liao, Non-fermentative pathways for synthesis of branched-chain higher alcohols as biofuels, Nature 451 (7174) (2008) 86.

[8] AFDC. Alternative fuels data center. Fuels and vehicles. Emerging fuels. https://afdc.energy.gov/fuels/emerging_biobutanol.html (Accessed November 28, 2019).

[9] A. Demirbas, Biomethanol production from organic waste materials, Energy Sources, Part A 30 (6) (2008) 565–572.

[10] H.J. Alves, C.B. Junior, R.R. Niklevicz, E.P. Frigo, M.S. Frigo, C.H. Coimbra-Araújo, Overview of hydrogen production technologies from biogas and the applications in fuel cells, Int. J. Hydrogen Energy 38 (13) (2013) 5215–5225.

[11] H. Li, D. Mehmood, E. Thorin, Z. Yu, Biomethane production via anaerobic digestion and biomass gasification, Energy Procedia 105 (2017) 1172–1177.

[12] R. Bhutani, K. Tharani, K. Sudha, Y. Tomar, Design of a cogeneration plant for sugar industries using renewable energy resources, J. Stat. Manag. Syst. 23 (1) (2020) 181–190.

[13] E.R. Cano, D. Banham, S. Ye, A. Hintennach, J. Lu, M. Fowler, Z. Chen, Batteries and fuel cells for emerging electric vehicle markets, Nature Energy 3 (4) (2018) 279–289.

[14] D. Humbird, R. Davis, L. Tao, C. Kinchin, D. Hsu, A. Aden, P. Schoen, J. Lukas, B. Olthof, M. Worley, D. Sexton, D. Dudgeon, Process design and economics for biochemical conversion of lignocellulosic biomass to ethanol—dilute-acid pretreatment and enzymatic hydrolysis of corn stover, National Renewable Energy Lab. (NREL), Golden, CO (United States), (2011).

[15] Sandia National Laboratories (SNL), 90-Billion gallon biofuel deployment study: executive summary, General Motors' R&D Center. https://digitalcommons.unl.edu/cgi/viewcontent.cgi?article=1083&context=usdoepub 2020 (Accessed January 12, 2020).

[16] G. Aranda, A. van der Drift, B.J. Vreugdenhil, H.J.M. Visser, C.F. Vilela, C.M. van der Meijden, Comparing direct and indirect fluidized bed gasification: effect of redox cycle on olivine activity, Environ. Prog. Sustain. Energy 33 (3) (2014) 711–720.

[17] M. Kraussler, F. Pontzen, M. Müller-Hagedorn, L. Nenning, M. Luisser, H. Hofbauer, Techno economic assessment of biomass-based natural gas substitutes against the background of the EU 2018 renewable energy directive, Biomass Convers. Bior. 8 (4) (2018) 935–944.

[18] E.M.W. Smeets, A. Faaij, Bioenergy potentials from forestry in 2050: as assessment of the drivers that determine the potentials, Climatic Change 81 (2007) 353–390.

[19] IEA, International Energy Agency, Good Practice Guidelines: Bioenergy Project Development and Biomass Supply, OECD/IEA, Paris, 2007.

[20] A. Eisentraut, Sustainable Production of Second-Generation Biofuels: Potential and Perspectives in Major Economies and Developing Countries, IEA ENERGY PAPERS, OECD/IEA, Paris, 2010.

[21] FAOSTAT. FAO Statistical Yearbook. Data. Crops and livestock products. https://www.fao.org/faostat/ (Accessed January 12, 2020).

[22] A. Gobin, P. Campling, L. Janssen, N. Desmet, H. Van Delden, J. Hurkens, P. Lavelle, S. Berman, Soil organic matter management across the EU—best practices, constraints and trade-offs, Final Report for the European Commission's DG, Environment 34 (2011).

[23] A.I. Magalhães Jr, J.C. de Carvalho, G.V. de Melo Pereira, S.G. Karp, M.C. Câmara, J.D.C. Medina, C.R. Soccol, Lignocellulosic biomass from agro-industrial residues in South America: current developments and perspectives, Biofuels Bioprod. Biorefin. 13 (6) (2019) 1505–1519.

CHAPTER SEVENTEEN

Semantic sustainability characterization of biorefineries: A logic-based model

Edgard Gnansounou, Catarina M. Alves, Elia Ruiz Pachón, Pavel Vaskan

Ecole Polytechnique Fédérale de Lausanne (EPFL), School of Environment, Civil Engineering, and Architecture, Institute of Civil Engineering, Bioenergy and Energy Planning Research Group, Lausanne, Switzerland

17.1 Introduction

In order to tap into the potential of innovation and contribute to the transition to a sustainable economy, early sustainability assessment must be undertaken at different levels ranging from strategic to design and products/systems levels. Ness et al. [1] reviewed existing tools of sustainability assessment by distinguishing three main analysis categories, (1) indicators/indexes, (2) product-related assessment, (3) integrated assessment. Several authors addressed the early sustainability assessment at the design phase that could be performed through different steps comprising comparison of design variants and in-depth characterization of the selected best option. Moncada et al. [2] developed a sustainability method to compare bio-processing to petrochemical processes for three cases: 16 systems for converting bio-based syngas into derivatives, 12 conversion systems of carbohydrates, and 8 processing systems of glycerol. The sustainability was assessed by combining into a single score a set of indicators using a weighted average. Patel et al. [3] developed an early sustainability assessment method using quantitative and qualitative indicators. The system to assess was chemical process including reaction and separation process. The indicators were grouped into five categories: (1) economic constraints, (2) environmental impact of raw materials, (3) process cost and environmental impact, (4) environmental Hazard And Safety Index, (5) risk aspects. Each category was divided into subcategories that were described using indicators. For example, for the environmental hazard and safety index, the subcategories were environment, hazard, and safety. Environmental indicators considered to describe the subcategory "Environment" were: persistency, air hazard, water hazard, and solid waste. The total score was the sum of the scores of the categories where each category was restrained to a maximum and the maximum total score is 1. Bertoni et al. [4] proposed a method for integrating value and sustainability into the design process, using Machine learning to address the simulation cost of design. For the sustainability assessment, they classified the variables into leading criteria. Each criterion was then characterized by indicators. The authors identified 10 leading criteria along the life cycle of the products, (1) critical materials, (2) recycled materials, (3) scrap recyclability, (4) risk of remanufacturing, (5) health and safety, (6) emissions, waste products and chemicals, (7) risk of being exposed to dangerous substances, (8) optimized product weight, (9) noise to the surroundings, (10) materials/components returned for remanufacturing and recycling. For example, in the case of the fifth leading criterion, the indicators are the following: number of injuries, risk of exposure, leakages. Gnansounou et al. [5] proposed an integrated method to compare the sustainability method of a set of design configurations of sugarcane biorefinery. The method was composed of five steps: (1) context definition, choice of cultural values and selection of the biorefineries to compare, (2) process design, flow sheeting, mass and energy balances, (3) sustainability metrics, (4) multicriteria comparison, and (5) In-depth analysis of the nondominated biorefineries. The authors developed the four first steps and applied them to a case study for comparative assessment of strategic choice between selected biorefinery configurations. The method used a multicriteria decision-making. This chapter aims at developing the fifth step of the methodology. The best option found after the fourth step was characterized using a Logic-Based Model (LBM) [6]. Multicriteria decision-making pertains to the integrated methods in the typology proposed by Ness et al. [1]. At a conceptual design phase, many of the assumptions and context elements of the biorefinery systems do not change from one design variant to another and, thus, in a comparative evaluation, some indicators become nondiscriminatory since they do not confer any difference among the design options. In a comparative assessment, the aspects showing greater differences between scenarios must be first identified and then, the sustainability metrics covering such aspects designed. Second, the most sustainable scenario can be determined using a multi-criteria comparison, given the indicators' weighting profile [5]. Furthermore, one can characterize the sustainability performance of the comparison evaluation best case, based on an extended set of indicators and benchmark against a reference system using support data. Sustainability characterization consists of an in-depth analysis which covers context-dependent indicators including both qualitative and quantitative type of metrics. This study proposes an integrated assessment methodology for sustainability characterization of biorefinery systems using the conventional concept of sustainability. A high sustainability performance can only be achieved if the economic, social and environmental conditions meet the set targets. The integrated assessment methodology consists of an LBM that differs from an arithmetic model due to the following arguments: it takes into account qualitative aspects of the sustainability concept; it relies on a hierarchical network of indicators based on their specificity levels; and, finally, it outlines a logic pathway that uses specific variables to define a more general variable, from the most specific one to the most general one—*sustainability grade*. Besides considering the three sustainability dimensions within the network, the logic-based method analyzes each particular indicator as per its impact beyond dimensions. In other words, an indicator classified in a certain dimension (e.g., economic) is related to the indicators in other sustainability dimensions (e.g., social, environmental), increasing the logic-based network reliability.

17.2 The problematic of sustainability characterization

A five-module methodology was proposed to analyze the sustainability of different schemes of biorefineries providing the same services to the community. The first four modules covered the comparative assessment of four biorefinery-centered systems [5]. The fifth module implies an in-depth analysis of the non-dominated biorefinery. The sustainability performance of a sugarcane-based ethanol distillery integrated into a lignocellulose First and Second Generation (1G2G) ethanol plant (Ethanol Distillery and Only Fuel production—ED OF) was higher than any other of the biorefinery systems considered by Gnansounou et al. [5]. Thus, ED OF was optimized in this module to improve the overall process efficiency. Then, the sustainability characterization of ED OF was completed according to the logic-based method used in this work.

17.2.1 Improvement of the energy efficiency

The design of ED OF was improved in this work. Then, the overall energy of the process was reduced through heat integration. The energy requirements of ED OF were reduced using Aspen Energy Analyzer V8.8 [7]. This software allows the user to analyze a process flowsheet for heating and cooling efficiency based on pinch technology [8]. Aspen Energy Analyzer constructs a pinch diagram for the design of an optimal heat exchanger network (HEN) to minimize utility consumption. Fixing the operating conditions of the process, the heat and cold streams of the simulated ED OF were analyzed. After evaluation, a new HEN providing maximum energy recovery and the minimum extra capital cost was proposed. The proposed HEN was implemented in the optimal design based on energy efficiency. The extra capital costs were considered in the techno-economic assessment.

17.2.2 Developing the logic-based model for sustainability characterization

17.2.2.1 Background of the model

As introduced by Gnansounou [6] and in chapter 8 of this book, the LBM for sustainability characterization consists of a logic network that contains several variables hierarchically disposed and interconnected. The variables specificity increases from the top to the bottom of the hierarchy. The top hierarchic level is the sustainability performance (with the lowest specificity) of the biorefinery system, which is the major output of the model. The sustainability performance is estimated based on three qualitative determinants—economic, social, and environmental. Each determinant is then calculated based on a set of general indicators that are more specific than the determinant itself. In turn, each general indicator is defined based on several specific indicators located at the bottom of the hierarchy network. A qualitative score expressed in linguistic terms—low, moderate or high—is attributed to each specific indicator as a result of a comparison with support data. Once converted into a numeric value (1, 2, or 3), the specific indicator scores are used to calculate the general indicator scores throughout a forward chain procedure by a rule-based system. Next, the general indicators are the inputs for estimating the determinants, whose scores are the inputs for estimating the final sustainability grade. The output scores of the general indicators and determinants, and the sustainability grade can be 0 (very low), 1 (low), 2 (moderate) or 3 (high).

17.2.2.2 Criteria for selecting the indicators

The first step of the development of the LBM is to build the hierarchical structure by selecting and designing the variables. The determinants were defined based on the three pillars of sustainability from the Brundtland report: economy, society, and environment [9]. Within each determinant, the authors selected a set of general indicators based on three criteria: relevance—the indicator is an important measure and captures a relevant issue in the context of the study; usefulness—the indicator suits to serve a certain purpose; and distinctiveness—the indicators are not redundant. For each general indicator, the specific indicators were designed to cover the main aspects related to a sugarcane biorefinery system and its impact on socioeconomic development and the environment in both short and long-term horizons.

17.2.2.2.1 Economic indicators

The indicators within the economic determinant covered aspects related to the biorefinery feasibility and viability of its products, affordability of feedstock cultivation and logistics, as well as sensitivity to product-market fluctuations. The economic indicators were calculated based on process simulation data (mass and energy balances) and techno-economic assessment, as in Gnansounou et al. [5]. The economics of the biorefinery system relies on the operational expenditure, including raw materials, utilities, maintenance, labor, among others; and the capital expenditure, including process equipment costs, construction, bulk materials and engineering. Several relevant assumptions were made such as product selling prices and the feedstock minimum selling price (MSP) at the Farm gate. Moreover, conversion costs were allocated to the different biorefinery products using the value-based method presented by Gnansounou et al. [5, 10].

17.2.2.2.2 Social indicators

The social indicators covered a wide range of aspects linked to sugarcane cultivation and bioenergy production from the perspective of Sao Paulo-Brazil and, more specifically, in the regional framework of Ribeirão Preto. The evaluation of the social indicators was performed based on the benchmarking of local data and common practices against reference support data. For most social-specific indicators, the authors identified the most relevant issues that can influence the value of the indicator. Additionally, given the integrative assessment, the authors used outputs of the life cycle analysis to assess issues related to toxicity levels and resources conservation, which can affect health and social well-being, food security, etc., and to benchmark against the corresponding reference system. It is worth noting that the attribution of the social scores unavoidably relies on context and assumptions regarding the biorefinery system boundaries, location, and practices.

17.2.2.2.3 Environmental indicators

Indicators related to land use, biodiversity, local and global ecosystems conservation were addressed within the environmental determinant. The environmental sustainability characterization was performed based on the life cycle analysis outputs and the benchmarking of biorefinery products against the corresponding reference systems. Like in the social indicators, part of the aspects discussed in the specific indicators involved the qualitative comparative judgment of biorefinery performance aspects against the state-of-the-art or a reference condition.

17.2.2.3 Rule-base

Once the scores were attributed to the specific indicators, a rule-base system was implemented to estimate the upper scores in the hierarchy. The representation of the variables hierarchic network is given by Fig. 17.1.

The rule-base system relies on distance formulas which use as reference the maximum possible difference between scores (d). The variables i, j, and k represent the determinant, general indicator, and specific indicator, respectively. Each general indicator score ($I_{i,j}$) is estimated based on the scores of its specific indicators ($I_{i,j,k}$) as per Eq. 17.1, n being the number of specific indicators within the general indicator. Likewise, the determinant score (D_i) is deducted based on the scores of its general indicators ($I_{i,j}$) as per Eq. 17.2, m being the number of general indicators within the determinant. Finally, the sustainability grade is calculated based on the scores of its determinants (D_i) as per

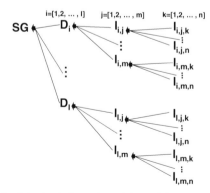

Fig. 17.1 Hierarchic network of the logic-based model variables.

Eq. 17.3, l being the number of determinants of the sustainability grade. The score calculation may be tuned manually should a semantic inconsistency be detected [6].

$$I_{i,j} = d\left[1 - \left(\sqrt{\frac{\sum_{k=[1,2,\ldots,n]}(I_{i,j,k} - d)^2}{nd^2}}\right)\right] \quad (17.1)$$

$$D_i = d\left[1 - \left(\sqrt{\frac{\sum_{j=[1,2,\ldots,m]}(I_{i,j} - d)^2}{md^2}}\right)\right] \quad (17.2)$$

$$\text{Sustainability grade} = d\left[1 - \left(\sqrt{\frac{\sum_{i=[1,2,\ldots,l]}(D_i - d)^2}{ld^2}}\right)\right] \quad (17.3)$$

An LBM was developed that involved 43 variables comprising 3 determinants, 11 general indicators, and 29 specific indicators. The variables are described and discussed in detail in the next section. The model is composed of 1475 rules.

17.3 Case study

17.3.1 Description of the case study

17.3.1.1 Process design

In total, 13,000 t/day of sugarcane entered the milling section of the 1G2G ethanol production plant. Table 17.1 shows the composition of sugarcane, sugarcane bagasse, and green harvesting residues used in the simulated plant.

Table 17.1 Composition of sugarcane, sugarcane bagasse, and green harvesting residues (GHR).

	Weight percent (%)		
	Sugarcane	Sugarcane bagasse	GHR (in plant)
H_2O	71	50	30
Glucose	0.5	0.1	0
Sucrose	13	3	0
Extract	1	2	6
Cellulose	6	19	25
Galactan	0.10	0.3	0.5
Mannan	0.05	0.15	0
Xylan	3	10	15
Arabinan	0.27	1	2
Lignin	4	12	15
Acetate	0	1	1
Ash	1	2	6

Since the old design was restrained to be energy-self-sufficient while producing electricity, bagasse was not used for cogeneration in this work [10]. The whole amount of bagasse produced in the sugarcane mill was pretreated according to the dilute acid technique. The process design in Gnansounou et al. [10] was adapted from the one performed by NREL [11]. The pretreated hydrolysate underwent a sequential hydrolysis and fermentation process where ethanol was obtained at 7% weight in the fermentation broth. This stream was distilled in the beer column. The number of stages of the distillation column was reduced to 32 operating at 48% efficiency. In such energy-optimized beer column, the mole reflux ratio required to release 99.95% ethanol mass through a vapor side-stream was 4.3:1. The side stream, mainly composed of ethanol and water, was conducted to a reduced rectification column using 40 stages at 76% efficiency. The mass reflux ratio was 3.78:1 to obtain 93% mass ethanol in the vapor overhead.

The whole bottom stream leaving the rectification column was recycled back to the pretreatment reactor as dilution water, whereas in the previous design part of this stream was sent to wastewater treatment. In the present case, the ethanol lost in bottoms was recycled entirely for ethanol production. Since cars in Brazil can run with hydrated ethanol, the dehydration process was avoided for saving both capital and operating costs. Thus, the ethanol product left the rectification column at 93% weight. Since the NREL design [11] did not include sucrose, new reactions were implemented in the reactors of the wastewater treatment and cogeneration areas for the residual sucrose not converted in the process. Moreover, the process water manifold was adjusted to the water requirements of the plant.

The overall energy of the simulated plant was evaluated with Aspen Energy Analyzer to propose a new HEN providing some energy savings. The pinch methodology results provided a saving potential of 255.15 GJ/h; up to 127.57 GJ/h could be saved from heating or cooling utilities. Based on the maximum energy recovery and minimum extra capital cost, an optimal HEN was designed by the addition of 4 heat exchangers in the design. With the proposed HEN, 122.47 GJ/h were removed from the heat (LP steam) required by the juice extraction area. In the pretreatment area, 46.4 GJ/h could be removed from the cold utilities (cooling water). In the ethanol recovery area, 76.5 GJ/h were saved for air conditioning. In total, 245 GJ/h were recovered from the cooling and heating utilities of the process through heat integration with an extra capital cost of 168,604 US$. The new HEN was implemented in the design and the mass and energy balances were calculated with this new configuration.

The main simulation results are listed in Table 17.2. Due to the conversion of the whole bagasse into ethanol, the amount of ethanol produced in the simulated ED OF increased by 28.6% compared to the design presented in Gnansounou et al. [10]. Regarding the energy balance, the electricity surplus produced increased by 12.4%, whereas the steam demand is reduced by 35.2%. The electricity consumption corresponds to 1.02 kWh/kg of ethanol produced, which corresponds to a 5.1% reduction when compared to the previous design. Thus, the energy efficiency of the process is improved.

Table 17.2 Process simulation results of the ED OF plant.

Ethanol production kg/t SC (w.b.)	Electricity surplus production kWh/t SC (w.b.)	Total electricity generated kWh/t SC (w.b.)	Electricity demand kWh/t SC (w.b.)	Steam demand GJ/t SC (w.b.)
115.9	28.1	146.5	118.4	1.05

tSC, tonne of sugarcane.

17.3.1.2 Context

The characterization of the sugarcane biorefinery system sustainability relies largely on the context in which the biorefinery plant is being implemented. Besides employing process data, life cycle assessment, and techno-economic analysis outputs, the reflection and benchmark of the indicators were completed based on the context and the assumptions described below.

System boundaries

The boundaries of the system were set as cradle-to-wheel for the bioethanol product, and cradle-to-gate for the surplus electricity. Cradle-to-wheel refers to the partial fuel product life cycle from the resource extraction (sugarcane cultivation) to the use phase. Though, for the surplus electricity, the assessment was performed uniquely until the biorefinery gate. The power distribution from the biorefinery to the end customer was performed over a complex national network that merges the power from hydro and other sources, whose emissions and impacts were not considered in the current study.

Location

The biorefinery plant is located in Sao Paulo due to the favorable climate conditions for cultivating sugarcane, the existence of know-how, infrastructure, and connectivity. Given the last decade's increase in the sugarcane demand, the production has been raising in Sao Paulo through cultivation land expansion to degraded and former pasture lands, or, remarkably, by improving agricultural practices and renovating existing plantations to increase productivities [12]. According to local sources, the suitable land for the expansion of sugarcane in the northeast dense production area of Sao Paulo is largely diminishing due to strict regulations [12, 13]. Therefore, a "renovation" scenario is assumed for the current biorefinery system, considering the scenario of occupation and renovation of an old sugarcane plantation. The selected location for the biorefinery is Ribeirão Preto situated in the northeast of Sao Paulo state, one of the regions presenting a large necessity for the renovation of sugarcane plantations and mills [14]. This area presents the highest density of sugar mills in the state, the majority built between 1975 and 1987, a period of abrupt expansion of sugarcane industry due to the increase of ethanol's demand as a consequence of the implementation of the Pro-Alcohol national program in 1975 [14].

Agro-practices

The sugarcane harvesting method has a large impact on the social and environmental performance of the biorefinery system. According to the Agro-Environmental Sugarcane Industry Protocol, sugarcane burning should be eradicated in Sao Paulo by

2017 [15, 16]. However, the 2017 scenario corresponds to an 85%–90% mechanical harvest, meaning that 15% of the sugarcane harvest is still performed by manual work and includes pre-harvest burning [17, 18]. For this work, 85% conservative choice is assumed. Furthermore, a practice that has been progressively implemented in the sugarcane fields in Sao Paulo is the use of the green harvesting residues as a soil cover/mulch to balance the nutrient content levels in the soil and decrease erosion [19, 20]. Therefore, 50% sugarcane residues were assumed to be left on the field. Finally, reduced tillage practices were considered to reduce the disruption of soil aggregates, which enhances the soil carbon levels and improves soil stability [21, 22].

17.3.2 Specific indicators
17.3.2.1 Economic specific indicators
17.3.2.1.1 Economic viability of the whole biorefinery

1. Economic affordability of the farm gate feedstock price $I_{1,1,1}$

The economic affordability of the farm gate feedstock price defines the ability to purchase the required sugarcane for running the facility. The indicator was assessed based on the Prospective Economic Performance, a concept developed by [23] which represents the difference between the maximum purchasing price (MPP) the biorefinery can pay for the feedstock and the MSP set by the feedstock supplier, both given in the US $/t of feedstock. While MSP is fixed, MPP relies on the economic performance of the biorefinery and represents the economic margin between profits and the total conversion costs excluding sugarcane feedstock. The higher the MPP, the higher the economic margin available for purchasing the feedstock at the farm gate, and the higher the affordability. MPP was estimated as 45.1 US $/t sugarcane. Given an MSP of 29.3 US $/t sugarcane in the context of Sao Paulo, the authors estimated a PEP of 15.8 US$/t sugarcane. The achieved PEP is considered reasonable given the stability of the sugarcane market in Sao Paulo. In addition, in contrast to other areas in Brazil, Ribeirão Preto/Sao Paulo region offers larger sugarcane cultivation know-how, as well as more advanced agro management practices. Such local features can lead to a more constant and long-term sugarcane supply, contributing to higher stability of the sector and, from the biorefinery perspective, higher affordability of the feedstock. Even so, a moderate score was attributed to the affordability indicator because the ethanol market is associated with uncertainties and there is a risk that the ethanol selling price drops reducing PEP.

2. Economic affordability of the feedstock logistic cost $I_{1,1,2}$

The affordability of the feedstock logistic cost reflects the ability of the biorefinery to meet the expenses related to the feedstock logistic activities such as the transportation from the farm gate to the biorefinery gate. Firstly, the authors have taken into consideration the PEP to evaluate the current indicator. As described in the previous indicator, the higher the PEP, the higher the margin to cover the feedstock total cost which includes the farm gate cost (set by the supplier) and the feedstock logistic cost.

Moreover, the feedstock logistic cost can be largely affected by the average distance between the farm gate (points of feedstock collection) and the biorefinery due to variations in the transportation expenses. In the current context, the biorefinery location was set in Ribeirão Preto, a region of a high density of sugarcane plantations, dense infrastructure network, and large implementation of agriculture management practices when compared to other regions of the country [24]. Therefore, transportation expenses were expected to be relatively reduced. Yet, due to the relatively large scale of the factory and requirements for sugarcane, the average distance from the biorefinery to the sugarcane supply can considerably increase in a scenario of sugarcane shortage, as well as the transportation expenses. A moderate score has been attributed to $I_{1,1,2}$ indicator due to the moderate PEP and to account for a potential scenario of sugarcane shortage in the target region.

3. Economic viability of the conversion cost $I_{1,1,3}$

The conversion cost represents the sum of the capital and operational costs necessary to convert one or more raw materials into many products. The estimation of the costs of conversion is important to study the economic viability of the whole biorefinery system. In the current biorefinery scenario the total conversion cost was estimated as 274.0 MUS$/y which represents about 82% of the total biorefinery sales revenues, 335.5 MUS$/y. The total conversion costs are considered moderate, reasonably lower than the sales revenues of the biorefinery, and, therefore, the score attributed to $I_{1,1,3}$ is moderate.

4. Economic sensitivity to product price volatility $I_{1,1,4}$

The volatility of product prices can impact the economic sensitivity of the whole biorefinery system. In the context of a comparative study, the current biorefinery was placed first in the ranking with the lowest economic sensitivity due to the reduced number of products and a higher profitability level, when compared to the remaining scenarios [5]. However, the present aim is reflecting upon the context where the biorefinery system is inserted. The biorefinery major revenues are associated with the 1G2G ethanol product, hence, the authors have looked into the volatility of the ethanol market in Brazil and assessed its potential impact on the viability of the biorefinery system. Ethanol market in Brazil depends on the crude oil market since ethanol is a current substitute for the gasoline petroleum product [25]. The crude oil price volatility is naturally high due to the low responsiveness or "inelasticity" of the supply and demand to price changes in the short run, which are a consequence of economic, geopolitical, and weather-related events disruptive for the oil market [26]. A direct positive relation was proved between the Brazilian ethanol and the world crude oil price level and volatility [25]. Thus, the economy of a biorefinery in Brazil producing mainly bioethanol is very sensitive to the crude oil market volatility, which justifies a low score for $I_{1,1,4}$.

17.3.2.1.2 Economic cross-subsidy between the products' value chains

In order to assess the level of cross-subsidization between the products' value chains, the viability of each biorefinery product was individually evaluated. For this purpose, the biorefinery operation costs and capital annuity were allocated to the different biorefinery products using a value-based approach, as performed by Gnansounou et al. [5].

a. Economic viability of 1G Ethanol $I_{1,2,1}$

The cost of production of 1G ethanol sums part of the costs of sugarcane handling, separation of the sugarcane juice from the bagasse, fermentation, and downstream. Regarding the costs associated with the shared operation units and process fixed expenses, the allocation to 1G ethanol was performed based on the sugar fraction, i.e., the fraction of soluble sugars (1G sugars) over the total amount of fermentable sugars in the cofermentation unit. The cost of conversion of 1G ethanol has been estimated at 0.434 US$/L. The conversion cost is significantly lower than the market price of bioethanol assumed as 0.586 US$/L (0.721 US$/kg) and the reference 1G ethanol production cost of 0.500 US$/L, reported by Junqueira et al. [27] (short term 1G scenario). Still, the viability of the 1G ethanol of the biorefinery system was evaluated as moderate due to a relatively reduced margin [27].

b. Economic viability of 2G Ethanol $I_{1,2,2}$

The economic viability of the 2G Ethanol produced in the biorefinery was evaluated based on the method described in the previous indicator. The allocation to the 2G Ethanol product was performed based on the ratio between the amount of hydrolyzed sugars (2G sugars) and the total fermentable sugars in the reactor. Besides sharing the costs of the common operations, the 2G ethanol product holds the costs associated with the pretreatment and hydrolysis stages. The estimated 2G ethanol production cost was 0.515 US$/L. Since the total production cost of 2G ethanol is not higher than the reference production cost of 2G ethanol, 0.660 US$/L, nor the selling price assumed for the bioethanol product, the economic viability of the 2G Ethanol was scored as moderate [27]. The results evidence cross-subsidization of the 2G ethanol given the lower price of the 1G ethanol when compared to 2G ethanol.

c. Economic viability of the electricity produced $I_{1,2,3}$

The economic viability of the electricity produced was assessed by comparing the cost of production of 1 kWh at the biorefinery with the reference selling price of electricity. As for the previous products, a value-based approach has been applied to allocate the costs of the biorefinery to the electricity. The authors have concluded that the electricity production cost rounds the 0.073 US$/kWh, a reasonably low cost compared to the assumed electricity selling price 0.088 US$/kWh. Closing, the economic viability of the electricity produced was scored as moderate.

17.3.2.2 Social specific indicators
17.3.2.2.1 Social acceptability

a. Social affordability $I_{2,1,1}$

Social affordability refers to the ability and willingness of a particular social group to pay for a minimum level of a defined service [28]. The ability to pay (if people can pay) is associated with aspects such as income levels, income distribution and employment.

In turn, the willingness to pay (if people want to pay) relates to the public opinion and social acceptance of the product or service, which can depend on aspects such as the level of education, urbanization, median farm size, and existence of land conflicts [29]. For assessing the "ability to pay," the authors have investigated the poverty level in the Ribeirão Preto region, when compared to Sao Paulo state and Brazil. The percentage of extremely poor and poor people sums 3.14% for Ribeirão Preto, which is far lower than the statistics for Sao Paulo (5.82%) and Brazil (21.82%). Moreover, the income distribution inequalities in the sugarcane sector in Sao Paulo-Brazil were analyzed against the agriculture sector, industry, and services sectors. Income distribution inequality can be measured in terms of the Gini coefficient. The Gini coefficient represents the inequality among values of frequency distribution and it ranges from zero to one, where 1 is the largest possible level of inequality [30]. According to de Carvalho Macedo [30] in Sao Paulo-Brazil, the sugarcane industry has lower income inequality—Gini coefficient was reported as 0.413- when compared to the agriculture sector in general or other economic sectors—Gini coefficient values were reported as 0.555, 0.493, and 0.537 for agriculture, industry, and services, respectively. Furthermore, women employment and empowerment can affect social affordability. In the past, mostly men would have access or could afford to buy the goods or services. Due to the increased level of mechanization in the sugarcane chain activities, women employment is largely increased and, consequently, women empowerment [31]. On the other hand, employment creation in a community will generally affect the ability of the community to purchase products and services. Therefore, even though jobs are created directly and indirectly due to the biorefinery implementation, the mechanization increase in the sugarcane harvest is expected to decrease the employment along the supply chain. Moreover, public opinion and acceptance are related to the "willingness to pay" the biorefinery products/services. The level of education affects public opinion since higher education most likely brings openness to accept innovative products and concepts [29]. According to IBGE, Ribeirão Preto municipality is in 1630th position in the basic education ranking over 5570 municipalities in the country, belonging to a range of the one third most educated municipalities [32]. The population of less urbanized areas is more likely to oppose new industrial constructions, given the fear of landscape change and shortage of resources [29]. Ribeirão Preto and, in a larger picture, the Sao Paulo state region present one of the highest degrees of urbanization of the country [32, 33] https://cidades.ibge.gov.br/brasil/sp/ribeirao-preto/panorama. Moreover, with the existence of land conflicts due to the biorefinery implementation it was feared that the willingness to pay would be affected [29]. The current biorefinery will be implemented in existing sugarcane plantations—renovation scenario—within a well-built sugarcane area in terms of farm development and infrastructures. Therefore, no expansion of sugarcane and no conflicts were expected with smallholders or indigenous communities. These arguments justify the attribution of a moderate score to the Social affordability indicator.

b. Transparency $I_{2,1,2}$

Transparency has increasing importance in today's businesses. The public opinion and social acceptance of a certain product/service are largely influenced by how a business is doing concerning social and environmental responsibility. In Brazil, several policies and instruments have been implemented at the country level to manage and regulate transparency in the public and private sectors in the social, environmental, and financial domains [34,35]. However, it takes time until the specific anticorruption laws are applied to corporate and government institutions at the national level. Corruption is a problem largely compromising transparency and, consequently, affecting people's trust and social acceptance [36]. Brazil occupies the 79th position in the Corruption Perception Index ranking, over a total of 176 countries under assessment [37]. Based on such ranking, the authors have attributed a moderate score to $I_{2,1,2}$ indicator.

c. Stakeholders' participation $I_{2,1,3}$

Stakeholders' participation increases with transparency. For direct intermediates with the biorefinery but also indirect actors and the community to engage, it is of high importance to be transparent regarding the sustainability performance of one biorefinery system. Data such as income distribution, women's employment, emissions, waste production, and financial reporting should be accessible to all stakeholders to induce trust, engagement, and participation. Moreover, employees' empowerment and, in particular, women empowerment, can increase trust and satisfaction in the local communities, leading to higher participation as part of the business model, for instance, through feedback links and advertising. Given the close association with the social affordability and transparency indicators, the moderate score has been equally attributed to the Stakeholders' participation indicator.

17.3.2.2.2 Social well-being and prosperity

1. Working conditions and rights $I_{2,2,1}$

Fair working conditions and respect for the employees' rights are key aspects of social well-being and prosperity. In the context of the sugarcane industry, the wages in sugar and ethanol industries in Sao Paulo state were identified. According to de Carvalho Macedo [30], the sugarcane and ethanol production industries in Sao Paulo pay higher salaries compared to the average in the rest of the country. The wages in the sugarcane industry are also larger than the average in the agro sector (average covering banana, coffee, citrus, manioc, corn, and soy). However, the wages earned may still be lower than the average expenses of rural families in the state and insufficient to allow a decent standard of living [38, 39]. Machado et al. [39] analyzed the socioeconomic and environmental issues of sugarcane cultivation in Sao Paulo and characterized the sensitivity regarding the wages in different state regions. The municipality of Ribeirão Preto has been classified in medium rank with regard to the wage of the workers in the sugarcane industry with recorded wages ranging from US$ 314.31 to 688.40 [39].

Moreover, the advance with the largest improvements in the working conditions of the sugarcane workers Brazil has been the mechanization of the sugarcane chain. The sugarcane chain mechanization has decreased the incidence of occupational accidents, improved average wages, increased educational level and the women immersion in the workforce [31, 40]. In Sao Paulo, mechanical harvesting was implemented at a large pace due to legal restrictions on the preharvest sugarcane burning [30, 41]. Sao Paulo aimed to eradicate sugarcane burning by 2020 [42]. In a renovation scenario, an 85% mechanical harvest will lead to reduced sugarcane burning and improved conditions in the harvest. Furthermore, the issues regarding child and forced labor practices were investigated in the context of the location chosen to build the biorefinery. The Brazilian government defines 18 years old as the minimum age for hard job workers and conducts inspections in the sugarcane sector to control child labor [43]. Child labor in the sugar and ethanol production industries in Brazil correspondents to <0.1% and 0.1%–0.4%, for 10–14 and 15–17-year-old children, respectively [38]. Brazil has been one of the most progressive and proactive nations in defending children's needs and rights [44]. For the US Department of Labor, sugarcane child labor in Brazil is not particularly concerning when compared to other countries such as India and Thailand [44]. Machado et al. [39] showed that, even though sporadic, there are still occurrences of child labor cases in Sao Paulo and, particularly, in Ribeirão Preto. On the other hand, the sugarcane industry in Sao Paulo is still linked to circumstances of forced and unfree labor, mainly due to the remaining sugarcane manual harvest [39, 44]. The manual harvesting in Sao Paulo sugarcane fields regularly employs migrant workers coming from north-eastern poor states who are seeking for an income [45, 46]. Often, the conditions provided to migrant workers are deplorable: unrecorded working hours, wage relies on the harvest personal productivity, wage deduction due to sickness, failure of payment adjustment due to holiday and weekends work [24,45,47]. There is also colossal discrimination and a wage gap between migrant and non-migrant workers, even though the latter usually does not perform extreme manual work, crossing the borderline of respect for labor rights [46]. Concluding, wages and working conditions have been largely enhanced in the last years due to the measures implemented in Sao Paulo state toward 100% mechanical harvest. However, the manual harvest and the linked unfree labor have not yet been fully eradicated, which justifies the attribution of moderate score to the current indicator.

2. Safety and health $I_{2,2,2}$

The safety and health of the employees and the local community are partly linked to the considerations of the previous indicator. Mechanical harvest implementation has a critical impact on the labor conditions, as well as in the workers' safety and health by reducing the number of accidents in the last decades [40]. One investigated the occurrence of work accidents in Sao Paulo and Brazil in the context of the ethanol production chain from cane cultivation to the harvest, postharvest, ethanol production, and fuel distribution activities. The Ministry of Labor and Employment estimated the work accidents incidence index based on the number of work accidents registered in one year over the exposed working population. Even though the sugar-ethanol chain index in Sao Paulo was reported to be lower than the average in Brazil, the index was greater than the ones of the majority of the economic activities in Brazil [48]. The observations induced the choice of a moderate score for the current indicator. Moreover, the current indicator should be in line with the evaluation of the local environment indicators analyzed further for the 3.2.3.2 Local environment indicator, since soil, air, and water conditions will impact the safety and health of the local community.

3. Food security $I_{2,2,3}$

Hunger and undernourishment in Brazil are mainly related to the low food purchasing power due to poverty, and not due to lack of food production capacity in the country [38]. In the biorefinery context, one started by analyzing the poverty levels in Sao Paulo in comparison with the figures at the country level. As explained for the (1) social affordability indicator, Sao Paulo state has one of the lowest poverty levels in Brazil being the state with the largest GDP per capita [97, 98]. In particular, the fraction of people considered poor in the region of Ribeirão Preto is lower than the average in Sao Paulo. Therefore, communities in the region have a relatively high food purchasing power when compared to the average in Brazil. Moreover, one investigated whether the sugarcane production associated with the biorefinery can impact food cultivation in the regions. Several authors agree that sugarcane cultivation for biofuel production does not constitute a threat to food availability and accessibility in the case of Brazil [38,39]. Brazil has plenty of lands available including degraded areas that can be renovated and used for both food and sugarcane production without conflict [39]. Naturally, the implementation of land use and landscape management practices is crucial to allocate the target lands to different cultivation purposes, as it will be further discussed for the 3.2.2.4 Resources conservation and 3.2.3.1 Land use indicators. Besides the occupation and renovation of degraded lands, increasing the cultivation productivities of existing crop area could contribute to enlarge the production without expanding to wildlands or competing with food production [38, 49]. Finally, food security can be compromised by the volatility of food prices. The crude oil price impacts on the ethanol price in Brazil as well as its volatility. Since ethanol is currently mostly produced from food crops in Brazil, fluctuations in the ethanol market can affect the feedstock market and, consequently, the sugar market [25]. The local industries can switch from sugar to ethanol production (and vice-versa) depending on how attractive it is to produce ethanol or sugar. Such practices lead to frequent changes in the sugar market, sugar availability and pricing, and, consequently, can impact food security. In the current case study, the system was designed to produce only 1G2G ethanol and, therefore, there will be no switches that can brutally affect the markets. Nevertheless, besides the positive scenario concerning food purchasing power in Brazil, the percentage of poor people is still high. Moreover, sugarcane expansion must be controlled to avoid competition with food, however, the management practices are not yet fully implemented and the outcomes of such actions may take a longer time. The given arguments in this paragraph justified the attribution of a conservative moderate score.

17.3.2.2.3 Energy security

1. Energy dependence $I_{2,3,1}$

Energy dependence was defined by Eurostat as the extent to which a country's economy relies on imports to satisfy its energy needs. The authors have estimated the impact of the biorefinery system on energy imports and exports in Brazil. Within the biorefinery industrial facility, the energy consumed is entirely self-produced. In turn, the diesel fuel consumed along the supply chain from the cultivation stage to the harvest, transport, and distribution activities, needs to be purchased. According to Brazil's National Agency of Petroleum, Natural Gas, and Biofuels, about 15% of the diesel oil fuel distributed in the country is imported [50, 51]. The estimated power imports associated with diesel consumption are 1.7 MW, based on a total diesel consumption of about 11.6 MW along the supply chain (calculations in 2) net nonrenewable energy). In this analysis, one has assumed that, besides diesel fuel, no other forms of energy were consumed along the chain. Moreover, biorefinery contributes to energy exports in the form of bioethanol. The percentage of Brazilian bioethanol market exported is 5% as per ANP [51,52]. The authors assumed that 5% bioethanol produced in the biorefinery was funneled to external markets, which corresponds to about 20.3 MW. Additionally, the production of bioethanol leads to currency savings since it avoids oil imports [53]. Thus, the biorefinery system contributes to reducing the energy dependence of the country with a net energy export of 18.6 MW, which justified a high score for the current indicator.

2. Energy affordability $I_{2,3,2}$

Energy affordability relates to how consumers bear the energy costs to meet their own needs [54]. The reflection was made based on major elements that can influence energy affordability: energy pricing, energy supply conditions, regulatory action, and the general economic situation of the target community [55]. 1G2G bioethanol and electricity are energy products, which are fuel and power supply for the local markets. The viability of the biorefinery products—previously assessed based on the margin between the market prices and the cost of production—has a direct relation with the prices established for these products, and, consequently, has an impact on the local energy pricing. A moderate score has been given to the viability of the biorefinery products, which should be coherent with the energy affordability score. Moreover, concerning the energy supply, the Sao Paulo region contains one of the most solid networks in the country in terms of road infrastructure for fuel transport and electricity distribution systems [56, 57]. Regarding regulatory action, the Brazilian Regulatory Electricity Agency (ANEEL) is the entity that regulates the national energy matrix and market [57,

58]. The energy sector in Brazil is considered highly centralized—98% of the energy produced in the country belongs to the National Interconnected System, a controlled interlinked power grid that serves all the states—and firmly regulated by the state [58]. The constant regulatory action balances the energy market and assures stable tariffs for consumers, resulting in continuous energy affordability. Furthermore, the economic performance of the state and wage levels in the target region can impact the energy affordability. Ribeirão Preto sugarcane wages are high, as per the (1) Working conditions and rights indicator. Concluding, the fairly regulated energy market, controlled energy tariffs and the high level of wages in the Ribeirão Preto region are arguments that justified a moderate score for the energy affordability indicator.

3. Net energy balance $I_{2,3,3}$

The net energy balance accounts for the difference between all the energy outputs and inputs throughout the boundaries of the biorefinery supply chain. The power outputs are in the form of bioethanol and surplus electricity which correspond to 440.0 MW and 15.2 MW, respectively. Moreover, the power input is the diesel load that corresponds to 11.6 MW. The net power balance of the biorefinery system supply chain is 443.6 MW, which is reflected in a high for the current indicator.

17.3.2.2.4 Resources conservation

1. Fossil fuels conservation $I_{2,4,1}$

Fossil fuel conservation was evaluated based on the fossil fuel depletion indicator which quantifies the contribution of one system to the global depletion of fossil fuel reserves [5]. The fossil fuel depletion (given in kg of oil equivalent) was estimated based on LCA methodology. An ethanol/gasoline fuel blend 85/15% (v/v) was considered for comparison with the reference system 100% gasoline Euro 4 since this is a major form of ethanol use in Brazil. The biorefinery surplus electricity product was benchmarked against the reference electricity mix grid in Brazil. The results have shown that fossil fuel depletion is reduced by 70% from the gasoline reference system to the biofuel blend. The biorefinery electricity fossil depletion is 74% lower than the grid. For both cases the improvements achieved are considered high, thus, to the authors' judgment, the indicator score was high.

2. Water conservation $I_{2,4,2}$

The biorefinery supply chain activities can impact the local water resources by affecting the availability and quality of freshwater. Therefore, at first, a reflection was carried out on the availability and quality of the local water resources in the region of the biorefinery and watershed conditions [39, 59, 60]. The criticality of Brazil's rivers and watersheds has been assessed by Machado et al. [39], who considers the water resources quantitative and qualitative aspects, and estimates its "criticality." Sao Paulo state presents a satisfactory qualitative-quantitative water resources balance (noncritical) since the state has implemented a sugarcane agroenvironmental zoning to manage water availability and quality [39]. Secondly, the authors looked into the freshwater requirements along the biorefinery chain. Irrigation water for crop cultivation can be neglected in the context of Sao Paulo sugarcane-producing regions, given the insignificant utilization of irrigation practices in this sugarcane region [18, 53]. Sugarcane crop presents the lowest water footprint among the biofuel crops (e.g. oil palm, cassava) and food crops (e.g. rice, soybean, maize, coffee) at the global level [61]. In biorefinery processing, a significant amount of water is provided by the sugarcane stalk itself since more than two-thirds of the cane's weight is water. Still, additional freshwater is required (550 t/h) in the various stages of the biorefinery processing. Finally, soil conservation is crucial in sugarcane cultivation to preserve the optimal soil characteristics (e.g., porosity, shear strength, stability) and keep high rates of infiltration, leading to more effective use of the rainfall and superior water conservation. Brazil was ranked as one of the best countries in the world in the adoption of soil conservation practices based on minimum soil disturbance by practicing reduced tillage, use of organic soil cover and use of crop rotations/associations [62]. Still, the cropland type of cover in the southeast region of Brazil (which includes Sao Paulo) presents high soil losses when compared to other regions in Brazil due to the higher rainfall levels and larger slope gradients in this region [62]. The high erosion levels are associated with the large periods during the preparation of the cane planting at the beginning of the rainy season that bare soils are exposed to erosion [62,63]. Thus, sugarcane experts largely encourage the adoption of best agriculture practices to reduce soil erosion. In the current evaluation, reduced tillage and 85/15% of mechanical/manual sugarcane harvest are assumed, as well as the use of green harvesting residues to cover the fields. Concluding, besides the satisfactory condition of the Sao Paulo watersheds and the application of best agro practices to avoid soil erosion, there are still considerable requirements of freshwater in the industrial facility which contributes to a reduction of local resources. Therefore, a moderate score has been given to the indicator.

3. Landscape and cultural heritage conservation $I_{2,4,3}$

Landscape and cultural heritage conservation consist of collective action of a diversity of stakeholders aiming to conserve ecological systems and their interconnections, cultural heritage, rural communities, urban areas, and other public and private lands being part of a fully integrated whole landscape [64]. The operation of a sugarcane biorefinery industry in a particular area can lead to changes not only in the landscape, but also in the lifestyles and local socioeconomic and cultural activities, and, thus, conservation studies and measures are required prior and during the project lifetime. In the context of Sao Paulo state, sugarcane agricultural regions have been the target of a large number of conservation studies given the importance of the sugar-ethanol industry [65,66]. Compared to other agro-regions in Brazil, the Sao Paulo sugarcane region presents a very reasonable level of landscape conservation and optimization given the long-term implementation of land use management practices, agro-ecological zoning, and policy making toward improved systems conservation. Moreover, sugarcane commercial agriculture has a long history in Sao Paulo and it is one of the most developed segments in terms of agro-practices in the country [18]. The majority of the plantations are medium-large size, with a reduced number of owners per cultivated hectare compared to other regions. Sao Paulo is also one of the most economically and socially developed states of Brazil, with reduced occupation by indigenous communities. Given such characteristics, there is a reduced risk of land grabbing from small farmers or invasion of indigenous cultures as a result of sugarcane expansion. To conclude the reflection, a moderate score has been given to the current indicator, justified by the levels of landscape and culture conservation in Sao Paulo.

17.3.2.2.5 Rural development and workforce

1. Job creation $I_{2,5,1}$

The creation of employment throughout the 1G2G ethanol value chain contributes to rural development. In the ethanol conventional supply chain, the largest fraction of employment corresponds to the work in the agricultural fields due to a large number of people necessary for performing the manual harvest. A strong decrease in employment in the sugarcane sector has been observed in the last decades as a result of the harvest mechanization [38]. Since the renovation scenario assumes only a 15% manual harvest, the number of people employed is expected to be largely reduced when compared to the prior situation. Moreover, the ethanol industry is shifting from basic 1G ethanol production to more optimized 1G and 1G2G scenarios, which are associated with lower levels of employment at the industrial unit [31]. The 1G ethanol conventional industry employs a higher number of people per liter of ethanol produced when compared to optimized scenarios due to the reduced efficiency [31]. Therefore, having as a benchmark the conventional harvest and 1G ethanol production systems, a low score was considered for the current indicator. However, compared to 1G, the integration of 1G2G would improve the ratio of qualified employments. Another comparison was performed based on the industrial employment/total sales ratio (jobs/M US$) in the current biorefinery scenario *versus* the average in the chemical industry. The estimated ratio for the current biorefinery scenario was 0.21, lower than the average employment ratio in the chemical industry ranging from 0.5–4 employees per MUS$ sales [67]. It was not easy to project in which extent 2G would improve the ratio related to ethanol biorefinery.

2. Rural livelihood improvement $I_{2,5,2}$

Rural livelihood can be defined as the capability of a rural community to fulfill its basic needs—housing, safety, access food, water and energy—and improve its well-being [68]. The rural livelihood indicator is transversal to socioeconomic and environmental aspects. Therefore, the evaluation of biorefinery impact on rural livelihood inevitably relies on other indicators of the current study such as social acceptability and affordability, social well-being and prosperity, energy security, and land use, whose average score is moderate. Moreover, the authors investigated two additional aspects that can strongly influence rural livelihood: geographical dispersion of the supply chain and percentage of migrant labor in the sugarcane harvesting activities. The geographical dispersion of the supply chain gives an indication of how concentrated the biorefinery activities are, and on how much the rural communities can profit from it in terms of accessibility to the products and employment. Reduced dispersion indicates more concentrated activities in the rural areas and, therefore, more opportunities. The biorefinery supply chain is geographically spread from the sugarcane fields to the bioethanol collection centers/fueling stations. In the context of Sao Paulo state, the sugarcane average traveling distance from the fields to the biorefinery gate is in the range 25–35 km [27, 69]. Further, the average required distance to flow the bioethanol product from the biorefinery to the collection centers is 100 km [70]. The geographical dispersion of the supply chain is considered reduced for the Sao Paulo state scale since the distances do not exceed 100 km. Moreover, the origin of the workers in the supply chain can have a great impact on rural livelihood. A common practice in the Sao Paulo sugarcane industry with an enormous social negative footprint is the employment of undereducated migrant workers for performing the physical manual harvesting tasks [24,71]. First, as mentioned for the (1) Working conditions and rights indicator, migrant labor is generally underpaid and the working and living conditions are often deplorable. Second, hiring migrant seasonal workers discourages investments in the education and training of potential local labor, compromising social progress in rural areas. The percentage of migrant labor in the sugarcane harvesting activities was estimated for the current case study. According to data in the context of Sao Paulo, it is fair assuming that the manual harvest is entirely performed by migrant labor [24, 47, 71]. About 369 migrant workers are required to harvest 585 kt of sugarcane (15% of the sugarcane required to feed the biorefinery), assuming manual harvest productivity of 12 t/worker/day and a period of the harvest of 132 days [40, 41]. The remaining 85% of the sugarcane (3315 kt) is harvested using machines with average productivity of 1 kt/machine/day [41]. Each harvesting machine generates about 20 vacancies for harvesting operators and tractor drivers [17]. A total of 502 local employees are necessary for the job. Therefore, about 42% of the workers in sugarcane harvest are migrants, which is rather disappointing for the supposed state of sugarcane industry development in Sao Paulo. In conclusion, the negative outcome regarding high migrant labor in the sugarcane harvest activities counterbalances the positive analysis of the concentration of the supply chain, which leads us to a moderate score.

3. Technological innovation and development of skilled workforce $I_{2,5,3}$

In the context of a renovation scenario for the 1G2G ethanol production supply chain, one expects considerable levels of technological innovation due to the increased mechanical harvesting and the enhanced production systems. Training courses and educational projects are required to allow the workers to operate with advanced machinery, which leads to the development of the skilled workforce in rural areas. Policy and new legislation have been introduced in Sao Paulo state to boost the professional requalification of former sugarcane cutters [53]. Furthermore, the implementation of a biorefinery system brings benefits to the local economy. Raw materials such as chemicals and fertilizers will be required from the agrochemical local industries. Likewise, the biorefinery will boost regional machinery manufacturing industries (e.g. sugarcane harvesters), transport vehicles, and construction industries, increasing technological innovation and employees' qualifications. Finally, the services industries associated with the biorefinery system activities will be also enhanced, such as storage, transportation, and trading. A high score was attributed to the current indicator.

17.3.2.3 Environmental specific indicators
17.3.2.3.1 Land use

1. Land use management $I_{3,1,1}$

    Managing land resource use and development is crucial for sustainable development. Land use management tools such as agro-ecological zoning have been introduced in Brazil as mechanisms for ecosystems and habitats protection and conservation [13,72]. Both federal and regional governments recognize the vital importance of such practices for a sustainable sugarcane industry expansion. In the last years, the policy making has been rather effective at zoning the territory suitable for sugarcane expansion and mapping the protection areas where sugarcane cannot be cultivated [72]. The expansion of sugarcane is likely to happen near existing plantations in the region of the southeast of Brazil due to the existing know-how and infrastructures, as well as the proximity to the markets [65]. Therefore, potential sugarcane expansion should have a reduced impact on the country's biodiversity and be no threat to protected ecosystems. Nevertheless, besides a large number of initiatives and policy making development, the success of land use management still relies on how effective the measures are, and on how fast they are implemented. For this reason, a conservative moderate score has been attributed to the current indicator.

2. Land productivity $I_{3,1,2}$

    Land productivity is an indicator of the land's ability to grow and sustain life. Land productivity becomes an important sustainability aspect when it comes to industrial crop production since it directly impacts the area of land required to produce the target tons, and, consequently, the land use and landscape management. Land productivity can be influenced by natural edaphic and climatic conditions, land use, and agricultural practices, ownership and environmental policy [65,73,74]. One major factor inducing sugarcane productivity is the use of irrigation. Yet, such practice is rarely applied due to its associated economic and environmental drawbacks. In the current work, one investigated the land productivity of different regions in Brazil to compare with Sao Paulo. Southeast Brazil and, more specifically, Sao Paulo state present the highest average productivity for sugarcane production, reported as 83 t/ha [65,75–77]. Further, a similar analysis was performed between Brazil and other sugarcane production regions in the world for a wider benchmark. The average sugarcane land productivity considering the entire territory of Brazil is close to Latin America, India, China, and Thailand productivities covering the range 60–73 t/ha, while the world average sugarcane productivity is 57 t/ha [65,75,78–79]. A high score was attributed to the current indicator given the difference between the land productivity in Sao Paulo and the average in Brazil's or other main producing regions in the world.

17.3.2.3.2 Local environment

1. Soil quality $I_{3,2,1}$

Soil quality comprises the physical, chemical, and biological characteristics of the soil. Factors such as climate, topography, type of soil use and practices, can largely influence soil quality and, thus, the local ecosystem dynamics and health. Higher levels of natural soil erosion may occur simply due to larger rainfall levels and slope gradients, which is the case in the southeast region of Brazil [62]. The intense agricultural activity can have a detrimental impact in the soil quality levels by increasing the soil exposure to erosion, decreasing the natural stability of the soil, and the nutrient and carbon contents [21,62–63]. Sugarcane intense cultivation can be particularly problematic to the soils due to the long periods that bare soils are exposed to water erosion during the period of preparation for cane planting [63]. In the context of Sao Paulo, Cherubin et al. [80] evidenced that both pasture and sugarcane cultivation activities (when compared to wild natural vegetation) lead to disturbances in the physical, chemical and biological properties of the soil. The implementation of suitable soil management practices is proposed has the only tactic to alleviate the negative impacts of sugarcane agriculture [19,22,63,77, 80–81]. In the cultivation of sugarcane, the greatest erosion and nutrient losses were linked to the areas where preharvesting burning has occurred [81]. Therefore, it is important to eradicate manual harvesting and its associated pre-harvest burning activities, as well as to promote the use of the green postharvesting residues for protecting from erosion, mulching, and enhancing the structure of the soil [19,22,77]. According to Martins Filho et al. [19], if 50%–100% of the residues are left in the sugarcane fields after harvest, one should be able to control erosion and increase the organic matter in the soil. Another important aspect is the reduction of soil tillage. Concerning this case study, best soil management practices will be implemented in the sugarcane site. Firstly, 85% of the sugarcane harvest is already performed mechanically, meaning that only 15% of the fields are subject to pre-harvest burning. Second, 50% of the green harvesting residues are left in the fields for soil mulching and erosion control (the remaining residues are required in the biorefinery facility to produce energy). Lastly, to avoid tillage, alternative sugarcane replanting techniques will be implemented aiming for reduced soil disturbances.

Two additional elements LCA outputs were analyzed to assess the impact of the biorefinery system: terrestrial acidification and ecotoxicity. Terrestrial acidification accounts for the changes in the soil's chemical properties such as changes in the soil pH. Soil acidification may lead to declining soil fertility and plant photosynthetic rates [82]. Due to the sugarcane cultivation, biomass-derived fuel and electricity have a much more negative impact in terms of terrestrial acidification than the reference systems gasoline and electricity mix grid. Similar conclusions were taken regarding terrestrial ecotoxicity, the indicator measuring the effects of soil chemical changes in the local fauna and flora. Even though the agricultural practices can be improved to promote soil quality, the biorefinery product impacts are very discrepant compared to the benchmark products. Therefore, a low score has been attributed to the soil quality indicator.

2. Air quality $I_{3,2,2}$

One of the air quality key impact factors in the context of the sugarcane industry is the level of combustion gases and particles released due to the preharvest burning of the fields. Typical pollutants from sugarcane burning are aerosols, particulate matter, carbon monoxide, aldehydes, nitrogen oxide, nitrous oxide, and polycyclic aromatic hydrocarbons [83]. Literature shows the effects of such pollutants in the respiratory and general health of humans [83,84]. Currently, in Sao Paulo state, more than 85% of sugarcane harvesting relies on mechanical cutting, meaning that less than 15% of sugarcane fields are subjected to pre-harvest burning [17]. Even though sugarcane pre-harvest burning has not yet been eradicated, those percentages are quite satisfactory when compared with the figures in the last decade, which justifies the attribution of a moderate score to the local air quality indicator. Likewise, sugarcane burning is linked to the formation and release of ozone precursor molecules that may lead to the generation of tropospheric ozone. Tropospheric ozone is a very reactive molecule damaging natural vegetation, forests, crops and, ultimately, human health [85]. One has investigated the levels of tropospheric ozone in Sao Paulo, the most intense sugarcane producing state, having as benchmark Brazil and other countries. Cooper et al. [85] have shown that, in the period 2004–2010, the levels in Sao Paulo corresponded to low-/mid-scale in the global distribution of tropospheric ozone. A similar outcome presented by NASA in 2016 showed relatively low levels of ozone in the troposphere of Sao Paulo/Brazil, compared to the maximum recorded at the global level [86]. In line with these observations, Francisco et al. [16] demonstrated that, even though sugarcane burning contributes to higher emissions of ozone precursor molecules, the seasonal emissions do not significantly impact the local tropospheric ozone concentration.

3 Water quality $I_{3,2,3}$

The quality of the local water resources can be affected by changes in the physical, chemical, biological, and radiological conditions. One preliminary factor is the minimum base flow level and watershed conditions in the region. Reduced base flow conditions can compromise the oxygen diffusion rates, which can lead to oxygen depletion and can compromise the respiration of countless aquatic organisms [21]. Intense agricultural exploitation can be responsible for decreasing base flow levels. As soil quality degrades due to intense use, infiltration rates are reduced which leads to unsteady and decreased base flows. The watershed conditions and base flow levels in Sao Paulo were studied by Martinelli et al. [87], covering the regions of most intense sugarcane production. Half of the largest watersheds in Sao Paulo are above the critical usage level, defined as Total/$Q_{7,10}$, which correspondents to the fraction between the total usage and the minimum weekly discharge over a 10-year return period. Mogi-Guaçu, Pardo, Baixo Grande-Pardo, and Sapucai Mirim-Grande watersheds, which cover and surround Ribeirão Preto plantations, present some of the most critical situations in terms of water availability. For Mogi-Guaçu watershed, the water use by sugarcane mills reaches the highest fraction in the state, which is a critical condition that must encourage water use optimization by the sugar mills in the region [87]. Another important aspect that compromises the quality of freshwater is the increase of the sediment loading levels, usually linked to the occurrence of stormflows, large runoffs, and intense soil erosion [21,62,88]. Unusual nutrient fluxes, high concentration of suspended matter, and increased water turbidity are common in high sediment loading conditions. Anache et al. [62] studied the runoff and erosion levels for several land covers in different regions of Brazil. Soil loss and runoff levels are much higher for cropland cover (including sugarcane) than shrubland (native vegetation). Against fallow type of land cover (bare soil), pasture and grassland, cropland presents also higher soil losses, even though with a lower gap. Moreover, within cropland type of cover, the Southeast region of Brazil (including Sao Paulo) shows higher soil losses than any other region, which can be justified due to the higher average slope gradient. Though, the authors proved that, with the application of the best agro-practices, the soil losses can be reduced to reach the shrubland natural erosion levels [62].

Moreover, two LCA outputs were analyzed to determine to which extent the biorefinery activities affect the local freshwater quality due to eutrophication and/or increase of ecotoxicity levels. Eutrophication is caused by an increase in the nutrient level in freshwater bodies, which induces aquatic biomass overgrowth and, ultimately, generates oxygen depletion and ecosystem disruption. It can occur due to point sources such as discharges of effluents and leachate from a specific site of waste disposal; or nonpoint sources such as runoffs from agricultural fields, pastures, and urban areas [89]. The results show that biofuels lead to higher freshwater eutrophication than the reference fossil-based product gasoline. The reason for such an impact is the agricultural exploitation of sugarcane to feed the biorefinery. The conclusions were similar for the freshwater ecotoxicity. In conclusion, a low score was attributed to the indicator due to the shreds of evidence of critical water availability and base flow conditions in the target region, higher soil losses in Sao Paulo cropland when compared to other regions and other land covers, and larger eutrophication and ecotoxicity impacts associated to the biorefinery products against the conventional products.

17.3.2.3.3 Biodiversity

1. Plant diversity $I_{3,3,1}$

Biodiversity is vital for the functioning of the ecosystems, therefore, it is essential to measure the impact of new economic activities in the conservation of species and habitats. Plant's health and diversity can be affected in a very direct way by soil and water conditions. The quality of the soil will determine the availability of nutrients that feed the vegetation, and water. In turn, the water should be proper to allow plant proliferation. Thus, this indicator is very interconnected to the performance of the (1) soil quality and (3) water quality indicators. Equally important is the application of measures for land use management (studied for 3.2.3.1 Land use indicator) and biodiversity conservation in the target areas. The creation of ecological corridors, buffer zones, and environmental conservation areas where construction and other activities are prohibited, as wells as the reduction of access to certain ecosystems for activities such as hunting and fishing, are some of the measures that can be adopted [59]. According to Machado et al. [39], Sao Paulo's most dense sugarcane cultivation areas are shown to be well covered regarding environmental conservation areas. Still, one of the most important biomes in Brazil—*Cerrado*—could be threatened in medium/long term due to the proximity of the sugarcane cultivation areas and potential sugarcane expansion [38]. Cerrado biome conservation is a concern for Brazil since is an enormous carbon sink with dozens of species threatened by extinction due to intense habitat fragmentation and inefficient preservation [38, 90–92], undervalued in the last decades in comparison to Amazon. Conservation measures have been insufficient and inefficient which leads to the deterioration of about 48% of the Cerrado native vegetation and consequently collapse of local habitats [91]. Ideas and programs do emerge from the academic world, NGOs, supply chain actors, and government, however, one notices inadequate implementation, lack of stakeholder's actions integration, and long-lasting policy making for regulating and monitoring the land-use in Cerrado and surrounding areas [91–92]. Even though in the context of the study there is no sugarcane expansion, a conservative low score is attributed due to the soil and water quality evaluations.

2. Animal diversity $I_{3,3,2}$

Animal diversity is closely related to plant diversity since plants are the base of the food chain, as well as to the soil and water conditions. McBride et al. [21] have proposed to screen the existence of taxa of special concern to measure the impact of bioenergy production on animal biodiversity. The main finding was the negative impact of sugarcane pre-harvest burning in the local insect and bird communities due to gases/particles release and increased air temperature [93]. Likewise, in the Sao Paulo Cerrado area and surroundings, priority should be given to the protection of insects and avifauna, which are the largest groups of organisms [94, 95]. Although nowadays sugarcane cultivation does not represent the main threat for Cerrado species, the situation can change if pre-harvest burning is not eradicated [90]. A low score has been given to the current specific indicator since 15% of the sugarcane is assumed to be still burned in the biorefinery fields, and for consistency with plant diversity, soil and water quality assessments.

3 Habitat and ecological systems $I_{3,3,3}$

The third specific indicator of 3.2.3.3 Biodiversity is related to the conservation of habitats and ecological systems. Ecosystem preservation aims to maintain natural resources and promote plant/ animal diversity. Cerrado is an example of a complex and remarkable ecological system, habitat for over 4800 plant and vertebrate species found nowhere else worldwide [92]. Not that the protection of habitats and ecological systems largely depends on the wide application of land use management and landscape zoning practices at the national or international level, aspects discussed for the 3.2.2.4 Resources conservation and 3.2.3.1 Land use indicators, rather than on unidirectional measures which target a particular plant/animal taxa. Still, a low score has been attributed to the current indicator, as it is very closely related to the two earlier indicators of 3.2.3.3 Biodiversity.

17.3.2.3.4 Global environment

1. Reduction of greenhouse gas emissions $I_{3,4,1}$

The contribution to the greenhouse gas (GHG) emissions was investigated using LCA methodology. As performed for the other LCA indicators, the biorefinery products were benchmarked against conventional reference systems: biofuel blend E85 *against* gasoline Euro 4 and biorefinery surplus electricity *against* electricity mix grid from Brazil. According to the LCA results, for each km traveled using biofuel instead of gasoline, GHG emissions are reduced by about 71%. In turn, for each kW of bioelectricity instead of grid electricity, GHG emissions are reduced by 80%. Brazil committed to reducing GHG emissions by 37% below 2005 levels in 2015 and by 43% in 2030 [96]. The current system offers a percentage of reduction far larger than the target percentage, contributing to achieving the objectives of the country. Additionally, the biorefinery system was benchmarked against relevant 1G2G ethanol production schemes in a similar context provided by Junqueira et al. [27]. Three schemes for three different timeframes are studied: short, medium, and long term. From short to long term, straw recovery and 2G ethanol yield were gradually increased, enzyme cost reduced, and energy cane was progressively introduced as additional feedstock. The scheme most comparable to the ED OF under is the short-term one, whose 1G2G ethanol emits 21.3 kg CO_2 eq. /GJ [27]. ED OF bioethanol GHG emissions are 12.4 kg CO_2 eq. /GJ, significantly lower than the reference case. The authors have attributed a high score to the indicator given the very large percentage of reduction of GHG emissions compared to gasoline and the reference case presented by Junqueira et al. [27].

2. Net nonrenewable energy $I_{3,4,2}$

The net nonrenewable energy indicator is defined as the ratio between the nonrenewable energy consumed and the total energy consumed along the biorefinery chain.

The biorefinery conversion plant is designed to be self-sufficient in terms of energy, meaning that all the energy consumed within the gates is produced in the cogeneration unit and is not derived from fossil fuels. The power requirements in the facility reached 66 MW. However, non-renewable diesel fuel is still required in agricultural and transport activities. In the context of the sugarcane industry in Sao Paulo state, an investigation was conducted to identify the common practices in terms of seedling application, fertilizer distribution, sugarcane handling and harvesting, type of transport vehicles, and average distances traveled by feedstock and products. One estimated the diesel consumption in sugarcane cultivation and transport activities correspond to the equivalent of 11.6 MW. The nonrenewable energy consumed along the supply chain corresponds to about 15% of the total energy consumed, which justifies a high score for the specific indicator net non-renewable energy.

17.3.3 Values of general indicators, determinants, and sustainability grade

The scores of the specific indicators were used as the inputs for the LBM to calculate the general indicators, determinants, and, finally, the biorefinery system sustainability grade (Table 17.3). The biorefinery system is characterized by a moderate score in both economic and social determinants. The economic viability of the whole biorefinery is moderate, even though it is sensitive to oil market volatility. The reflection on the social aspects of the biorefinery in the context of Ribeirão Preto shows balanced social impacts, where negative aspects such as lower employment creation are outweighed by positive outcomes such as improved working conditions. The score of the environmental determinant is low mainly due to the harmful local impacts in soil and water quality, and the concerns about biodiversity and ecosystems conservation in the sugarcane production region. Finally, the proposed LBM estimates a Moderate sustainability grade for the biorefinery system.

17.4 Conclusions and perspectives

An LBM was developed to characterize the sustainability of a sugarcane biorefinery. This chapter presents a detailed description of the variables and the justification of the qualitative values attributed to the specific indicators, the inputs for the LBM. The model assessed the biorefinery as moderately sustainable (moderate sustainability grade). However, this result depends on the design of the rule-base and the qualitative evaluation of the indicators. Contrary to the possible expectation of a unique and quantitative assessment of sustainability as a criterion of rigor, sustainability is fundamentally a qualitative concept. Its assessment must be transparent, relevant, consistent, and contestable. By emphasizing these requirements, the LBM is a promising method for characterizing biorefineries' sustainability.

Table 17.3 LBM hierarchic inputs and output network.

Sustainability grade																														
2																														Outputs
D_1					D_2													D_3												
2					2													1												
$I_{1,1}$				$I_{1,2}$			$I_{2,1}$			$I_{2,2}$			$I_{2,3}$			$I_{2,4}$			$I_{2,5}$			$I_{3,1}$			$I_{3,2}$		$I_{3,3}$			$I_{3,4}$
2				2			2			2			3			2			2			3			1		0			3
1	2	3	4	1	2	3	1	2	3	1	2	3	1	2	3	1	2	3	1	2	3	1	2	1	2	3	1	2	3	1	2	Inputs
2	2	2	1	2	2	2	2	2	2	2	2	3	2	3	3	3	2	2	1	2	3	2	3	1	2	1	1	1	1	3	3	

Possible scores of specific indicators: 1(low), 2 (moderate), 3(high). Possible scores of general indicators and determinants: 0 (very low), 1(low), 2(moderate), 3(high).

References

[1] B. Ness, E. Urbel-Piirsalua, S. Anderbergd, L. Olssona, Categorising tools for sustainability assessment, Ecol. Econ. 60 (3) (2007) 498–508.

[2] J. Moncada, J.A. Posadam, A. Andrea Ramírez, Early sustainability assessment for potential configurations of integrated biorefineries. Screening of bio-based derivatives from platform chemicals, Biofuel, Bioprod. Bioref. 9 (2015) 722–748.

[3] A.D. Patel, K. Koen Meesters, H. den Uil, Ed. de Jong, K. Bloka, M.K. Patel, Sustainability assessment of novel chemical processes at early stage: application to biobased processes, Energy Environ. Sci. 5 (2012) 8430.

[4] A. Bertoni, S.I. Hallstedt, S.K. Dasari, P. Andersson, Integration of value and sustainability assessment in design space exploration by machine learning: an aerospace application, Des. Sci. 6 (2020) 1–32.

[5] E. Gnansounou, C.M. Alves, E. Ruiz Pachón, P. Vaskan, Comparative assessment of selected sugarcane biorefinery-centered systems in Brazil: a multi-criteria method based on sustainability indicators, Bioresour. Technol. 243 (2017) 300–610.

[6] E. Gnansounou, Assessing the sustainability of biofuels: a logic-based model, Energy 36 (2011) 2089–2096.

[7] N. Brownrigg, J. Zhang, Jump Start: Aspen Energy Analyzer V8. A Brief Tutorial (and supplement to training and online documentation), Aspen Technology, Burlington, VT, 2013.

[8] M. Ebrahim, Pinch technology: an efficient tool for chemical-plant energy and capital-cost saving, Appl. Energy 65 (2000) 45–49.

[9] WCED, Brundtland Report, Our common future. United Nations World Commission on Environment and Development, Oxford University Press, New York, 1987.

[10] E. Gnansounou, P. Vaskan, E. Ruiz Pachón, Comparative techno-economic assessment and LCA of selected integrated sugarcane-based biorefineries, Bioresour. Technol. 196 (2015) 364–375.

[11] D. Humbird, R. Davis, L. Tao, C. Kinchin, D. Hsu, A. Aden, P. Schoen, J. Lukas, B. Olthof, M. Worley, D. Sexton, D. Dudgeon, Process Design and Economics for Biochemical Conversion of Lignocellulosic Biomass to Ethanol—Dilute-Acid Pretreatment and Enzymatic Hydrolysis of Corn Stover, NREL, Golden, CO (United States), 2011.

[12] R. de Oliveira Bordonal, R. Lal, D.A. Aguiar, E.B. de Figueiredo, N. La Scala, Greenhouse gas balance from cultivation and direct land use change of recently established sugarcane (*Saccharum officinarum*) plantation in south-central Brazil, Renew. Sust. Energy 52 (2015) 547–556.

[13] E. Verde, Zoneamento Agroambiental para o Setor Sucroalcooleiro, Sao Paulo State Government, 2017. http://www.ambiente.sp.gov.br/etanolverde/zoneamento-agroambiental/ Accessed February 15, 2020.

[14] G. Kohlhepp, Analysis of the ethanol and biodiesel production in Brazil, Estudos Avançados 24 (2010) 223–253 (in Portuguese).

[15] Unica, Agro-environmental protocol of Sao Paulo's sugar-energetic sector: data from 2007/2008 to 2013/2014 harvesting. http://www.iea.sp.gov.br/Relat%F3rioConsolidado1512.pdf, 2014 (Accessed February 15, 2020).

[16] A.P. Francisco, D.D.S. Alvim, L.V. Gatti, C.R. Pesquero, J.V.D. Assunção, Tropospheric ozone and volatile organic compounds on a region impacted by the sugarcane industry, Química Nova 39 (2016) 1177–1183 (in Portuguese).

[17] A. C. Fábio, Rural workers give way to a cane harvester operator that earns up to 2.6 k R$. São Paulo. https://economia.uol.com.br/agronegocio/noticias/redacao/2013/11/29/boia-fria-da-lugar-a-operador-de-colhedora-de-cana-que-ganha-ate-r-26-mil.htm#fotoNav=2, 2013 (Accessed February 15, 2020).

[18] Sugarcane, Best Cultivation Practices. http://sugarcane.org/sustainability/best-practices, 2018 (Accessed February 15, 2020).

[19] M.V. Martins Filho, T.T. Liccioti, G.T Pereira, J. Marques Júnior, R.B. Sanchez, Soil and nutrients losses of an alfisol with sugarcane crop residue, Eng. Agríc. 29 (2009) 8–18.

[20] D.A. Ferreira, H.C. Franco, H.R. Otto, A.C. Vitti, P.C. Trivelin, Contribution of N from green harvest residues for sugarcane nutrition in Brazil, GCB Bioenergy 8 (2016) 859–866.

[21] A.C. McBride, V.H. Dale, L.M. Baskaran, M.E. Downing, P.E. Storey, J. M. Indicators to support environmental sustainability of bioenergy systems, Ecol. Ind. 11 (2011) 1277–1289.

[22] E.F. Luca, V. Chaplot, M. Mutema, C. Feller, H.T.Z. Couto, Effect of conversion from sugarcane pre-harvest burning to residues green-trashing on SOC stocks and soil fertility status: results from different soil conditions in Brazil, Geoderma 310 (2018) 238–248.

[23] E. Gnansounou (Ed.), Economic Assessment of Biofuels, in Biofuels—alternative Feedstocks and Conversion Processes for the Production of Liquid and Gaseous Biofuels, 2nd Edition, Elsevier Inc., 2019, pp. 95–121.

[24] L.A. Rosa, V.L. Navarro, Trabalho e trabalhadores dos canaviais: perfil dos cortadores de cana da região de Ribeirão Preto (SP), Cad. Psicol. Soc. Trab. 17 (2014) 143–160 (in Portuguese).

[25] T. Serra, D. Zilberman, J. Gil, Price volatility in ethanol markets, Eur. Rev. Agr. Econ. 38 (2010) 259–280.

[26] Eia.gov, What drives crude oil prices: spot prices? US Energy Information Administration, 2018. https://www.eia.gov/finance/markets/crudeoil/spot_prices.php Accessed February 15, 2020.

[27] T.L. Junqueira, M.F. Chagas, V.L. Gouveia, M.C. Rezende, A. Bonomi, Techno-economic analysis and climate change impacts of sugarcane biorefineries considering different time horizons, Biotechnol. Biofuels 10 (2017) 1–12.

[28] S. Fankhauser, S. Tepic, Can poor consumers pay for energy and water? An affordability analysis for transition countries, Energy Policy 35 (2007) 1038–1049.

[29] G. E. Lee, S. Loveridge, S. Joshi, Local Acceptance and Heterogeneous Externality of biorefineries: A Case Study from the State of Michigan. https://www.canr.msu.edu/afre/uploads/files/Gi-Eu_Lee_Job_Market_Paper_-_Biorefinery_01.pdf, 2015 (Accessed February 15, 2020).

[30] I. de Carvalho Macedo, Sugar cane's energy: twelve studies on Brazilian sugar cane agribusiness and its sustainability, UNICA—Sugar Cane Agriindustry Union—Benedis Editions Lida São Paulo Brazil, 2005.

[31] L.S.V.D. Souza, L.F.C. Nascimento, Air pollutants and hospital admission due to pneumonia in children: a time series analysis. Revista Associação Médica Brasileira 62 (2016) 151–156.

[32] IBGE, Ribeirão Preto statistics, Brazilian Institute of Geography and Statistics. https://cidades.ibge.gov.br/brasil/sp/ribeirao-preto/panorama, 2017 (Accessed February 15, 2020).

[33] L.A. Martinelli, R. Garrett, S. Ferraz, R. Naylor, Sugar and ethanol production as a rural development strategy in Brazil: Evidence from the state of São Paulo, Agric. Syst. 104 (2011) 419–428.

[34] E.D.M. Rico, Corporate social responsibility and the state: an alliance for sustainable development, São Paulo Perspec 18 (2004) 73–82 (in Portuguese).

[35] ibase. Social balance model. http://www.balancosocial.org.br/cgi/cgilua.exe/sys/start.htm, 2017 (Accessed February 15, 2020).

[36] R. Sanyal, Determinants of bribery in international business: the cultural and economic factors, J. Bus. Ethics 59 (2005) 139–145.

[37] Transparency International, Corruption perceptions index 2016. https://www.transparency.org/news/feature/corruption_perceptions_index_2016, 2017 (Accessed February 15, 2020).

[38] E. Smeets, M. Junginger, A. Faaij, A. Walter, W. Turkenburg, The sustainability of Brazilian ethanol—an assessment of the possibilities of certified production, Biomass Bioenergy 32 (2008) 781–813.

[39] P.G. Machado, N.A.M. Rampazo, M.C.A. Picoli, C.G. Miranda, K.R.E. de Jesus, Analysis of socioeconomic and environmental sensitivity of sugarcane cultivation using a Geographic Information System, Land Use Policy 69 (2017) 64–74.

[40] J.G. Baccarin, J.J. Gebara, B.M. Silva, Acceleration of mechanical harvesting and its effects on sugarcane formal occupation in the state of São Paulo, 2007–2012, Informações Econômicas 43 (2013) 19–31 (in Portuguese).

[41] L.M. Moreno, 2011. Transition from manual to mechanical harvest of sugarcane in the Sao Paulo: scenarios and perspectives. Dissertation from Sao Paulo University (in Portuguese). Sao Paulo, Brazil.

[42] I.C. Macedo, J.E.A. Seabra, J.E.A.R. Silva, Greenhouse gases emissions in the production and use of ethanol from sugarcane in Brazil: The 2005/2006 averages and a prediction for 2020, Biomass Bioenergy 32 (2008) 582–595.

[43] A. Barber, G. Pelloy, M.A. Pereira, The sustainability of brazilian sugarcane bioethanol: a literature review, AgriLINK NZ (2008). https://www.bioenergy.org.nz/ (Accessed February 15, 2020).

[44] ILAB, List of Goods Produced by Child Labor or Forced Labor, Bureau of International Labor Affairs, 2016. https://www.dol.gov/ilab/reports/child-labor/list-of-goods/goods/?q=sugarcane (Accessed February 15, 2020).

[45] S. McGrath, Fuelling global production networks with slave labour? Migrant sugar cane workers in the Brazilian ethanol GPN, Geoforum 44 (2013) 32–43.

[46] N. Phillips, Unfree labour and adverse incorporation in the global economy: comparative perspectives on Brazil and India, Econ. Soc. 42 (2013) 171–196.

[47] J.R.P. Novaes, Productivity champions: pains and fevers in the sugarcane plantations in Sao Paulo, Estudos Avançados 21 (2007) 167–177 (in Portuguese).

[48] A.Z. Fernandes, Statistical Yearbook of Occupational Accidents (in Portuguese). Brasilia (2015).

[49] F.R. Marin Jr, G.B. Martha, K.G. Cassman, P. Grassini, Prospects for increasing sugarcane and bioethanol production on existing crop area in Brazil, BioScience 66 (2016) 307–316.

[50] ANP, Petroleum fuel derivatives sales. Brazil's National Agency of Petroleum, Natural Gas Biofuels (2017). http://www.anp.gov.br/wwwanp/images/DADOS_ESTATISTICOS/Vendas_de_Combustiveis/Vendas_de_Combustiveis_m3.xls Accessed February 15, 2020.

[51] ANP, Petroleum fuel derivatives imports and exports statistics. Brazil's National Agency of Petroleum, Natural Gas Biofuels (2017). http://www.anp.gov.br/wwwanp/images/DADOS_ESTATISTICOS/importacao_exportacao/Importacoes_Exportacoes_m3.xlsx Accessed February 15, 2020.

[52] ANP. Ethanol production statistics. Brazil's National Agency of Petroleum, Natural Gas and Biofuels. http://www.anp.gov.br/wwwanp/images/DADOS_ESTATISTICOS/Producao_etanol/Producao-de-Etanol-m3.xls, 2017 (Accessed February 15, 2020).

[53] J. Goldemberg, S.T. Coelho, P. Guardabassi, The sustainability of ethanol production from sugarcane, Energy Policy 36 (2008) 2086–2097.

[54] A. Moors, J. DeCicco, Energy affordability: an overview, Energy Instit. Michigan. (2016).

[55] U. Dubois, H. Meier, Energy affordability and energy inequality in Europe: implications for policy-making, Energy Res. Social Sci. 18 (2016) 21–35.

[56] FIESP, State of Sao Paulo overview, Department of Trade and Foreign Affairs, 2017. https://www.mzv.cz/file/1737832/Apresentac_o_Estado_de_S_o_Paulo.pdf Accessed February 15, 2020.

[57] M. Teixeira. The Brazilian energy distribution system. TechninBrazil. https://techinbrazil.com/the-brazilian-energy-distribution-system, 2016 (Accessed February 15, 2020).

[58] Export. Brazil—Electrical Power Systems. Brazil Country Commercial Guide. https://www.export.gov/article?id=Brazil-Electrical-Power-Systems., 2017 (Accessed February 15, 2020).

[59] N. Rettenmaier, G. Hienz, A. Schorb, R. Diaz-Chavez, D. Rutz, R. Janssen, Strategies for the harmonization of environmental and socio-economic sustainability criteria, Global-Bio-Pact. Deliverable D 5 (2012).

[60] G.M. Souza, M.V.R. Ballester, C.H. de Brito Cruz, H. Chum, R. Maciel Filho, The role of bioenergy in a climate-changing world, Environ. Devel. 23 (2017) 57–64.

[61] P. Kaenchan, S.H. Gheewala, A review of the water footprint of biofuel crop production in Thailand, J. Sustain. Energy Environ. 4 (2013) 45–52.

[62] J.A. Anache, E.C. Wendland, P.T. Oliveira, D.C. Flanagan, M.A. Nearing, Runoff and soil erosion plot-scale studies under natural rainfall: a meta-analysis of the Brazilian experience, Catena 152 (2017) 29–39.

[63] S. Filoso, J.B. do Carmo, S.F. Mardegan, S.R.M. Lins, S.R.M.L.A. Martinelli, Reassessing the environmental impacts of sugarcane ethanol production in Brazil to help meet sustainability goals, Ren. Sustain. Energy Rev. 52 (2015) 1847–1856.

[64] NLC. What is landscape conservation? Network for landscape conservation. http://www.largelandscapenetwork.org/about/landscape-conservation-what/, 2017 (Accessed February 15, 2020).

[65] D.M. Lapola, J.A. Priess, A. Bondeau, Modeling the land requirements and potential productivity of sugarcane and jatropha in Brazil and India using the LPJmL dynamic global vegetation model, Biomass Bioenergy 33 (2009) 1087–1095.

[66] C.M. Kennedy, D.A. Miteva, L. Baumgarten, P.L. Hawthorne, J. Kiesecker, Bigger is better: improved nature conservation and economic returns from landscape-level mitigation, Sci. Adv. 2 (2016) 1–9.

[67] AIChE, Social responsibility, AIChE Sustain. Index (2017). https://www.aiche.org/ifs/resources/sustainability-index/social-responsibility Accessed February 15, 2020.

[68] F.A. Mphande, Rural livelihood, in: F.A. Mphande (Ed.), Infectious diseases and rural livelihood in developing countries, Springer, Singapore, 2016, pp. 17–34.

[69] R.F. Françoso, A. Bigaton, H.J.T. da Silva, P.V. Marques, Ratio of the cost of transportation of sugarcane due to the distance, Revista iPecege 3 (2017) 100–105.

[70] Novacana. Ethanol logistics: infrastructure and transportation for the export of ethanol. https://www.novacana.com/etanol/logistica-infraestrutura-transporte/, 2017 (Accessed February 15, 2020).

[71] D.M.P. Nunes, M.S.D. Silva, R.D.L.M. Cordeiro, Work and risk experience among Northeastern migrant works in sugarcane plantations in the state of São Paulo, Brazil, Saúde Soc. 25 (2016) 1122–1135.

[72] C. S. Gurgel. Legal-environmental approach to the agroecological zoning of sugarcane in Brazil. https://jus.com.br/artigos/23964/abordagem-juridico-ambiental-do-zoneamento-agroecologico-da-cana-de-acucar-no-brasil, 2013 (Accessed February 15, 2020).

[73] G. Kaschuk, O. Alberton, M. Hungria, Quantifying effects of different agricultural land uses on soil microbial biomass and activity in Brazilian biomes: inferences to improve soil quality, Plant Soil 338 (2011) 467–481.

[74] J. Assunção, J. Chiavari, Towards efficient land use in Brazil, Climate Policy Initiative, 2015, pp. 1–28.

[75] C. Monfreda, N. Ramankutty, J.A. Foley, Farming the planet: 2. Geographic distribution of crop areas, yields, physiological types, and net primary production in the year 2000, Global Biogeochem. Cycles 22 (2008) 1–19.

[76] A.D.C. Teixeira, J.F Leivas, C.C. Ronquim, D.D.C Victoria, Sugarcane water productivity assessments in the São Paulo state, Brazil. Int. J. Remote Sens. App. 6 (2016) 84–95.

[77] F.C Dalchiavon, M.D.P. Carvalho, R. Montanari, M. Andreotti, E.A. Dal Bem, Sugarcane trash management assessed by the interaction of yield with soil properties, Rev. Bras. Ciência Solo 37 (2016) 1709–1719.

[78] L. Peng, P.A. Jackson, Q.W. Li, H.H. Deng, Potential for bioenergy production from sugarcane in China, Bioenergy Res. 7 (2014) 1045–1059.

[79] T. Silalertruksa, S.H. Gheewala, Environmental sustainability assessment of bio-ethanol production in Thailand, Energy 34 (2009) 1933–1946.

[80] M.R. Cherubin, D. L.Karlen, C.E. Cerri, A.L. Franco, C.C Cerri, Soil quality indexing strategies for evaluating sugarcane expansion in Brazil, Plos One 11 (2016) e0150860 1-26.

[81] N. S. de Andrade, M.V. Martins Filho, J. L.Torres, G.T. Pereira, J. Marques Júnior. Economic and technical impact in soil and nutrient loss through erosion in the cultivation of sugar can. Eng. Agrícola (2011) 31, 539-550.

[82] L.B. Azevedo, P. Roy, F. Verones, R. van Zelm, M.A.J. Huijbregts. Terrestrial acidification. Report project LC-Impact methodology, 2013. https://lc-impact.eu/doc/deliverables/Terrestrial_acidification.pdf (Accessed February 15, 2020).

[83] M.L.D.S. Paraiso, N. Gouveia, Health risks due to pre-harvesting sugarcane burning in São Paulo State, Brazil, Revista Brasileira de Epidemiologia 18 (2015) 691–701.

[84] P.R.D.S. Pestana, A.L.F. Braga, E.M.C. Ramos, A.F. de Oliveira, D. Ramos, Effects of air pollution caused by sugarcane burning in Western São Paulo on the cardiovascular system, Rev. Saúde Pública 51 (2017) 1–8.

[85] O.R. Cooper, D.D. Parrish, J. Ziemke, N.V. Balashov, V. Naik, Global distribution and trends of tropospheric ozone: an observation-based review, Elem. Sci. Anth 2 (2014) 1–28.

[86] J.R. Ziemke, N.A. Kramarova, P.K. Bhartia, D.A. Degenstein, M.T. Deland. Highlights from the 11-year record of tropospheric ozone from OMI/MLS and continuation of that long record using OMPS (2016). In European Geosciences Union General Assembly 2016.

[87] L.A. Martinelli, S. Filoso, C. de Barros Aranha, S.F. Ferraz, P.B. de Camargo, Water Use in Sugar and Ethanol Industry in the State of São Paulo (Southeast Brazil), J. Sust. Bioenergy Syst. 3 (2013) 135–142.

[88] K. Edelmann, R.L. Nóbrega, G. Gerold, Stormflow influence on nutrient dynamics in microcatchments under contrasting land use in the Cerrado and Amazon Biomes, Brazil, Geophys. Res. Abstr. 19 (2017) EGU2017–EG16094.

[89] V.H. Smith, G.D. Tilman, J.C Nekola, Eutrophication: impacts of excess nutrient inputs on freshwater, marine, and terrestrial ecosystems, Env. Pollution 100 (1999) 179–196.

[90] G. Durigan, M.F.D. Siqueira, G.A.D.C. Franco, Threats to the Cerrado remnants of the state of São Paulo, Brazil, Scientia Agricola 64 (2007) 355–363.

[91] R.S. Ganem, J.A. Drummond, J.L.D.A. Franco, Conservation polices and control of habitat fragmentation in the Brazilian Cerrado biome, Ambiente Sociedade 16 (2013) 99–118.

[92] B.B. Strassburg, T. Brooks, R. Feltran-Barbieri, A. Iribarrem, B. Soares-Filho, Moment of truth for the Cerrado hotspot, Nat. Ecol. Evol 1 (2017) 1–3.

[93] R.A. Araújo, A.H. Gonring, R.N. Guedes, Impact of controlled burning of sugarcane straw on a local insect community, Neotrop. Entomol 34 (2005) 649–658.

[94] L.T. Manica, M. Telles, M.M. Dias, Bird richness and composition in a Cerrado fragment in the State of São Paulo, Braz. J. Biol 70 (2010) 243–254.

[95] A. Consorte-McCrea, E.F. Santos (Eds.), Policy Intervention in the Cerrado Savannas of Brazil: Changes in the Land Use and Effects on Conservation, Ecology and Conservation of the Maned Wolf: Multidisciplinary Perspectives, CRC press, Brasilia, Brazil, 2014, pp. 293–308.

[96] FRB. Intended nationally determined contribution towards achieving the objective of the United Nations Framework Convention on Climate Change. Federative Republic of Brazil (FRB). http://www4.unfccc.int/submissions/INDC/Published%20Documents/Brazil/1/BRAZIL%20iNDC%20english%20FINAL.pdf, 2015 (Accessed February 15, 2020).

[97] BBC. Brazil: Key Facts and Figures. http://news.bbc.co.uk/2/hi/americas/8702891.stm, 2010 (Accessed February 15, 2020).

[98] IBGE. A quarter of the population lives with less than R$387 per month. Brazilian Institute of Geography and Statistics. https://agenciadenoticias.ibge.gov.br/agencia-noticias/2012-agencia-de-noticias/noticias/18825-um-quarto-da-populacao-vive-com-menos-de-r-387-por-mes.html 2017 (Accessed February 15, 2020).

CHAPTER EIGHTEEN

Solid biofuels

Ashish Manandhar, Seyed Hashem Mousavi-Avval, Jaden Tatum, Esha Shrestha, Parisa Nazemi, Ajay Shah

Department of Food, Agricultural and Biological Engineering, The Ohio State University, Wooster, OH, United States

18.1 Introduction

Solid biofuels are derived from nonfossil, organic materials, including plant biomass, animal waste, and municipal waste, which can be used to produce energy for heating, cooking, or electricity generation. Solid biofuel in the form of wood has been used since fire was discovered by humans for cooking, heating, and other applications. In recent years, solid biofuels are obtained from multiple sources including forest products, animal and agricultural residues, and dedicated energy crops.

Renewable energy sources provided approximately 17.7% of global energy consumption in 2017 [1]. Approximately 70% of the renewable energy was contributed from bioenergy sources, equivalent to 55.6 EJ of biomass [1]. Solid biofuels contributed almost 86% of the bioenergy use in the form of wood chips, wood pellets, and fuelwood for cooking and heating [1]. In 2017, 596 TWh of electricity was generated from biomass-based sources, making it the 3rd largest renewable electricity source after hydropower and wind [1]. Most of this electricity is produced using solid biofuels. Globally, biomass is a dominant source of renewable energy. Biomass contributed to approximately 96% of the renewable energy supply in Africa, 65% in Asia, and 59% in the Americas and Europe (Table 18.1).

Solid biofuels are expected to be a dominant energy source in the future as well. The IEA (2012) predicts a global solid biofuel use of 160 EJ by 2050, of which 100 EJ would be for heat and power generation [3]. This energy would provide 7.5% of global power demand, 15% of global industrial heat demand, and 20% of residential heating. Depending on different critical issues such as competition for land and water resources, use of fertilizers, pest control measures, and competition with food and feed production, the global potential for biomass-derived energy in 2050 can vary from 0 to 700 EJ for energy crops on agricultural lands, <60–110 EJ for biomass on marginal land, 15–70 EJ for agricultural residues, 30–150 EJ for forest residues, 5–55 EJ for dung, and 5–50 EJ for organic wastes [4]. The use of solid biofuels for energy currently as well as in the future will also depend on policies implemented for solid biofuels.

Biomass, Biofuels, Biochemicals
DOI: https://doi.org/10.1016/B978-0-12-819242-9.00017-8

Table 18.1 Share of renewables and biomass in primary energy supply in 2017 [2].

	Total energy (EJ)	Renewables (EJ)	Biomass (EJ)
Africa	24.5	16.1	15.4
Americas	92.1	18.3	10.8
Asia	170.0	33.2	21.6
Europe	77.0	12.7	7.5
Oceania	3.9	0.7	0.3
World	370.0	81.0	55.6

18.2 Solid biofuel types

Solid biofuels can be obtained from primary or secondary sources (Table 18.2). Primary sources include biomass obtained directly for the purpose of producing solid biofuel (e.g., energy crops and woody biomass from forests and plantations). Secondary sources include biomass obtained as a byproduct from other economic activities, such as agricultural and forest residues, organic fractions of municipal wastes, and animal dung. Final forms of solid biofuels include unprocessed biomass (e.g., fuelwood), minimally processed biofuel (e.g., wood chips and sawdust), densified biofuel (e.g., briquettes and pellets), or thermochemically upgraded biomass (e.g., biochar, torrefied biomass, and torrefied pellets). Biomass can be upgraded to achieve desired density, energy content, size, and composition. The final biofuel forms have specific properties which impact their handling, storage, and use. Biomass upgrading can help provide a homogeneous feedstock for consistent energy production, increase feedstock density which can reduce transportation and storage costs, and decrease plant maintenance costs [5].

Table 18.2 Types of feedstock for solid biofuels [6–8].

Feedstock category	Feedstock type	Examples
Primary	Dedicated energy crops	Woody crops	Small round wood, willow, poplar
		Herbaceous energy crops	Switchgrass, miscanthus, common reed, giant reed
	Forestry	Forestry byproducts	Bark, wood blocks, wood chips from tops and branches, wood chips from thinning, logs from thinning
Secondary	Agricultural residues	Lignocellulosic agricultural residues	Corn stover, wheat straw, sugar beet leaves
		Dry livestock waste	Solid manure
	Industry wastes	Wood industry residues	Bark, sawdust, wood chips, slabs, off-cuts
		Food industry residues	Wet cellulosic material (beet root tails), tallow, slaughterhouse wastes
		Contaminated waste	Demolition wood, municipal waste

18.2.1 Unprocessed solid biofuels

Unprocessed solid biofuels include forestry products, agricultural residues, and livestock wastes. They are usually in easy handling sizes and forms and can be burned without any prior processing. However, for forestry products and agricultural residues, as their size increases, the need for processing gains more importance.

Unprocessed solid biofuels are used without any additional processing in the form obtained from the sources. Out of the total biomass used for bioenergy applications in the world, 65% is used as traditional unprocessed biomass such as fuelwood, charcoal and animal dung [9]. Unprocessed biofuels are used for cooking and heating in rural and peri-urban areas and represent the main source of energy for many people in developing countries.

18.2.1.1 Dry animal manure

Animal manure is known to be a rich source of nutrients for soil amendment, however, in rural regions of developing countries where energy resources are limited, it plays an important role in supplying energy mainly for cooking. Dry animal manure is defined as dung cake having less than 30% moisture content which makes it easier to burn. Depending on the animal, their food, and the season, energy content of this resource can vary. A study [10] indicated the average lower calorific value of cow dung over different seasons to be 16.34 MJ/kg. It is estimated that more than 2.5 billion people in developing countries burn dry animal dung as one of their energy sources for cooking, and this number is expected to increase to 2.7 billion by 2030 in a scenario with no change in policies and available technologies [11–14].

18.2.1.2 Fuelwood

Fuelwood, or forestry resources are in high demand in developing countries. The availability of fuelwood made it one of the major energy resources. However, studies show an increase in the use of these resources which later will cause severe deforestation in developing countries and an increase in the price of fuelwood. It is estimated that fuelwood consumption will decrease to less than 50% by 2040; however, there will still be around 650 million people in sub-Saharan Africa using such resources for heating and cooking [15]. Thus, many studies are focusing on finding solutions to make a balance in supply-demand chain of fuelwood and prevention of forest depletion [15–18].

18.2.1.3 Wood and agricultural industry residues

Residues from wood and agricultural industries are considered nonprocessed solid biofuels as they are used directly as residues without processing. Saw dust and bark residues from wood processing industries are examples of this type of solid biofuels. They are produced by cutting or chipping larger forms of wood like forest woods. Wood residues are directly burned in combustion chambers to produce energy. However, compared to wood pellets they have higher moisture content and lower energy content [19]. As mentioned earlier, there is an increasing gap between the supply and demand of wood resources, thus, use of agricultural wastes like rice husk and bagasse can decrease this gap. Rice husk is a byproduct of the rice production industries and the production rate for it was estimated to be more than 10 million tons in 2012 [20]. An increase in use of agricultural resources will help prevent forest depletion [21]. Bagasse is the fibrous content left after processing sugarcane. It can play a crucial role in fulfilling the global energy demand due to the worldwide production of sugarcane and its high calorific value of 8–18 MJ/kg which makes it an efficient source of energy [22–26].

18.2.2 Minimally processed solid biofuels

This group of solid biofuels are those which are almost ready to use. However, they are subjected to minor processes to increase the efficiency as fuels. Wood chips and municipal solid waste (MSW) are categorized in this group.

18.2.2.1 Wood chips
Wood chips are smaller pieces of wood formed after cutting larger wood logs and residues. To have a uniform chip size and an increased surface area, larger pieces of wood go through a wood-chipper to make smaller chips with dimensions of 20–50 mm. Wood chips with high moisture content may undergo an additional step of drying [27,28].

18.2.2.2 Municipal solid waste
MSW refers to waste generated by household, commercial, or light industrial sources. The major components in MSW are plastics, food, paper, metals, and garden wastes [29,30]. MSW is considered as a solid biofuel when it is used for energy recovery in a process known as incineration, which involves removal of glass and metal components from MSW and combustion. The share of different waste types varies based on the income of countries and their location (Table 18.3) [30,31]. Hoornweg and Bhada-Tata [31] sorted different countries around the world based on their GDP income and classified them into four groups of lower, lower-middle, upper-middle, and high income. In countries with lower income, the major fraction of wastes is organic contents

Table 18.3 Composition of MSW [30,31].

Waste type	Composition (%)
Organic	28.0–72.0
Paper	4.0–31.0
Plastic	4.0–13.0
Glass	1.0–7.0
Metals	0.2–5.0
Other	10.0–47.0

(up to 72%), while the share of glass and metal are minimum (1% and 0.2%, respectively). However, in high-income countries, while paper is the most dominant content of wastes (31%), the average share of organic content is 28% and the shares of glass and metal are 7% and 5%, respectively. The glass and metal fractions cannot be processed and burned during incineration. Thus, prior to using MSW as a solid biofuel, it usually goes through separation steps including use of a magnetic separator and manual selection to remove components undesired for use as solid biofuels.

18.2.3 Processed solid biofuels

Processed solid biofuels have gained a lot of interests since the final products will have upgraded and uniform characteristics after the process compared to unprocessed biomass. Characteristics like lower moisture content, higher surface area, higher density, higher carbon content, result in fuel with higher calorific value, the lower energy requirement for grinding, and improved transportability and storability.

18.2.3.1 Torrefied biomass and hydrochar

Torrefied biomass is pyrolyzed biomass, created using the torrefaction. Torrefaction is the process of heating the biomass at 200–300°C in absence of oxygen. This process includes devolatilization, which is the production of noncondensable gases like CO_2 and O_2 as well as condensable gases like water vapor, volatile organic compounds and lipids, followed by depolymerization, and carbonization of hemicellulose, lignin and cellulose [32,33]. After torrefaction, about 70% of solid biomass is retained as solids with higher surface area, higher calorific value, higher carbon content, and improved grindability and pelletability [34,35].

Hydrothermal carbonization is a thermochemical process performed at temperatures ranging between 180–250°C in presence of water [36]. This process yields hydrochar together with gases including CO_2, and aqueous extracted compounds like inorganic salts, organic acids, and depolymerized and degraded sugars. Produced hydrochar has lower ash content, higher calorific value, higher carbon content, and less hydrogen content than raw biomass [37–43].

18.2.3.2 Briquettes and pellets

Briquettes and pellets are the most common densified forms of solid biofuels. To produce both forms, the feedstock is compressed using external forces. This external force compacts the biomass by increasing pressure. Sengar et al. [44] tested a variety of feedstocks and showed the applied pressure decreases moisture content from 60%–70% to a range of 41.9%–59.5% and thus increases their density and calorific value. The main advantage of briquettes and pellets over the raw biomass is higher density and uniform shape and size, which makes it easier to handle and burn. Considering the large distance between the production sites and the end use, the high density significantly improves the logistics and decreases the required storage capacity [45]. The properties of raw biomass used for briquetting and pelleting—particle size, moisture content, chemical composition, and shape of the particles—determine the final quality of the products. Both woody and nonwoody biomass can be used to produce pellets and briquettes. Feedstock sources can be in the form of sawdust, corn stover, miscanthus, rice straw, wheat straw, and many other biomass wastes and energy crops. Briquettes and pellets can have differences in size, shape, and moisture content due to different processing technologies.

18.3 Solid biofuel properties

Properties of solid biofuels can vary significantly depending on their source and the processing steps implemented to obtain the final delivered forms. Such variation can influence the biofuel quality and impact biofuel handling, storage, and thermochemical conversion. Low-quality biofuel can also emit compounds that can have health and environmental impacts. Thus, there is a high demand of solid biofuels that meet quality standards. Different standards are developed to standardize procedures to determine biomass properties as well as maintain biofuel quality for commercial use in boilers in household and industries (Table 18.4).

18.3.1 Physical properties

18.3.1.1 Moisture

Moisture content of the solid biofuel is determined as the proportion of water present in the biofuel (which can be removed from the biomass under defined conditions) to the total weight of biofuel [62]. Different standards are developed to determine the moisture content of the fuel (Table 18.4). Moisture content of the solid biofuel can vary depending on the feedstock type, harvest conditions, weather, and time between biofuel delivery and use. Moisture content lower than 10% is recommended for solid biofuels [63]. Fuel with high moisture content provides less heat per unit mass as much of the

Table 18.4 Standard for determination of different properties of solid biofuel.

Fuel properties	Standards
Moisture content	EN ISO 18134–1:2015 [46], ASTM E 871 [47]
Bulk density	ASAE S269.5: 2012 [48]
Particle size	EN 15150: 2012 [49]
Mechanical durability	ASAE S269.5: 2012 [48], CEN/TS 15210 [50]
Volatile matter	EN ISO 18123:2015 [51], ASTM D-3175–20 [52]
Ash content	EN ISO 18122:2015 [53], ASTM E 1755 [54]
Fixed Carbon	ASTM 3172 [55]
Ash melting behavior	CEN/TS 15370 [56], ASTM D-3174–04 [57]
Calorific value	EN ISO 18125:2017 [58], ASTM D240–19 [59]
Carbon, hydrogen	EN ISO 16948:2015 [60], ASTM E777–17 [61]

energy is used up to heat and vaporize water in the fuel. Moisture contents higher than the optimum level increases microbial activity leading to biomass degradation and less efficient storage.

18.3.1.2 Bulk density

Bulk density of solid biofuel is defined as the ratio of the mass of the biofuel particles that occupy the unit volume in bulk form and can be determined following different standards (Table 18.4). Bulk density of the solid biofuels can determine the space requirement for storage, handling and transportation, which can impact the final delivered cost of biofuels. Bulk density can also influence the mechanical durability and thermal characteristics [64]. Most of the biomass inherently has low bulk density which can limit their use. Thus, bulk density of biomass is increased using techniques such as baling (for agricultural biomass), briquetting, and pelletization (for fine wood particles such as sawdust) before it is delivered for use (Table 18.5).

18.3.1.3 Particle size and density

Particle size and particle density refer to the actual size and density of the particles without considering the space between the particles. Particle size and density can affect

Table 18.5 Typical bulk densities, energy densities and particle sizes of different biomass forms

Densified form for feedstocks	Bulk density (kg/m³)	Energy density (MJ/m³)	Size range (mm) Input	Size range (mm) Output
Unprocessed biomass				
Agricultural residues [65,66]	36.1–52.1	444–900	Variable	Variable
Energy crops [65–67]	67.5–85	1200–1500		
Woody crops [68,69]	170–310	3100–6350		
Forest wood residue [69,70]	150	2735		
Coal – for comparison [71]	750	13,600		
Minimally processed biomass				
Miscanthus chopped [68]	90	1600	Variable	30–40
Willow chip [68]	150	2700		3–80
Wood chips [69,70]	220–265	2693–3244		3–80
Processed biomass				
Briquettes [72,73]	350	6400	6–30	Dia: 50–100 Length: 60–150
Pellets [72,74,75]	450–700	8,200–12,700	0.2–3	Dia: 6–8 Length: 18–32
Torrefied biomass [76]	230	4600	Variable	
Torrefied pellets [76]	750–850	13,600–15,500	0.2–3	

the flow, volumetric requirements, and stacking, and influence the fuel handling and combustion equipment. Higher particle density also indicates increased material hardness and lower risk of damage or fine material loss.

18.3.1.4 Mechanical durability

Mechanical durability of solid biofuels indicates its ability to resist abrasion, breakage and material loss during handling, storage, and transport. The particle density, lignin, and moisture content impact the mechanical durability of the solid biofuels. Higher particle density improves the hardness and thus, improves the mechanical durability of solid biofuels. Higher lignin content can improve the binding of other components of solid biofuel [77]. Mechanical durability improves with the increase of moisture content to an optimum level (8% to 10%) and decreases as the moisture content exceeds above optimum level [78,79].

18.4 Chemical properties

18.4.1 Volatile Matter

Volatile matter is the measure of organic matter producing gases, exclusive of water vapor, formed during heating. Volatile matter can be used to establish burning characteristics, rank solid biofuels, and provide basis for purchasing and selling of solid biofuels. Volatile matter is determined by heating the biomass sample at 950°C for 6 min in an oxygen-free environment, strictly adhering to the standards (Table 18.4). Fuels with high volatile matter content contain more tar and release most of the calorific value as vapor. Biomass materials have higher volatile matter content (agricultural residue: 63%–80%, wood: 72%–78%, peat: 70%, coal: up to 40%) compared to charcoal (3%–30%) [80,81].

18.4.2 Ash content and melting properties

Ash is the residue remaining after combustion of solid biofuel. It is the inorganic material bound in the physical structure of the biomass or the soil remaining in the biomass. Ash can be calculated as the residue remaining after heating the sample in the furnace following the standards (Table 18.4). Ash content varies significantly depending on feedstocks from around 0.3%–5% for biofuels from woody crops and forest residues, and up to 16% for agricultural residues [80,82]. The ash content for agricultural residues varies depending on the crop processing, collection, and weather conditions. For comparison, ash content of coal can vary from 5%–17% [80]. Higher ash content in solid biofuels can cause problems like slagging and fouling during combustion. Presence of certain mineral components in solid biofuel, primarily silica, potassium and chlorine, can cause slagging and fouling at relatively low temperatures. Depending on the temperature slag formation or even a complete meltdown of ash can occur, leading to severe technical disadvantage or unit shutdowns. Ash content and its melting behavior can impact the design and operation of boilers and emissions from the system. Thus, higher ash content decreases the efficiency of the system and increases the cost of cleaning of combustion systems.

18.4.3 Fixed carbon

Fixed carbon is the combustible solid residue remaining after the release of moisture and volatile matter from the solid biofuels. Fixed carbon is used as an estimation of the amount of coal that will be generated from solid biofuels and estimated by proximate analysis. Fixed carbon content of solid biofuels range between 11% and 26% [83]. For comparison, fixed carbon for coal is 56%, and biomass char range between 55% and 89% [83].

18.4.4 Heating value (calorific value)

Heating or calorific value represents the total amount of energy that is available in solid biofuels and is defined as the amount of heat released when it undergoes complete combustion with oxygen under standard conditions. It can be determined following the standards (Table 18.4). Heating value of solid biofuels is mostly a function of their chemical composition and can be expressed as higher heating value (HHV) or lower heating value (LHV).

HHV also known as gross heating value, is defined as the total amount of heat energy that is available in the fuel, including the energy contained in the water vapor in the exhaust gases. LHV, also known as a net heating value, does not include the energy embodied in the water vapor. LHV includes the energy required to evaporate the water from the fuel and thus has reduced energy available compared to HHV of biofuels. Most of the biomass feedstocks used for solid biofuels have HHV between 15–21 MJ/kg, with HHV of agricultural residues in the range 15–20 MJ/kg and woody biomass in the range 18–22 MJ/kg [84].

18.4.5 Carbon and hydrogen contents

Carbon and hydrogen are the important components in solid biofuels which help in the combustion process. Carbon and hydrogen are oxidized during combustion process by exothermic reaction, which has a positive effect on HHV. Carbon content in the fuel influences the release of heat via oxidation, calorific value, oxygen demand and particle emission. Woody biomass has a higher carbon content (47–50%) compared to the herbaceous biomass (43%–48%) fuels, and therefore generates HHV under the same ignition conditions [62]. Hydrogen content in solid biofuels is between 5% and 7% and influences the water formation and LHV. Carbon and hydrogen content can be determined by ultimate analysis following the standards (Table 18.4).

18.4.6 Sulphur content

Higher sulfur content in solid biofuels affects the concentration of SO_2 in the flue gas. The SO_2 concentration causes sulfation of alkali and earth-alkali forming deposits and corrosion on the surface of heat exchanger tubes. Solid biofuels with sulfur content above 0.2 wt% can have environmental problems related to emissions [82].

18.4.7 Heavy metals

Heavy metals determine the quality and the usability of the ash formed and collected during combustion and gasification. During combustion process heavy metals like Pb, Zn, Hg, remain in the residue and can cause pollution in the soil when reused as fertilizer. Some heavy metals like copper and iron chlorine can act as a catalyst for the formation of dioxin and furan which are toxic to human and the environment [85]. Presence of heavy metals in the biofuel depends on biomass origin. Heavy metals are accumulated during the growth stage from external sources, such as pollution of the soil. Wood fuels from forest trees have higher heavy metal concentrations than annually harvested crops because of the longer growing time.

18.5 Costs of solid biofuels supply

Costs of solid biofuels vary depending on the type of feedstock, logistics operations and conversion processes, as well as the regional, sectoral and market demand. Further details are provided in *sections 18.4.1* and *18.4.2*, followed by an example in *section 18.4.3*.

18.5.1 Cost of feedstock production/acquisition

The cost of feedstock acquisition is an important factor affecting the economics of the solid biofuels. Feedstock acquisition cost depends on the type of feedstock. For feedstocks categorized as residuals or wastes, the acquisition cost is defined differently compared to dedicated energy crops. Since production of residuals and wastes is not the purpose, the acquisition cost cannot be defined in the same way as that of dedicated energy crops, which is calculated by assessing the growing costs. For some feedstocks like MSW, the acquisition cost is negative. It means a producer in the United States pays a cost between 53–55 $/t MSW called tipping fee [86]. This cost is attributed to expenses related to landfilling of MSW. Increasing tipping fees may improve the economic viability of conversion of MSW to solid biofuels [86]. However, for other types of feedstocks like agricultural residuals and forest resources an average cost of 20–40 $/t and 33–50 $/t, respectively, need to be paid [87–90]. The included costs for estimation of this data are payment to farmer, harvesting, shredding, and baling.

The production cost of dedicated energy crops depends on different factors including land rent, inputs requirement (seeds, fertilizer, pesticide, and water), field operations (land preparation, planting, fertilizer and pesticide applications, cultivating, labor, fuel, and machinery), maintenance, and harvest costs. The acquisition cost for this type of feedstock ranges between 70–90 $/t [90]. In recent years, the cost of dedicated energy crops has been increased in response to the further development and expansion of the biofuel and bioenergy industries. In this regard, commercial production of several bioenergy crops has been developed to answer the predicted future demand for this type of feedstock [91].

18.5.2 Costs of feedstock logistics and conversion to solid biofuels

Solid biofuels are burned directly for domestic heating and cooking in most developing countries, and for heat and electricity generation in most developed countries. The purpose of solid biofuels production is important for identifying the downstream processes needed for solid biofuels. Different components of logistics and conversion which affect the economics of solid biofuels include harvest, postharvest logistics, and processing and conversion to solid biofuels. Harvesting cost includes the costs of labor, harvesting equipment, as well as the working capital. Storage costs significantly impact the total solid biofuels production costs. The biomass storage costs can vary depending on the storage duration and the biomass properties and qualities, which can determine the storage area condition (open or roofed area) and location (within the growing land or at a separate storage place). The conversion cost of solid biofuels is affected by different factors, such as the type of processes needed, which depends on the purpose of use of solid biofuels.

18.5.3 An example of costs of solid biofuels

The cost of solid biofuels is dependent on different costs associated with biomass production and delivery, and facility, labor, energy, and consumables used during biomass processing. Total production cost for torrefied wood pellets in the United States was estimated to be ~$200/t [92], from which biomass supply cost was the main contributor (~29%), followed by facility and equipment related cost (~25%), binder cost (~13%), and labor cost (~10%) (Fig. 18.1).

The facility location affects the economic viability of establishing solid biofuels production plants. In addition, the uncertainties on sustainable biomass supply have significant influences on identifying the best locations for the facilities. Considering the uncertainties associated with biomass supply and the high cost of transportation of low bulk density biomass, some recent studies have focused on using portable systems to convert forest biomass to solid biofuels after harvest. In addition to reducing the transportation costs, portable systems reduce the investment on large-scale facility establishment by integrating feedstock location and conversion platforms, and they reduce the uncertainty associated with biomass supply uncertainties [93]. In a recent study, total costs of portable systems to produce woodchips briquettes, torrefied-woodchips briquettes, and biochar from forest residue were estimated to be $580,000, $790,000, and $720,000, respectively [94]. Feedstock contributed to 7%–10% of total cost, and the contribution of capital costs was low (17%–32%) compared to operational

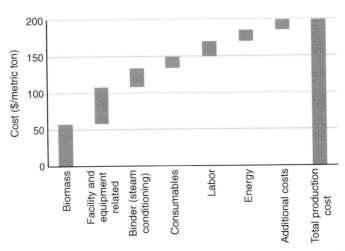

Fig. 18.1 Contribution of different sources to torrefied pellet production cost (data from [92]).

costs (68–83%) (Fig. 18.2). The fixed operating cost includes annual insurance, property taxes, repair and maintenance costs, and total cost of relocations. In addition, the minimum selling price of woodchips briquettes, torrefied-woodchips briquettes and biochar has been reported as $162/t, $274/t, and $1044/t, respectively [94], which shows the economic viability of using portable systems for conversion of forest residues to solid biofuels.

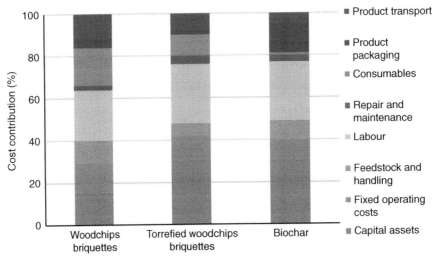

Fig. 18.2 Percentage contribution of different cost components of solid biofuels production from forest residues using portable system (adapted from [94]).

18.6 Life-cycle environmental impacts

Solid biofuels are considered an environmentally advantageous energy source. So, it is important to consider the environmental impacts of solid biofuels over the course of their life-cycles. This includes understanding the impact of biofuels on the carbon cycle and the steps of the biofuel production processes that majorly impact the life-cycle greenhouse gas (GHG) emissions of solid biofuels.

18.6.1 Carbon cycle of solid biofuels

The carbon cycle describes how carbon moves between the atmosphere, organisms, soil bodies, and water bodies [95]. Large carbon stores exist in the atmosphere, in the soil, in the ocean, and in fossils and sediments. Part of this cycle moves very slowly (i.e., exchanges with fossils and reactive sediments), but part of the cycle moves relatively quickly, (i.e., carbon taken up from the atmosphere by plants and stored in plant biomass or soil carbon). Some carbon remains stored while a portion is returned to the atmosphere through the plant and soil respiration and combustion. Increasing production of biofuels from agricultural and forestry residues may reduce the natural replenishing of carbon stores in the soil and further increase atmospheric carbon.

Compared to fossil fuels, which release carbon from long-term stores to the atmosphere during combustion, solid biofuels are often considered to be "carbon neutral" because the carbon released during combustion of this type of fuels is the same amount of CO_2 captured by the plant during the growth (Fig. 18.3). However, additional carbon emissions are incurred by fossil fuel consumption during the planting, harvesting, transporting, and treatment/processing of solid biofuel. The use of a life cycle approach in tracking the solid biofuel systems usually estimates the environmental impacts and determines the energy use of solid biofuel systems.

18.6.2 Energy use and GHG emissions of solid biofuel systems

While solid biofuels are considered as a source of energy, there are different types of environmental impacts during the production and use phases. Energy and GHG balances of solid biofuel systems depend on the type of feedstock, conversion technologies, end-use technologies, and system boundaries [96]. Land-use for feedstock production and type of coproducts also affect the energy and GHG balances of systems. The biomass production system may largely contribute to the GHG emissions in solid biofuel production, while on the other hand production of useful coproducts decreases the share of GHG emissions allocated to solid biofuels. The system boundary of solid biofuels (Fig. 18.4) can be considered as fuel-to-combustion or fuel-to-energy. Fuel-to-combustion system boundary includes feedstock growing, logistics operations (collection, handling, transportation and storage), conversion of feedstock to solid biofuels, and transportation of solid biofuels. Fuel-to-energy system boundary, in addition to

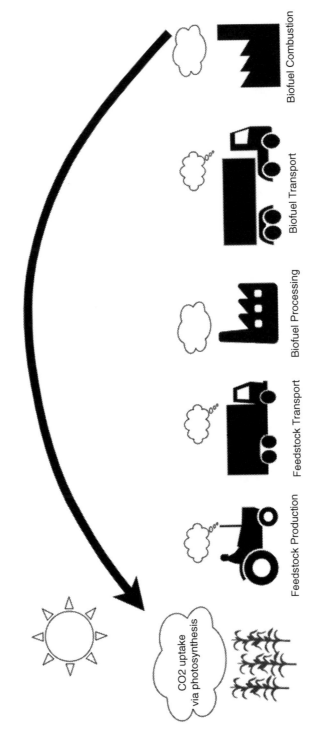

Fig. 18.3 Carbon cycle of solid biofuels.

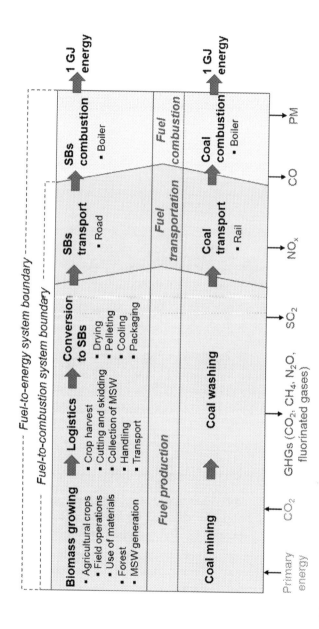

Fig. 18.4 Primary energy use, CO_2 uptake and environmental emissions of solid biofuels life-cycle compared to fossil-based coal (adapted from [101]).

fuel-to-combustion system boundary, includes conversion of solid biofuels to energy, usually in a boiler. Environmental emissions associated with solid biofuels life cycle are classified as direct and indirect emissions. Direct emissions include those associated with on-site use of materials, energy and fuel during the production as well as emissions during the combustion of solid biofuels. Indirect emissions include those associated with background processes for the production and transportation of equipment, material, and other inputs to the end-user. The main GHGs during the life cycle of solid biofuels include CO_2, CH_4, and N_2O. The CO_2 equivalent emissions are estimated by considering the global warming potential of CH_4 (i.e., 23 kg CO_2eq/CH_4) and N_2O (i.e., 296 kg CO_2eq/ N_2O). Other air pollutants include SO_2, NO_x, CO, and particulate matter (PM) (Fig. 18.4). Types of emissions vary depending on the operations needed during the life cycle of solid biofuels, which are explained in more detail below.

Feedstock growing and logistics operations: emissions associated with feedstock growing vary depending on the source of feedstock and allocation method. Feedstock supply from dedicated energy crops, such as switchgrass and miscanthus, need different field operations, fertilizers application, harvest and postharvest logistics. Usually, the main product, to which the emissions are allocated, is considered for energy generation. Since crop residue is the by-product after harvest, use of fertilizers and pesticides do not usually have significant contribution to the environmental life cycle of solid biofuels, as they are mainly allocated to the main-product which has higher economic value. However, crop residue usually needs different logistics operations, such as windrowing, baling, collecting, loading, transportation, unloading and storage. Considering the low bulk density of crop residue, transportation is the main contributor to GHG emissions [97,98]. The use of forestry biomass avoids the feedstock production emissions, as they are grown with no use of inputs. However, logistics operation for cutting, screening, loading, transportation and unloading emits GHGs due to fuel combustion and use of equipment. GHG emissions of forestry biomass supply have been reported less than GHG emissions associated with feedstock supply from dedicated energy crops or agricultural residue [97]. MSW management also needs operations for collection, segregation, and transportation which contribute to the life cycle GHG emissions of solid biofuels.

Conversion of biomass to solid biofuels: Depending on the type of solid biofuels, some of the conversion processes, such as drying, pelleting, cooling, and packaging, are needed. Electricity and natural gas are the main sources of energy during conversion processes. For example, pellets need more energy-intensive processes for conversion compared to other types of solid biofuels, which consequently, cause more emissions during the conversion processes [99].

Use of solid biofuels: combustion is the main mode of the utilizing solid biofuels. During the combustion process, there are different types of potential pollutants depending on the type of biomass, contaminants in the fuel, type of technology used for combustion, and completeness of combustion process [99]. Principal emissions during the combustion of solid biofuels include (1) PM including salts, soot, condensable and volatile organic compounds, (2) oxides of nitrogen (NO_x) including NO, NO_2 and N_2O, (3) oxides of carbon (CO_x) including CO and CO_2, (4) oxides of sulfur (SO_x) including SO_2 and SO_3, and (5) dioxins/furans [100]. Among these, PM and NOx are considered as pollutants of the highest concern during the combustion of solid biofuels. Nevertheless, type of feedstock and upgrading processes highly affect the emission levels. For example, wood pellets typically emit lower levels of PM and NOx than wood chips and wood logs [100]. In addition, while pellet production has more emissions during the processing, combustion of pellets has less emissions relative to other types of wood; this is primarily due to their consistent quality, lack of bark (white premium pellets) in their composition, less ash content (around 0.5%), as well as low and consistent moisture content (around 8%). Thus, pellets may be considered as a low-emissions type of solid biofuels [99]. Emissions of solid biofuels combustion also depend on the technology used for combustion. Use of appropriate combustion systems specifically designed for the type of solid biofuels can minimize combustion emissions. The considerations for designing the combustion system include the solid biofuels moisture content, ash, and chlorine contents, and physical size and characteristics [100].

Life-cycle GHG emissions of solid biofuels are evaluated by taking into account the indirect emissions during the raw material extraction and processing, direct emissions during the feedstock production, logistics, conversion, and use, as well as CO_2 uptake during biomass growth (Fig. 18.4). Compared to the reference system of energy generation, e.g., coal production and combustion (Fig. 18.4), the net amount of emissions of solid biofuels depends on fossil carbon displacement efficiency. Usually, the energy use and life cycle emission of solid biofuels are less than those of fossil-based fuels. As an example, the energy use, GHG emissions in terms of CO_2 equivalent, and air pollutant emissions of wood pellet and coal are presented in Table 18.6.

18.6.3 Case studies of environmental impacts of solid biofuels

Feedstock type and their associated logistics operations are the main factors affecting the overall GHG emissions during the solid biofuel's life-cycle. Previous studies have

Table 18.6 Energy use, GHG emissions and air pollutants of wood pellet compared to fossil-based coal.

Fuel	Unit	Energy use (MJ)	GHG (g CO_2eq)	CO	NO_x	SO_2	PM
Coal	1 kg	3.3	866.0	0.1	0.5	4.3	1.2
	1 MJ	0.16	41	0.01	0.02	0.2	0.1
Wood pellet	1 kg	1.4	137.5	0.2	0.7	0.8	0.2
	1 MJ	0.08	8	0.01	0.04	0.04	0.01

Criteria emissions (g) span CO, NO_x, SO_2, PM columns.

Source: [101].

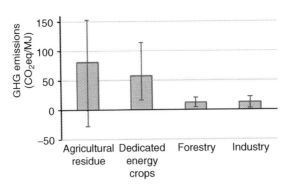

Fig. 18.5 GHG emissions of electrical energy generation from different types of biomass [97].

reported GHG emissions of electrical energy generation from different types of feedstocks, by considering the system boundary from fuel to heat. Average GHG emissions of electrical energy generation from agriculture residues, dedicated energy crops, forestry, and industry are 81, 58, 12, and 13 gCO$_2$e/MJ, respectively (Fig. 18.5) [97].

Lower GHG emissions of forestry and industry-based feedstocks are due to using the byproduct resources as feedstock, compared to feedstocks from dedicated energy crops and agricultural residue which use chemical fertilizers and other inputs for production. Nevertheless, energy generation from dedicated energy crops have lower GHG emissions compared to fossil-based energy generation systems, which shows the significant potential of these crops to reduce the GHG emissions in a global perspective. In addition, GHG emissions of energy generation from dedicated energy crops is mainly (62%–82%) attributed to chemical fertilizers [102]. Accordingly, use of organic manure in dedicated energy crop production is a potential strategy to further reduce the life cycle GHG emissions of energy generation from this type of feedstock. In addition, strategies, such as best management practices to maximize the yield and minimize resource use, are recommended for the sustainable establishment of energy generation from dedicated energy crops. Use of agricultural residue, such as corn stover and rice straw, as feedstock for solid biofuels and energy generation contributes to GHG emissions mainly due to the involved transportation of feedstock [103].

In addition to the type of feedstock, the supply chain significantly affects the environmental impacts of energy generation from solid biofuels. In a previous study, the global warming potential (GWP) of 1 MJ energy from forest-based firewood was investigated [104]. The system boundary included forest management, felling trees, full tree extraction, processing tree at landing, off-road transport, road transport, processing (sawing and splitting), distribution, and combustion. The main contribution to environmental impacts was the transportation of forest biomass, so an analysis was performed for two scenarios of short- and long-supply chains. The short supply chain considered harvest of forest biomass from local forests, while the long supply chain considered the harvest of biomass from neighboring countries. GWP of energy generation from forest-based firewood for short and long supply chains was estimated to be 4.3 and 9.8 gCO$_2$eq/MJ, respectively [104]. The contribution of different elements of the supply chain to the GWP are presented in Fig. 18.6. In the short supply chain scenario, combustion is the main contributor to total GWP, while in the long supply chain, on-road transport has the highest contribution.

18.7 Solid biofuel policies

Some international and national policies have been enacted to promote renewable energy consumption and production. The United Nations Sustainable Development Goals (SDGs) were adopted by all member states in 2015 and included two goals related directly to renewable energy. Goal 7 is to "ensure access to affordable, reliable, sustainable and modern energy for all" and Goal 13 is to "Take urgent action to combat climate change and its impacts." The 2019 progress report toward SDGs found that while renewable energy share of total energy consumption had increased to 17.5% in 2016, atmospheric GHG emissions concentration has continued to rise [105]. Some of the existing policies help promote renewable energy by providing targets for the share of renewable energy in total energy production, financial incentives for developing and using clean energy, and regulating GHG emission levels while promoting "emissions trading." Such policies together with solid biofuels standards promote the safe use of solid biofuels for residential and industrial purposes and facilitate solid biofuels trading in markets throughout the world.

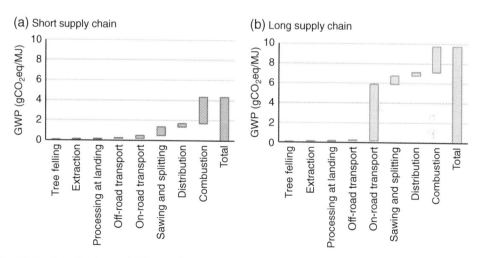

Fig. 18.6 Contributions of different elements of supply chain to the GWP of energy generation from forest-based firewood in short supply chain (A), and long supply chain (B) (adapted from [104]). *GWP*, global warming potential.

18.7.1 Policies promoting renewable energy
18.7.1.1 Demand-side policies
Demand-side policies include regulations that set quotas and requirements for renewable energy production or consumption. An example includes the European Union's Renewable Energy Directive 2009/28/EC, which set targets for renewable energy sources to make up 20% of final energy consumption and a minimum of 10% for biofuels by 2020. Member countries were on track for this target in 2017, with 17.52% of its energy consumption coming from renewable sources (European Commission, 2019). In the United States, most of the requirements for renewable energy supply occur at the state level through Renewable Portfolio Standards. As of 2019, 37 states, the District of Columbia, and four US territories have enacted Renewable Portfolio Standards [106]. These demand-side policies have significantly stimulated international trade in biofuels.

18.7.1.2 Supply-side policies
Supply-side policies provide direct or indirect stimulation of biofuel systems. Indirect stimulation includes funding for research and development related to all steps along the biofuel production and distribution chain. Direct support comes through financial incentives for collection or processing equipment or through tax reductions/exemptions for biofuels. One example is the United States Renewable Electricity Production Tax Credit, which grants tax credits of approximately $0.012/kWh for renewable electricity production, with full credit for wind energy and "open-loop biomass" using waste or by-products, and half credit for "closed-loop" biomass including dedicated energy crops [107]. These supply-side policies have primarily stimulated biofuel production in domestic markets but have not had noticeable increases in international trade. It is worth noting that most solid biofuel production and export comes primarily from countries with existing export-focused forestry and wood processing infrastructure.

18.7.2 Policies regulating GHG emissions and promoting "emissions trading"
Increased awareness of the negative effects of GHG has prompted some institutional, national, and international actions to establish an "emissions trading" system. One such system is termed "cap-and-trade," where nations or specific industry sectors are allotted a GHG emissions cap. If they exceed their allowed emissions, they can purchase carbon credits, which is a standard unit of one metric ton of carbon dioxide equivalent. These credits can be sold by countries or entities producing lower GHG than their allotted level. Within the Kyoto Protocol, these credits can also be earned by countries for sponsoring clean energy projects to reduce GHG emissions in other countries through Clean Development Mechanisms, which establish the potential project types and monitoring system. Currently, biofuel production is an existing Clean Development Mechanism, but none of the solid biofuel projects have previously been approved.

18.8 Opportunities for using solid biofuels

Solid biofuels are versatile source of energy which can be effectively acquired, stored, and used for different applications. Different studies have shown that biomass could contribute to 200–400 EJ of energy generation by 2050 [4]. However, most of the increment is expected to be in the form of modern bioenergy use, such as transportation fuels, rather than traditional use, e.g. burning for domestic heat. Solid biofuels also contribute toward diversification of energy sources and reduced dependence on imported fossil fuels in most countries with limited fossil reserves. This is even more critical in the future as the fossil fuel reserves are currently depleting at an alarming rate.

Solid biofuels provide opportunity for the production and utilization of carbon-neutral biomass available in local resources. Energy production from solid biofuels using locally available biomass resources can be particularly relevant for sparsely populated rural areas, where deployment of centralized energy infrastructure is prohibitive due to high investment to provide energy to these areas. Biomass is also considered to be carbon-neutral as the atmospheric carbon is sequestered during the biomass growing phase, which is finally emitted during conversion and use of solid biofuel. Thus, solid biofuels are expected to provide an alternative cleaner energy than the fossil fuels.

The feedstock for solid biofuels is available locally or can be grown in most of rural areas. Thus, compared to fossil-based fuels, long transportation is not needed. This reduces the overall cost and emissions due to long transportation.

In addition, there is a growing interest in strategies to reduce landfilling and incineration of MSW. Use of MSW as feedstock for solid biofuels helps is more appealing than converting purpose-grown agricultural products. while solving the MSW disposal problems, this will reduce the emissions due to landfilling of MSW and reduce the costs associated with MSW management. High landfilling tipping fees result in rapidly improving of the economics of use of MSW as feedstock for solid biofuels.

Increased use of solid biofuels is also expected to help growth of the biobased industries and jobs. It would also help diversify the markets for different energy and products that can be obtained from the conversion of solid biofuels. Use of solid biofuels can also increase revenue for the communities in developing countries from carbon credits. Thus, solid fuels can have positive economic, environmental, and social impacts in both developed and developing countries.

18.9 Challenges for solid biofuels

Main challenges to the growth of solid biofuel use are technological limitations, uncertainties due to feedstock supply, and questions of environmental, social, and economic impacts.

18.9.1 Technological limitations

Technological aspects of solid biofuels vary from feedstock growing, pretreatment and conversion technologies, and end-use applications. Within feedstock growing, possible new energy-dense feedstocks may be identified, or the energy density of existing feedstock sources may be increased. New technologies may improve the cultivation, harvesting, and transport of feedstocks and improve pretreatment, conversion, and end-use technologies. Any technological improvements in biomass energy conversion technology will make solid biofuel systems more feasible by lowering their associated GHG emissions or costs [96]. One limitation for technological improvements, especially in biomass production system optimization, is that research findings may only be relevant regionally because of the dependence on different soils, climates, and farming systems [108]. Universal application of research findings is not possible because of the impacts of these localized components, and many research findings may not apply in developing countries where climates are semi-arid, arid, or tropical [108]. Reducing these technological limitations relies on increased research, which requires action and interest from funding agencies and national governments.

18.9.2 Uncertainty of feedstock supply

Feedstock availability varies geographically and seasonally, and properties of biomass vary widely. Properties of biomass can vary with the species of plant, the region where it was grown, growing conditions, and harvest and storage conditions. Additionally, the growth characteristics of different feedstocks may shift with the changing climate. The ability to measure biomass properties consistently and accurately is critical to designing bioprocessing operations. Large operations using solid biofuels year-round would need to increase their storage capacities and ensure that storage practices minimize losses of quality and quantity over time. Additionally, biomass feedstocks have low bulk density and may not be cost-effective to transport before some form of pre-treatment or processing.

18.9.3 Environmental impacts

Considering the environmental impacts of solid biofuel systems include their impacts on land use, water use, air quality, carbon sinks, and GHG emissions. Land has competing uses for housing, food, feed, fiber, chemicals, and energy, and some biomass feedstocks may require land conversion for their cultivation. Growing dedicated energy feedstocks requires water and energy-intensive inputs for their cultivation, harvesting, and transport, while other biomass feedstocks (e.g., crop and forestry residues and municipal wastes) are considered wastes or byproducts of other systems and are typically not attributed with the associated emissions used to produce the primary product. Solid biofuel combustion has impacts on air quality, especially for solid wastes incineration plants. Solid biofuel combustion is often thought of as carbon neutral because it releases the CO_2 taken up by the plant, but the removal and combustion of the plant prevent the replenishment of a "carbon sink" whereby a portion of the carbon not decomposed would have been stored in the soil [96]. Overall, determining the GHG emissions from bioenergy cannot be generalized, and is impacted by the feedstock, conversion method, and end-use applications [96].

18.9.4 Social and economic impacts

Solid biofuels are associated with social and economic impacts and increasing solid biofuel production and utilization must be carefully considered against these impacts. In developing countries, solid biofuel combustion for home cooking is often associated with negative health impacts from indoor air pollution. Improving education and access to more efficient stoves can decrease these risks. On a global scale, increasing the cultivation of biofuel feedstocks could displace some food crops and impact the price of food. Another potential issue for the widespread adoption of solid biofuels is the current lack of regulation and standardization in trade. Moving forward with regulations and land conversion should include involvement with farmers and other stakeholders.

18.10 Conclusions and perspectives

Solid biofuels can utilize biomass from multiple sources including biomass grown for bioenergy production as well as agricultural and municipal wastes. Some of these biomass needs to be processed before it can be used as a solid biofuel. The quality of solid biofuels varies considerably depending on the type of feedstock used, processing and logistics of the feedstocks and biofuels. Different standards have been developed to maintain the quality of solid biofuels so that it can be traded at fair price around the world. High quality solid biofuels can be produced at prices competitive to other sources with minimal environmental impacts. Different policies related to improving energy security, improving health impacts, and reducing environmental impacts can further promote the use of solid biofuels. Such initiatives can improve the bioeconomy and provide sustainable energy using renewable biomass.

References

[1] World Bioenergy Association. Global Bioenergy Statistics. 2019. https://worldbioenergy.org/uploads/191129 WBA GBS 2019_LQ.pdf. (Accessed February 21, 2020).
[2] International Energy Agency. Energy Data and Statistics 2019. https://www.iea.org/statistics/ (Accessed February 12, 2020).
[3] International Energy Agency. Technology Roadmap, Bioenergy for Heat and Power, EA, Paris. 2012. www.iea.org/reports/technology-roadmap-bioenergy-for-heat-and-power. (Accessed February 15, 2020).
[4] International Energy Agency. Potential Contribution of Bioenergy to the World's Future Energy Demand. 2007. www.iea.org/reports/technology-roadmap-bioenergy-for-heat-and-power (Accessed February 15, 2020).
[5] S. Van Loo, J. Koppejan. The Handbook of Biomass Combustion and Co-firing. Earthscan, London 2008.

[6] E. Virmond, R.L. Schacker, W. Albrecht, C.A Althoff, M. de Souza, R.F.P.M. Moreira, et al., Organic solid waste originating from the meat processing industry as an alternative energy source, Energy 36 (2011) 3897–3906.
[7] I. Hamawand, P. Pittaway, L. Lewis, S. Chakrabarty, J. Caldwell, J. Eberhard, et al., Waste management in the meat processing industry: conversion of paunch and DAF sludge into solid fuel, Waste Manag. 60 (2017) 340–350.
[8] P. Zhao, Y. Shen, S. Ge, K. Yoshikawa, Energy recycling from sewage sludge by producing solid biofuel with hydrothermal carbonization, Energy Convers. Manag. 78 (2014) 815–821.
[9] REN 21. Renewables 2017: Global Status Report 2017. www.ren21.net/wp-content/uploads/2019/05/GSR2017_Full-Report_English.pdf (Accessed March 12, 2020).
[10] A. Szymajda, G. Łaska, The effect of moisture and ash on the calorific value of cow dung biomass, Proceedings 16 (4) (2019). https://doi.org/10.3390/proceedings2019016004.
[11] L. Marufu, J. Ludwig, M.O. Andreae, F.X. Meixner, G. Helas, Domestic biomass burning in rural and urban Zimbabwe: Part A, Biomass Bioenergy 12 (1997) 53–68.
[12] I.S. Mudway, S.T. Duggan, C. Venkataraman, G. Habib, F.J. Kelly, J. Grigg, Combustion of dried animal dung as biofuel results in the generation of highly redox active fine particulates, Part Fibre Toxicol. 2 (2005) 6.
[13] K. Geremew, M. Gedefaw, Z. Dagnew, D. Jara, Current level and correlates of traditional cooking energy sources utilization in urban settings in the context of climate change and health, Northwest Ethiopia: a case of Debre Markos town, Biomed. Res. Int. 2014 (2014).
[14] G.T. Tucho, S. Nonhebel, Bio-wastes as an alternative household cooking energy source in Ethiopia, Energies 8 (2015) 9565–9583.
[15] P.J. Mograbi, E.T.F. Witkowski, B.F.N. Erasmus, G.P. Asner, J.T. Fisher, R. Mathieu, et al., Fuelwood extraction intensity drives compensatory regrowth in African savanna communal lands, L Degrad Dev 30 (2019) 190–201.
[16] B. Bowonder, S.S.R. Prasad, N.V.M. Unni, Dynamics of fuelwood prices in India: Policy implications, World Dev. 16 (1988) 1213–1229.
[17] B.P. Bhatt, S.S. Rathore, M. Lemtur, B. Sarkar, Fuelwood energy pattern and biomass resources in Eastern Himalaya, Renew. Energy 94 (2016) 410–417.
[18] A.M. Swemmer, M. Mashele, P.D. Ndhlovu, Evidence for ecological sustainability of fuelwood harvesting at a rural village in South Africa, Reg. Environ. Chang. 19 (2019) 403–413.
[19] S.R. Chandrasekaran, P.K. Hopke, L. Rector, G. Allen, L. Lin, Chemical composition of wood chips and wood pellets, Energy Fuels 26 (2012) 4932–4937.
[20] S.H. Chang, Rice husk and its pretreatments for bio-oil production via fast pyrolysis: a review, BioEnergy Res 13 (2020) 23–42.
[21] M. Ahiduzzaman, A.K.M.S. Islam, Assessment of rice husk briquette fuel use as an alternative source of woodfuel, Int. J. Renew. Energy Res. 6 (2016) 1602–1611.
[22] N. Ahmed, M. Zeeshan, N. Iqbal, M.Z. Farooq, S.A. Shah, Investigation on bio-oil yield and quality with scrap tire addition in sugarcane bagasse pyrolysis, J. Clean Prod. (2018). https://doi.org/10.1016/j.jclepro.2018.06.142.
[23] K. Jayaraman, I. Gokalp, S. Petrus, V. Belandria, S. Bostyn, Energy recovery analysis from sugar cane bagasse pyrolysis and gasification using thermogravimetry, mass spectrometry and kinetic models, J. Anal. Appl. Pyrolysis (2018). https://doi.org/10.1016/j.jaap.2018.02.003.
[24] F.O. Centeno-González, E.E. Silva Lora, H.F. Villa Nova, L.J. Mendes Neto, A.M. Martínez Reyes, A. Ratner, et al., CFD modeling of combustion of sugarcane bagasse in an industrial boiler, Fuel (2017). https://doi.org/10.1016/j.fuel.2016.11.105.
[25] E.V.S. Costa, C.F. Pereira MP de, C.M.S. da Silva, B.L.C. Pereira, M.F.V. Rocha, Carneiro A de C.O., Torrefied briquettes of sugar cane bagasse and eucalyptus, Rev Arvore (2019). https://doi.org/10.1590/1806-90882019000100001.
[26] M. Arshad, S. Ahmed, Cogeneration through bagasse: a renewable strategy to meet the future energy needs, Renew. Sustain. Energy Rev. 54 (2016) 732–737.
[27] L.J.R. Nunes, R. Godina, J.C.O. Matias, J.P.S. Catalão, Evaluation of the utilization of woodchips as fuel for industrial boilers, J. Clean Prod. 223 (2019) 270–277.
[28] G. Glock, Method and Device for Drying Wood Chips, U.S. Patent. US 2020/0200384 A1. 2020.
[29] S. Farmanbordar, K. Karimi, H. Amiri, Municipal solid waste as a suitable substrate for butanol production as an advanced biofuel, Energy Convers. Manag. 157 (2018) 396–408.

[30] T.V. Ramachandra, H.A. Bharath, G. Kulkarni, S.S. Han, Municipal solid waste: generation, composition and GHG emissions in Bangalore, India. Renew. Sustain. Energy Rev. 82 (2018) 1122–1136.

[31] Hoornweg D., Bhada-Tata P. What a waste: a global review of solid waste management. Urban development series; Knowledge papers no. 15. 2012. World Bank, Washington, DC. © World Bank. https://openknowledge.worldbank.org/handle/10986/17388. License: CC BY 3.0 IGO.

[32] Y. Uemura, W. Omar, N.A. Othman, S. Yusup, T. Tsutsui, Torrefaction of oil palm EFB in the presence of oxygen, Fuel 103 (2013) 156–160.

[33] J.S. Tumuluru, S. Sokhansanj, J.R. Hess, C.T. Wright, R.D. Boardman, A review on biomass torrefaction process and product properties for energy applications, Ind. Biotechnol. 7 (2011) 384–401. https://doi.org/10.1089/ind.2011.7.384.

[34] L. Wang, E. Barta-Rajnai, Ø. Skreiberg, R. Khalil, Z. Czégény, E. Jakab, et al., Effect of torrefaction on physiochemical characteristics and grindability of stem wood, stump and bark, Appl. Energy 227 (2018) 137–148.

[35] C. Zhang, S-H. Ho, W-H. Chen, Y. Xie, Z. Liu, J-S. Chang, Torrefaction performance and energy usage of biomass wastes and their correlations with torrefaction severity index, Appl. Energy 220 (2018) 598–604.

[36] D.S. Bajwa, T. Peterson, N. Sharma, J. Shojaeiarani, S.G. Bajwa, A review of densified solid biomass for energy production, Renew. Sustain. Energy Rev. 96 (2018) 296–305.

[37] M. Wilk, A. Magdziarz, Hydrothermal carbonization, torrefaction and slow pyrolysis of Miscanthus giganteus, Energy 140 (2017) 1292–1304.

[38] H.S. Kambo, A. Dutta, Strength, storage, and combustion characteristics of densified lignocellulosic biomass produced via torrefaction and hydrothermal carbonization, Appl Energy 135 (2014) 182–191. https://doi.org/10.1016/j.apenergy.2014.08.094.

[39] J. Lee, K. Lee, D. Sohn, Y.M. Kim, K.Y. Park, Hydrothermal carbonization of lipid extracted algae for hydrochar production and feasibility of using hydrochar as a solid fuel, Energy 153 (2018) 913–920.

[40] Z. Liu, A. Quek, S.K. Hoekman, M.P. Srinivasan, R. Balasubramanian, Thermogravimetric investigation of hydrochar-lignite co-combustion, Bioresour. Technol. 123 (2012) 646–652.

[41] Z. Liu, A. Quek, S.K. Hoekman, R. Balasubramanian, Production of solid biochar fuel from waste biomass by hydrothermal carbonization, Fuel 103 (2013) 943–949.

[42] J. Minaret, A. Dutta, Comparison of liquid and vapor hydrothermal carbonization of corn husk for the use as a solid fuel, Bioresour. Technol. 200 (2016) 804–811.

[43] Teng H., Wang G., Zhang J., Zhang N., Wang C. Conversion of maize straw to blast furnace injected fuel by hydrothermal carbonization 2019, Volume 02: Proceedings of 11th International Conference on Applied Energy, Part 1, Sweden, 2019. Paper ID: 895.

[44] S.H. Sengar, A.G. Mohod, Y.P. Khandetod, S.S. Patil, A.D. Chendake, Performance of briquetting machine for briquette fuel, Int. J. Energy Eng. 2 (2012) 28–34.

[45] A.A. Rentizelas, A.J. Tolis, I.P. Tatsiopoulos, Logistics issues of biomass: the storage problem and the multi-biomass supply chain, Renew. Sustain. Energy Rev. 13 (2009) 887–894. http://dx.doi.org/10.1016/j.rser.2008.01.003.

[46] ISO. BS EN ISO 18134-1:2015. Solid biofuels. determination of moisture content. Oven dry method. Total moisture. Reference method 2015.

[47] ASTM. ASTM E871-82, Standard Test Method for Moisture Analysis of Particulate Wood Fuels, 2019. https://doi.org/10.1520/E0871-82R19.

[48] ASABE. Densified products for bulk handling—definitions and Method. Asae S2695 2012.

[49] CEN. Solid biofuels—determination of particle density. DIN EN 151502012 2012.

[50] CEN. Solid biofuels—methods for the determination of mechanical durability of pellets and briquettes—Part 1: pellets. CEN/TS 15210–1 JULY 2006 2006.

[51] ISO. Solid biofuels. determination of the content of volatile matter. EN ISO 181232015 2015.

[52] ASTM. Standard test method for volatile matter in the analysis sample of coal and coke. ASTM D3175–20 2020. https://doi.org/10.1520/D3175-20.

[53] ISO. Solid biofuels. determination of ash content. EN ISO 181222015 2015.

[54] ASTM. Standard test method for ash in biomass. West Conshohocken, PA: 2015.

[55] ASTM. Standard practice for proximate analysis of coal and coke. ASTM D3172–13 2013. https://doi.org/10.1520/D3172-13.

[56] CEN. Solid biofuels—method for the determination of ash melting behaviour—part 1: characteristic temperatures method. CEN/TS 15370-1 2006.
[57] ASTM International. D3174-12 standard test method for ash in the analysis sample of coal and coke from coal 2011;14:1–6. https://doi.org/10.1520/D3174-12.2.
[58] ISO. Solid biofuels. Determination of calorific value. EN ISO 181252017 2017.
[59] ASTM. Standard test method for heat of combustion of liquid hydrocarbon fuels by bomb calorimeter. ASTM D240–19 2019. https://doi.org/10.1520/D0240-19.
[60] ISO. Solid biofuels. Determination of total content of carbon, hydrogen and nitrogen. EN ISO 169482015 2015.
[61] ASTM. Standard test method for carbon and hydrogen in the analysis sample of refuse-derived fuel. ASTM E777 – 17a 2017. https://doi.org/10.1520/E0777-17A.
[62] H. Hartmann, Solid Biofuelssolid biofuel, Fuels and Their Characteristics editor, in: R.A. Meyers (Ed.), Solid Biofuelssolid biofuel, Fuels and Their Characteristics, Encyclopedia of Sustainability Science and Technology (2012) 9821–9851. https://doi.org/10.1007/978-1-4419-0851-3_245.
[63] K. Rupar-Gadd, J. Forss, Self-heating properties of softwood samples investigated by using isothermal calorimetry, Biomass Bioenergy 111 (2018) 206–212.
[64] J. Shojaeiarani, D.S. Bajwa, S.G. Bajwa, Properties of densified solid biofuels in relation to chemical composition, moisture content, and bulk density of the biomass, BioResources 14 (2019) 4996–5015.
[65] N. Chevanan, A.R. Womac, V.S.P. Bitra, C. Igathinathane, Y.T. Yang, P.I. Miu, et al., Bulk density and compaction behavior of knife mill chopped switchgrass, wheat straw, and corn stover, Bioresour. Technol. 101 (2010) 207–214. https://doi.org/10.1016/j.biortech.2009.07.083.
[66] P. McKendry, Energy production from biomass (part 1): overview of biomass, Bioresour. Technol. 83 (2002) 37–46. https://doi.org/10.1016/S0960-8524(01)00118-3.
[67] Sustainable Energy Authority of Ireland. Fact sheet Miscanthus, 2007. www.doc-developpement-durable.org/file/Culture-Culture-fourrages/herbe_a_elephant/miscanthus_g%C3%A9ant/Miscanthis_Factsheet.pdf (Accessed May 26, 2021).
[68] B. Caslin, J. Finnan, C. Johnston, A. McCracken, L. Walsh, Short rotation coppice willow - Best practice guidelines. 2015. www.teagasc.ie/media/website/publications/2011/Short_Rotation_Coppice_Best_Practice_Guidelines.pdf. (Accessed May 26, 2021).
[69] J. Krajnc, Wood fuels handbook. Food and Agriculture Organization of the United Nations, 2015. ISBN 978-92-5-108728-2. https://doi.org/10.1017/CBO9781107415324.004.
[70] S. Hoyne, A. Thomas, Forest Residues: Harvesting, Storage and Fuel Value, 2001. ISBN: 190269614X, p. 26.
[71] Engineering Toolbox. Bulk density of products 2016. http://www.engineeringtoolbox.com/classification-coal-d_164.html (Accessed July 29, 2016).
[72] S. Sokhansanj, J. Fenton, Cost benefit of biomass supply and pre-processing, A BIOCAP Research Integration Program. Synthesis Paper. BIOCAP Canada, (2006) 1–33 www.cesarnet.ca/biocap-archive/rif/report/Sokhansanj_S.pdf. (Accessed May 26, 2021).
[73] Forest Research. Biomass briquettes 2016. http://www.forestry.gov.uk/fr/beeh-9ujpmn. (Accessed December 10, 2016).
[74] S. Mani, L.G. Tabil, S. Sokhansanj, Effects of compressive force, particle size and moisture content on mechanical properties of biomass pellets from grasses, Biomass Bioenergy 30 (2006) 648–654. https://doi.org/10.1016/j.biombioe.2005.01.004.
[75] P. Lehtikangas, Quality properties of pelletised sawdust, logging residues and bark, Biomass Bioenergy 20 (2001) 351–360. https://doi.org/10.1016/S0961-9534(00)00092-1.
[76] Bergman PCA. Combined torrefaction and pelletisation, The TOP process. 2005. https://publications.tno.nl/publication/34628560/S7JA61/c05073.pdf. (Accessed May 26, 2021).
[77] J.M. Castellano, M. Gómez, M. Fernández, L.S. Esteban, J.E. Carrasco, Study on the effects of raw materials composition and pelletization conditions on the quality and properties of pellets obtained from different woody and non woody biomasses, Fuel 139 (2015) 629–636.
[78] N. Kaliyan, R.V. Morey. Densification characteristics of corn stover and switchgrass, Trans ASABE. 52 (3) (2009) 907-920. https://doi.org/10.13031/2013.27380.
[79] J.C. Serrano-Ruiz, R. Luque, A. Sepúlveda-Escribano, A. Sepulveda-Escribano, Transformations of biomass-derived platform molecules: from high added-value chemicals to fuels via aqueous-phase processing, Chem. Soc. Rev. 40 (2011) 5266. https://doi.org/10.1039/c1cs15131b.

[80] C. Turare. Biomass Gasification - Technology and Utilization. Impact of fuel properties on gasification, 2020. http://ftpmirror.your.org/pub/misc/cd3wd/1002/_engy_biomass_gasification_artes_chandrikant_en_lp_131210_.pdf. (Accessed May 26, 2021).

[81] A. Demirbas, Combustion characteristics of different biomass fuels, Prog. Energy Combust. Sci. 30 (2004) 219–230.

[82] I. Obernberger, T. Brunner, G. Bärnthaler, Chemical properties of solid biofuels—significance and impact, Biomass Bioenergy 30 (2006) 973–982.

[83] S. Gaur, T.B. Reed, Thermal data for natural and synthetic fuels, Marcel Dekker, New York, NY, 1998.

[84] Biorenewables Education Laboratory, Biomass Properties and Handling, Biomass Prop. Instr. Guid. (2011). www.engineering.iastate.edu/brl/files/2011/10/brl_biomassppop_instructor.pdf Accessed March 20, 2020.

[85] G. Cagnetta, M.M. Hassan, J. Huang, G. Yu, R. Weber, Dioxins reformation and destruction in secondary copper smelting fly ash under ball milling, Sci. Rep. 6 (2016) 22925. https://doi.org/10.1038/srep22925.

[86] D. Kanter, L. Staley BF. Analysis of MSW landfill tipping fees, Environmental Research & Education Foundation, 2019. www.erefdn.org. (Accessed May 26, 2021).

[87] L.D. Mapemba, F.M. Epplin, R.L. Huhnke, C.M. Taliaferro, Herbaceous plant biomass harvest and delivery cost with harvest segmented by month and number of harvest machines endogenously determined, Biomass Bioenergy (2008). https://doi.org/10.1016/j.biombioe.2008.02.003.

[88] F. Zhang, D.M. Johnson, J. Wang, C. Yu, Cost, energy use and GHG emissions for forest biomass harvesting operations, Energy 114 (Nov 1) (2016) 1053-1062. https://doi.org/10.1016/j.energy.2016.07.086.

[89] S. Sokhansanj, S. Mani, S. Tagore, A.F. Turhollow, Techno-economic analysis of using corn stover to supply heat and power to a corn ethanol plant—Part 1: cost of feedstock supply logistics, Biomass Bioenergy 114 (2010)Nov 1; 1053-1062. https://doi.org/10.1016/j.biombioe.2009.10.001.

[90] H.J. Huang, S. Ramaswamy, W. Al-Dajani, U. Tschirner, R.A. Cairncross, Effect of biomass species and plant size on cellulosic ethanol: a comparative process and economic analysis, Biomass Bioenergy 114 (2009)Nov 1; 1053-1062. https://doi.org/10.1016/j.biombioe.2008.05.007.

[91] W. Jiang, M.G. Jacobson, M.H. Langholtz, A sustainability framework for assessing studies about marginal lands for planting perennial energy crops, Biofuels, Bioprod. Bioref. 114 (2019)Nov 1; 1053-1062. https://doi.org/10.1002/bbb.1948.

[92] A. Pirraglia, R. Gonzalez, D. Saloni, J. Denig, Technical and economic assessment for the production of torrefied ligno-cellulosic biomass pellets in the US, Energy Convers. Manag. 66 (2013) 153–164. https://doi.org/10.1016/j.enconman.2012.09.024.

[93] A. Mirkouei, K.R. Haapala, J. Sessions, G.S. Murthy, A mixed biomass-based energy supply chain for enhancing economic and environmental sustainability benefits: a multi-criteria decision making framework, Appl. Energy 114 (2017)Nov 1; 1053-1062. https://doi.org/10.1016/j.apenergy.2017.09.001.

[94] K. Sahoo, E. Bilek, R. Bergman, S. Mani, Techno-economic analysis of producing solid biofuels and biochar from forest residues using portable systems, Appl Energy 235 (2019) 578–590. https://doi.org/10.1016/j.apenergy.2018.10.076.

[95] Riebeek H. The Carbon Cycle. NASA Earth Observatory. Https://EarthobservatoryNasaGov/Features/CarbonCycle (Accessed October 5, 2020) 2011.

[96] F. Cherubini, N.D. Bird, A. Cowie, G. Jungmeier, B. Schlamadinger, S. Woess-Gallasch, Energy- and greenhouse gas-based LCA of biofuel and bioenergy systems: Key issues, ranges and recommendations, Resour. Conserv. Recycl. 53 (2009) 434–447. https://doi.org/10.1016/j.resconrec.2009.03.013.

[97] A. Kadiyala, R. Kommalapati, Z. Huque, Evaluation of the life cycle greenhouse gas emissions from different biomass feedstock electricity generation systems, Sustainability 8 (2016) 1181. https://doi.org/10.3390/su8111181.

[98] N. Yang, Bao, Zhao, Xie, Potential reductions in greenhouse gas and fine particulate matter emissions using corn stover for ethanol production in China, Energies 12 (2019) 3700. https://doi.org/10.3390/en12193700.

[99] P.A. Beauchemin, M. Tampier, Emissions from wood-fired combustion equipment, Envirochem. Serv. Inc., North Vancouver, BC (2008). https://www2GovBc.ca/Assets/Gov/Environment/Waste-Management/Industrial-Waste/Industrial-Waste/Pulp-Paper-Wood/Emissions_report_08Pdf Accessed July 3, 2020.

[100] IRBEA (Irish BioEnergy Association). Study on biomass combustion emissions: a report completed by Fehily Timoney & Company and kindly supported by the sustainable energy authority of Ireland, 2016. www.seai.ie/resources/publications/2016_RDD_108._Biomass_Combustion_Emissions_Study_-_IrBEA.pdf. (Accessed on May 26, 2021).

[101] C. Wang, Y. Chang, L. Zhang, M. Pang, Y. Hao, A life-cycle comparison of the energy, environmental and economic impacts of coal versus wood pellets for generating heat in China, Energy 120 (2017) 374–384. https://doi.org/10.1016/j.energy.2016.11.085.

[102] I. Butnar, J. Rodrigo, C.M. Gasol, F. Castells, Life-cycle assessment of electricity from biomass: Case studies of two biocrops in Spain, Biomass Bioenergy 34 (2010) 1780–1788. https://doi.org/10.1016/j.biombioe.2010.07.013.

[103] S.M. Shafie, H.H. Masjuki, T.M.I. Mahlia, Life cycle assessment of rice straw-based power generation in Malaysia, Energy 70 (2014) 401–410. https://doi.org/10.1016/j.energy.2014.04.014.

[104] F. Pierobon, M. Zanetti, S. Grigolato, A. Sgarbossa, T Anfodillo, R. Cavalli, Life cycle environmental impact of firewood production—a case study in Italy, Appl. Energy 150 (2015) 185–195. https://doi.org/10.1016/j.apenergy.2015.04.033.

[105] UNESC, Special edition: progress towards the sustainable development goals, Rep. Secr. (2019). Https://DoiOrg/101163/Ej9789004180048.i-962115 Accessed October 5, 2020.

[106] NCSL, State renewable portfolio standards and goals, Natl. Conf. State Legi. (2019). Https://WwwNcslOrg/Research/Energy/Renewable-Portfolio-StandardsAspx Accessed October 5, 2020.

[107] M.F. Sherlock, The renewable electricity production tax credit, Renew. Energy Tax Incent. Sel. Issues Anal. (2018). Https://FasOrg/Sgp/Crs/Misc/R43453Pdf Accessed October 5, 2020.

[108] C. Panoutsou, Supply of solid biofuels: potential feedstocks, cost and sustainability issues in EU27, Green Energy Technol. 28 (2011) 1–20. https://doi.org/10.1007/978-1-84996-393-0_1.

CHAPTER NINETEEN

Potential value-added products from wineries residues

Prasad Mandade, Edgard Gnansounou

Ecole Polytechnique Fédérale de Lausanne (EPFL), School of Environment, Civil Engineering, and Architecture, Institute of Civil Engineering, Bioenergy and Energy Planning Research Group, Lausanne, Switzerland

19.1 Introduction

Currently, the world is confronting significant challenges primarily connected with the substitution of fossil-based materials; therefore, there has been developing enthusiasm for bio-based materials, fuels, and products in the last two decades [1]. Interest to improve the sustainability of resources is increasing due to the thrust to reduce dependence on fossil oil, reduction in greenhouse gas (GHG) emission, jobs and wealth creation, etc. The uncontrolled deterioration of organic waste from the food and agricultural sector results in a negative effect on human and environmental health due to increased contamination of air, soil, and water. The disintegration of 1 metric ton of organic waste releases 50–110 m^3 of CO_2 and 90–140 m^3 of CH_4 [2]. Integrated solutions and strategies for the valorization of these wastes/residues are needed for environmental and economical and sustainable management [3].

Viticulture produces different types of wastes/residues which is one of the leading rural activities across the world. Because of raising environmental concerns valorization of waste/residues produced in the winery industry is becoming one of the most crucial fields of research to recover the value-added compounds in the high demand energy economies [4]. It is imperative to move from the current model of waste generation by producing the desired product consuming the available resources to the circular model of wealth creation by improving the management of winery waste. The current management practices of handling the residues and wastes lead to the lowering of their pollution load without valorization which has negative economic, environmental, and economic impacts on the winery industries [5].

As indicated by the Food and Agriculture Organization Corporate Statistical Database, the grape production for 2017 was higher than 77 million *t* representing one of the major horticultural crops in the world [6]. As stated in Statistical Report on World Viticulture as stated by IOV (International Organization of Vine and Wine), 39% of the total world grape production contributed by Europe, 34% Asia, and 18% America. Countries like Spain (14%), China (11%), France (10%), Italy (9%), and Turkey (7%) indicating a significant part of the world vineyard [7]. France is the main wine-delivering nation on the planet which produces 41.4 million hectoliters (16.4%) followed by Italy (15.9%) and Spain with (12.1%) of the worldwide production for the year 2014 [8]. Winemaking is a multistage process generating substantial quantities of wastes and residues such as grape marc, wine lees, wastewater, etc. depending on the production capacity of the winery. It also produces high quantities of vine shoots during the cultivation and harvesting of grapes per hectare per year. Significant efforts are needed for safe waste treatment and discharge considering the increasing cost of conventional treatment of winery waste [9]. This triggers the possibilities of potential recovery of value-added compounds such as chemicals, fuels, and products from winery waste/residues and wastewater [10]. Economic, environmental, and social drivers will help for further development of new chemicals, material, and energy from the waste, to reduce the climate change impact of their production [8].

Increasing environmental regulation and disposal cost of winery wastes and residues pushes to recover useful products and holds great potential for future prospective production, with applications including pharmaceutical, phytochemicals, nutraceuticals, biofuels, enzymes, flavors, and pigments [11]. Currently, several research efforts regarding winery waste valorization focusing are being taken worldwide. Composting the winery waste is a preferred way of traditional winery waste disposal. Extraction of polyphenols as a food additive due to its high antioxidant and anti-inflammatory properties, recovery of ethanol and tartaric acid (TA) from wine residues but the use of extracted antioxidant as a nutritional supplement, has an additional economic advantage over other recovered products. Oil extraction from seeds, production of protein, and biosurfactants are also being proposed as novel ways of valorization of grape seeds [10]. The wine industry produces a huge amount of wastewater so the treatment is becoming more challenging. Most of the wastewater treatment frameworks have been planned with huge air circulation tanks thus, wastewater treatment plants are larger and very hard to oversee. High organic load in the winery wastewater advances extreme biomass development, which creates problems of sludge production and its disposal, and thus the adoption of sustainable management strategies by reducing sludge production to improve the quality of water could be the key to the success of the winery wastewater treatment plants but it is still at an early stage of development [12]. In the winemaking procedure, the time between the waste age and its valorization directly affects the concentration of phytochemicals and their bioactive potential. Valorization incorporates extra strides for the advancement of novel, separation, refinement, and recuperation methods to improve the quality and amount of the recovered material [7]. Execution of suitable waste management is a difficult task for the advancement of inventive and successful valorization methods for winery wastes. Developing interest in biobased products and the desperation in the reduction of the environmental footprint of agro-industrial activities have pushed an intense legal framework system for the recovery and reuse methods [13]. Wastes created by the agriculture and food industry is the perfect feedstock for valorization to deliver low-value high-volume (LVHV) items to satisfy the worldwide need along with the creation of high-value low-volume items considering the criteria's such as size, nutritive content, and supply of the material [8].

19.2 A large diversity of wastes/residues of grape

19.2.1 Wastes VS residues

The Renewable Energy Directive 2009/28/EC amended through Directive (EU) 2015/1513 (RED) includes definitions of and specific incentives for the promotion of biofuels or bioliquids made from waste and residues. The verification and assessment of material and its waste or residue status is important, as there are no EU-wide harmonized lists defining waste or residues.

In the case of waste and residues, the main task at the origin is to verify the type of raw material used and its status as a genuine waste or residue. Waste and processing residues other than those directly derived from agriculture, aquaculture, fisheries, and forestry do not need to comply with the sustainability requirements for sustainable cultivation of biomass. "According to Renewable Energy Directive and Fuel Quality Directive (RED-FQD) wastes are "considered to have zero life-cycle GHG emissions up to the collection of those materials indicating no land-use emissions." Waste Framework Directive (WFD), defines "waste" as any substance which the holder discards or intends or is required to discard [14].

RED-FQD provides differential treatment to sub-categories of residues and coproducts. Residues are divided into four main sub-categories: agricultural, forestry, aquaculture, and fisheries and processing [15]. Determining the subcategory of a residue and whether it is a coproduct has significant implications on the life cycle sustainability impacts. Depending on the current utilization of produced material/substance it can be treated as waste or residue. Most of the materials/substances generated in viticulture and grape processing except wastewater, holders intend to valorize hence these substances/materials are considered as residues.

19.2.2 Viticulture waste

Viticulture is the cultivation and harvesting of grapes. The global wine and table grape industry generate substantial quantities of pruning waste each year which is conventionally composted or burned as waste, often incurring related cost to the winery. Since agricultural byproducts/waste are barely used comparing with highly valuable products, they are ignored which is available in large quantities [16, 17]. Vine shoots a grape waste that is produced from pruning in high quantities from 2 to 4 t of residues per hectare. Vine shoots are consisting of three major fractions: cellulose, hemicelluloses, and lignin as a suitable raw material for the food, pharmaceutical, and chemical industries by fractionation processes. Several kinds of research have been carried out to use vine shoots as a source of high value-added compounds like polyphenols, preparation of cellulose pulp, and as a source of renewable sugars that could be converted into bulk products.

Fractionation of lignocellulosic biomass is the center of the biorefinery facility, aiming to produce bulk chemicals such as lactic acid (LA), xylitol, or ethanol and further process them into marketable products [18].

19.2.3 Winery wastes and residues

Wine production is one of the major agro-industrial activities in representing the biggest agrarian part of numerous European nations. Food and beverage products make important contributions to a range of environmental impacts considering the scale of production and the wine sector is not an exception [19]. During winemaking, enormous amounts of wastes such as grape marc/pomace (the residue obtained after pressing for white wines or vinification for red wines) and stalks are produced as by-products. Grape marc constitutes pulps, skins, and seeds that are obtained after the crushing of grapes to get the fermentation must. Wet wastes are delivered while processing of grape into wine, and that comprises 62% marc, 14% lees, 12 % stalk, and 12% water waste [16]. One ton of grape processing approximately gives 0.13 t grape marc, 0.06 t wine lees, 0.03 t of stalks, and 1.65 m^3 of winery wastewater [11]. These waste and residues have different origins and can be characterized by relying upon the vinification procedure where they are formed [18].

Grape stalks were obtained by grape stripping operation. It has a high degree of filaments (lignin and cellulose) and a high nutritive mineral element like nitrogen and potassium. Grape stems and dewatered sludge represent 12% of total wet waste. Grape stems represent the woody skeleton of the grape packs having lignocellulosic synthesis and are acquired after the destemming procedure, before the crushing of grapes [4]. One of the generally utilized practices is spreading them legitimately on fields or in the wake of treating the soil, to recuperate this substrate as manure. The second most abundant residue is wine lees constitute 14% of total wet waste that is settled at the base of the vessels after the wine maturation process. Wine lees contain high organic matter, having high acid pH values and low electrical conductivity. The organic matter consists of many soluble compounds (ethanol, malic and TA along with sugars), substances partially adsorbed on particles such as polyphenols, and microbial biomass (mainly yeasts). The quantity of wastes produced depends on the grape cultivar and winemaking process. The fermentation stage for red wine fermenting must is in contact with the skins, seeds, and stems, while solid parts are not involved in the white wine production [8].

19.2.4 Estimation of the potential of wastes and residues

Wine production represents one of the largest agricultural sectors of many European countries in terms of cultural heritage and economy. Increasingly expensive conventional treatments of winery waste necessitate significant effort, resources, and energy for the safe discharge of waste. For the valorization of winery waste, it is important to know the potential availability of waste/residues to explore the different biorefinery options.

This chapter illustrates the estimation of the potential of different types of residues and wastes based on the assumptions of the crop to residue ratios for the different cantons of the Switzerland context to explore the different valorization options [20]. Data for the area of grape cultivation with white and red variety along with the wine production in each canton is taken from the wine statistics for Switzerland. The Vine shoot is the major sub-product of vineyards with a production of 2–4 t/ha. Vine shoot to the grape ratio of 3 t/ha is used for the estimation of the potential of vine shoot biomass [18].

Along with the vine shoot as viticulture waste, different types of wastes and residues in the winery are being produced. The quantification of these wastes provides important information about the scale of the waste generation that helps to explore the different waste treatment, such as reuse, recovery options from waste considering multiple factors focusing on social, environmental, and economic aspects of valorization. Table 19.2 gives the estimated potential of residues and wastes from the winery for the different cantons of Switzerland. The residues to crop ratio from Table 19.1 are used for estimation.

The primary objective of this work is to assess feasible options that could be applied to valorize winery wastes utilizing the cascading approach of biorefinery to produce chemicals, fuels, and energy. As the current winery waste is not utilized to its full potential and has the potential to recover the important value-added products along with the conventional utilization of lignocellulosic waste. This offers potential advantages of winemaking waste and residues to produce new products by exploring different options focusing on newer technologies and their novel applications in pharmaceutical, horticultural, domesticated animal fields, and in energy recovery systems. This chapter also illustrates the estimation of the potential of different types of winery waste and residues for Switzerland context that provides the basis for decision making and policy design by considering the multiple aspects of sustainability such as social, environmental, and economic.

19.3 Valorization of the residues and wastes

Plant-based biomass has been intensively grown to produce chemicals, fuels and power that prompts land-use conflicts for crop development. Winery wastes and residues contain different high-valued chemical substances including carbohydrates, proteins, and other value-added compounds depending on its origin [1]. Recovered materials have

Table 19.1 Residues/waste to crop (grape) ratios used in the assessment.

Type of waste	Residues/waste to the crop ratio
Grape stalks	0.03
Grape marc	0.13
Wine Lees	0.06
Wastewater	3.21

Cinzia et al. [19]; Oliveira and Duarte [10].

Table 19.2 Estimation of waste/residues for the different cantons of Switzerland.

Name of the canton	Vine shoots (t)	Grape marc (t)	Grape stalk (t)	Wine lees (t)	Wastewater (L)
Aargau (AG)	1168.52	425.57	98.21	196.42	78,926.50
Appenzell (Appenzell Ausserrhoden (AR)/ Appenzell Innerrhoden (AI))	14.35	4.52	1.04	2.09	839.03
Bern/Berne (BE) (sans lac de Bienne)	744.78	280.23	64.67	129.34	51,971.52
Basel Land (BL))	356.49	128.02	29.54	59.09	23,743.53
Glarus (GL)	5.88	2.15	0.50	0.99	398.62
Graubünden (GR) (San Misox)	1357.24	484.61	111.83	223.66	89,875.65
Luzern (LU)	191.68	79.54	18.36	36.71	14,752.11
Sankt Gallen (SG)	634.73	196.33	45.31	90.62	36,412.53
Schaffhausen (SH)	1440.71	598.42	138.10	276.19	110,983.15
Solothurn (SO)	31.46	8.39	1.94	3.87	1555.90
Schwyz (SZ)	114.68	47.46	10.95	21.90	8801.76
Thurgau (TG)	751.39	309.42	71.40	142.81	57,385.01
Uri (UR) Unterwalden (Obwalden (OW)/ Nidwalden (NW))	17.62	6.64	1.53	3.06	1231.22
Zug (ZG)	10.96	3.81	0.88	1.76	707.23
Zürich (ZH)	1830.29	777.99	179.54	359.07	14,4287.10
Ticino (TI)	3277.27	933.87	215.51	431.02	173,196.60
Freiburg / Fribourg (FR)	348.59	155.90	35.98	71.95	28,912.71
Genève (GE)	4229.59	1747.84	403.35	806.70	324,157.34
Jura (JU)	53.50	12.27	2.83	5.66	2275.98
Neuchâtel (NE)	1819.13	612.92	141.44	282.89	113,673.83
Vaud (VD)	11,324.49	5172.53	1193.66	2387.32	959,304.75
Valais/Wallis (VS)	14,411.37	7278.16	1679.58	3359.15	1,349,819.25
Total	44,134.72	19,266.59	4446.14	8892.27	3,573,211.30

potential applications in different sectors which will help in the reduction of volume and concentration of waste [8]. The development of an integrated methodology of winery waste and residues management is of considerable interest to different actors of the supply chain such as growers, consultants, contractors, policy regulators, equipment providers, and the general public. The valorization of winery waste and residue provides the opportunity to design and produce chemicals and fuels that in turn helps in improving the profitability of the wine sector [21].

19.3.1 Composition of the wastes/residues

To address waste management, it necessitates having the details of the wastes/residues along with the composition so that specific value-added products can be extracted from it. As the different types of wastes/residues are generated from the various stages of grape processing, the composition of these wastes/residues is different. Winemaking is seasonal and occurring from January to April in the south half of the globe and among August and October on the north side of the equator. Geographical, edaphic, seasonal, and weather-related factors along with the type of grape variety affect the composition of grapes [22]. Typically vine shoots are the lignocellulosic residues produced from pruning contains cellulose (20%–48%), hemicellulose (4%–36%), lignin (16%–27%). Max et al. [23] provide vine shoots composition that constitutes the main component cellulose (34%) followed by lignin (27%) and hemicellulose (19%). Bhaskar et al. [21] pointed out the residual chemical composition of vine shoots with cellulose 32%–49%, lignin 21%–26.4%, and hemicellulose 11.2%–13.6%.

Winery wastes can be classified into two fundamental categories such as solid waste that composed of mainly up to 45% grape pomace, up to 7.5% grape stalks, about 6% grape seeds, and liquid wastes representing a major portion of winery wastewater and other liquid wastes. Grape pomace mainly consists of seeds, skin, and remaining pulp and its composition depending on the grape variety and winemaking technology. After the extraction of the majority of phenolic compounds during the winemaking process, its by-products and wastes also contain significant amounts of phenolic compounds due to their incomplete extraction. Red grape pomace contains approximately 60%–65% of phenolic compounds after red wine production [22]. The typical chemical composition of grape pomace represents up to 72% of moisture content including cellulose (9%–48%), hemicellulose (4%–35%), lignin (11%–24%), and phenolic compounds along with other compounds like tartrate acid, pigments, etc. Bhaskar et al. [21] indicated the chemical composition for grape pomace consists of cellulose 20.8%–48.2%, lignin 16.8%–24.2%, and hemicellulose 4.8%–35.3%.

Zacharof [7] pointed out grape pomace with moisture from 50% to 72% and have lignin content from 17% up to 24%, cellulose ranging from 27% to 37%, and less than 4% protein content. Grape pomace contains up to 15% sugars, 0.9% pigments, and phenolics while red grape pomace constitutes up to 1 % tartrate acid and about 40% fiber. Jin et al. [24] estimated the chemical composition of grape pomace as cellulose 9.2%–14.5% hemicellulose 4.0%–10.3%, lignin 11.6%–17.2%, pectin 5.4%–5.7%, and protein 7.0%–14.5%. The most prominent polyphenols found in grape pomace are (1) phenolic acids, such as caffeine, gallic, and syringic acids (2) phenolic alcohols (3) flavonoids, for example, catechin, epicatechin, procyanidins, quercetin, and luteolin (4) stilbenes, and (5) proanthocyanidins. Phenolic mixes are disseminated in the mash (10%), seeds (60%–70%), and the skin (28%–35%) of the grape part [8].

Grape seeds represent a rich source of antioxidant compounds, such as vitamin E, containing saturated or unsaturated isoprenoid. Grape seeds also contain polyunsaturated fatty acids majorly linoleic acid, followed by α-linolenic and oleic acids [2]. Grape seeds (w/w) constitutes 60%–70% of carbohydrates 13%–19% essential oil, 11% protein,

7% phenolic compounds, and other substances minerals [25]. Total extracts from grape seed constitute 60% to 70% polyphenol compounds [26]. Grape seeds represent up to 5% of the grape weight and contributing about 38%–52% of solid wastes generated by wine industries representing about 40% fiber, 10%–20% lipid, 10% protein, complex phenolics as well as sugars and minerals [27]. Wine lees, composed of solid and liquid fractions and their composition depends on their agronomic and environmental conditions, type of grape variety, and the time of maturing in the wood barrels [28]. The solid part represents the remains precipitated consisting of bacterial biomass, undissolved carbohydrates, phenolic compounds, lignin, proteins, inorganic salts, organic acid salts, and other materials at the bottom of the tank. The liquid phase constitutes spent fermentation broth rich in organic acids and ethanol. Typically, wine lees contain 62.9% (w/w) water, 5.7% (w/w) ethanol and 31.4%(w/w) solids on a dry basis [9]. Vinasse is a by-product of the wine lees, it is liquid fraction waste derived from the refining of the wine lees, which is carried out to recover ethanol and elaborate distilled beverages [8].

Grape stalk represents 14% of total winemaking solid wastes and its composition contains 17%–26% lignin, 20%–30% cellulose and 3%–20% hemicellulose, and 6%–9% ash [29]. Grape stems contain flavan-3-ols, hydroxycinnamic acids, monomeric and oligomeric flavonols and stilbenes which is around 5.8% on a dry weight basis [13]. Prozil et al. [30] evaluated the chemical composition of grape stalks of red grape constituting mainly of cellulose (30.3%), hemicelluloses (21%) (xylan 12% as a most abundant component), lignin (17.4%), proteins (6.1%) and tannins (15.9%). Ping et al. [31] estimated the composition of the grape stalk with 34% lignin, 36% cellulose, 26% hemicellulose, and 6% tannins. Bhaskar et al [21] indicated the chemical composition for grape stalk that contains cellulose (20.1%–37%), lignin (2%–56.1%), and hemicellulose (4.8%–35.3%). Table 19.3 presents average quantities of winery waste and byproduct concentrations for winery wastes and residues.

Grape processing also generates a huge amount of wastewaters during the different stages of processing, including fermentation, storage, and maturation clarification, decanting, and bottling [12]. Cleaning generates wastewater by use of water along with solvents, detergents, and chemical agents, like sodium hydroxide at different process stages. Winery wastewater produced has been estimated as 0.2-48.2 L/L of wine produced. Winery wastewater is mostly acidic with high organic content with high biochemical oxygen demand (BOD) therefore it necessitates treatment before final release in the environment. Typical composition of winery waste wastewater is given in Table 19.4 [8].

19.3.2 Conventional valorization of the wastes/residues

Winery wastes and residues are considered as a valuable resource that could potentially be used as a nutrient-rich organic soil amendment. But due to higher scale of production of these wastes at the local level prompts removal of the material on rural land that can harm crops by discharging extreme measures of phytotoxic polyphenols to soils

Table 19.3 Winery waste and by product concentration (adapted and modified [28]).

Type of winery waste	% weight of grapes
Grape stalks	2.5%–7.5%
Grape pomace	~15% dry (wet up to 25%–45%)
– Sugars	~ (up to 150 g/kg)
– Phenolics/Pigments:	~9 kg/t (red grape pomace)
– Tartrate	~50 to 75 kg/t
– Fibers	~ 30 % to 40 %
Grape seeds	~ 3 % to 6 %
– Grape seed oil	~Oils 12%–17% (76% linoleic (omega-6 fatty acid)
– Phenolics	~4%–6%
Wine lees	~ 3.5%–8.5 % (compared to initial grape quantity)
– Pigments/colorants:	~ 12 kg/t (red wine lees)
– Tartrate	>100–150 kg/t
– Ethanol	~ 50 % of 10%–12% vol wine
– Beta-1,3-glucans:	~ 6%–12 % of dry weight
Vine pruning	~2–4 t/ha/yr
Winery wastewater	0.2–48.22 L/L of wine

Table 19.4 Typical winery wastewater composition (adapted from [8]).

Parameters	Unit	Min.	Mean	Max.
pH	mg/L	2.5	5.3	12.9
Total solid (TS)		190	8660	18,332
Total suspended solids (TSS)		66	1700	8600
Total volatile solids (TVS)		661	5625	12,385
Chemical oxygen demand (COD)		320	11,886	49,105
Biological oxygen demand (BOD)		181	6750	22,418
Total organic carbon (TOC)		41	1876	7363
Total Phosphorous (TP)		2.1	53	280
Total nitrogen (TN)		10	118	415
Total phenolic compound (TPh)		0.51	205	1450
Electrical conductivity (EC)	mS/cm	1.1	3.46	7.2

and causes groundwater pollution. Earthworms can digest polyphenols that potentially eliminates the agronomic problems associated with the application of the winery waste to the soil by vermicomposting [32, 33]. The quantity and composition of the residues rely upon the type of grape variety. Conventionally, wine lees and grape pomace have been used as a supplement in animal feeding as well as fermentation nutrient supplements. These wastes and residues are also incinerated, which demands high costs of operation and release of toxic gases which can cause harm to human health and the environment [4].

In some studies, it can be found grape marcs use for compost, as a soil conditioner, and even for the production of pellets [10]. Conventional use of grape seeds is for oil recovery and as a source of tannins and protein for animal feed or burning [2, 22]. Vine shoots are traditionally burned in the field to avoid the proliferation of phytopathogens on soil or even left in the field to recover nutrients as organic fertilizer since their economic value is very small [17]. Traditionally grape stalks are used as fertilizer but due to the presence of polyphenols makes them incompatible with agricultural requirements and therefore, they must be conditioned before use. Lin et al. [33] reported that grape stalks are burnt in the field releasing toxic compounds like polycyclic aromatic hydrocarbons that causes significant environmental problems. Composting and co-composting became one of the common important applications for all types of winery wastes to stabilize of vineyards with poor soil represents a method for nutrients recycle [19]. Kontogiannopoulos et al. [34] pointed out the traditional use of wine lees for animal feeding [34]. Due to the presence of different polyphenolic compounds in the matrix of winery wastes incineration or landfilling may be harmful to the environment. These compounds can cause a decrease in pH and increased resistance to biological degradation. Surface and groundwater pollution poses a major environmental problem due to conventional methods of winery waste management [10].

19.3.3 Potential valorization from viticulture and viniculture wastes/residues

The present chapter explores potential options of valorization of different winery wastes such as vine shoots, grape marc, wine lees, vinasses, and winery wastewater as a feedstock to produce platform chemicals and chemical intermediates, such as polyphenols, TA, LA, ethanol, enzymes, and fuels for energy recovery from pyrolysis and anaerobic digestion. Therefore, valorization of these wastes would be interesting for the wine market for additional economic stimulus and it would diminish the mandatory removal expenses, which are expected to pay to prevent contamination of the environment [4,35].

Industrial production of numerous platform chemicals along with the value-added compounds can be produced in biorefineries with significant commercial potential Table 19.5. However, significant efforts are needed to overcome the challenges in valorization to produce chemicals with high selectivity and yields [7]. Various industries like BASF, DuPont, Dow Chemical, etc., are aggressively pursuing biomass valorization. Integrated biorefinery producing both bio-based products and energy such as fuel, power, and heat is expected to increase substantially in the coming years with increased market demand. Biorefinery products delivered ought to be market competitive due to their economic sustainability and should reduce GHG emissions. This stimulates innovation in the biorefining processes and helps in the expansion of a biobased economy leading to a more environmentally benign chemical manufacturing industrial sector. To build up a sustainable integrated biorefinery, it is important to deliver high worth included bioproducts alongside bio-energies in a proficient way [36].

Table 19.5 Potential value-added biorefinery products from grape waste.

Type of waste	Treatment or technology used	Product	References
Vine shoots	Hydrolysis, fermentation of hemicellulosic sugars by L. pentosus	Lactic acid, biosurfactants	[71,83]
	Hydrolysis, delignification and simultaneous saccharification and fermentation of cellulosic fraction	Lactic acid	[84]
	Hydrolysis and fermentation of hemicellulosic sugars using Lactobacillus and *Debaryomyces hansenii*	Lactic acid; xylitol; biosurfactants	[37,79]
	Pulping process	Cellulose pulp	[80]
	Carbon dioxide activation	Activated carbon	[81]
	Hydrolysis and séparation	Prebiotic oligosaccharides, lignin, and glucose	[77,82]
Grape marc	Extraction	Polyphenols	[76,99]
	Extraction	Tartaric acid	[33]
	Ultrasound-assisted extraction	Proanthocyanidins	[100]
	Extraction using a mixture of water–sodium hydroxide	Tannin for use of wood adhesive	[30,101]
	Extraction of red grape pomace	Polyphenols and dietary fibers	[34,102]
	Grape skin extraction using chromatography	Polyphenols	[103]
	Extraction of grape seed	Proanthocyanidins and other phenolic compounds	[104]
	Pelletization	Pellets	[22]
	Hydrolysis, fermentation with L. pentosus	Bio-emulsifiers	[79]
	Solid-state fermentation	Hydrolytic enzymes	[51]
	Hydrolysis, fermentation with L. pentosus	Lactic acid biosurfactants	[79]
	Gasification, combustion, pyrolysis and anaerobic digestion	Energy recovery in terms of electricity, biogas, gaseous, liquid and solid fuel	[39]
	Grapeseed oil Fermentation with *Pseudomana aeruginosa*	Biosurfactants	[105]
	Fractionation of grape seed and skin extracts from grape waste	Polyphenol	[106]
	Extraction of grapeseed oil	Skin moisturizer (gel formulation)	[107]
	Anaerobic digestion of grape marc	Methane production	[43]
	Pre-treatment of lignocellulose using acid hydrolysis, followed by enzymatic saccharification and then fermentation	Ethanol	[47,108]
	Anaerobic digestion of grape marc	Methane	[48,52]
	Esterification of oil using fruit seeds	Biodiesel	[61]

(continued)

Table 19.5 (Cont'd)

Type of waste	Treatment or technology used	Product	References
Wine lees	Cation exchange resin	Tartaric acid recovery	[34]
	Fermentation, fractionation and bacterial conversion	Recovery of antioxidant, tartaric acid, ethanol and nutrients for PHB production	[9,68]
	Extraction using different methods	Recovery of phenolic compounds	[72,109]
	Extraction	Flavonols and anthocyanins	[69,73]
Winery wastewater	Codigestion of grape waste with wastewater	Biogas	[88–90]
	Electrodialysis at 60°C	Tartaric acid	[93,110]
	Removal of polyphenols using porous carbons developed from exhausted grape pomace	Biogas production	[12]

19.3.3.1 Potential utilization of grape marc (pomace)

Grape marc constitutes hemicellulosic sugars mainly that, after hydrolysis produces mixtures of xylose and glucose that could be converted to different chemicals in the presence of microorganisms. Researchers have used L. pentosus and L. rhamnosus to produce biosurfactants that induce the simultaneous production of LA [37]. Several efforts to recover polyphenols from grape marc, using different methods and solvents such as ethanol, methanol, etc., have been made with a considerable success rate [22]. Grape marcs are being extracted as a source of phytochemicals including phenolics, antioxidants, and pigments for a wide range of applications [38]. After extraction of useful chemical constituents its application using thermochemical and biological treatment to value-added products, agricultural and environmental applications such as composting, soil amendments, bio sorbent, etc., have been demonstrated in the literature [39].

Spigno et al. [40] extracted dried red grape marc to evaluate the antioxidant activity of phenolic compounds using the effects of solvent composition, time, and temperature (45 and 60 °C) on total phenols, anthocyanins (pigments), etc. The study also compared different alternative treatments such as ultrasounds (US), pulsed electric fields (PEF), and high voltage electric discharges (HVED) on solvent-free extraction of high value-added components from fermented grape pomace and found HVED to be the most efficient method to recover higher phenolic compounds with lower energy requirement than PEF and US [40]. Watson [41] discussed the extraction of polyphenols using different techniques such as US assisted extraction, microwave-assisted extraction, solid–liquid extraction, enzyme assisted extraction, etc. to recover the polyphenols from grapes. Currently, grape pomace is utilized as a source of resveratrol and flavonoids, which are utilized as food supplements [8].

De la Cerda-Carrasco et al. [42] showed a comparatively higher yield of extraction of phenolics and proanthocyanidins in white grape marc compared to red grape marc using methanol and acetone as a solvent. Zabaniotou et al. [3] proposed closed-loop integrated solutions for the concerns of a winery waste (pomace) to produce bio-based products, energy carriers, and biochar with the production of six products, along with wine from grapes. Martinez et al. [43] demonstrated multi-purpose biorefinery to valorize red grape pomace to obtain natural antioxidants, volatile fatty acids, biopolymers, and biomethane. Beres et al. [44] suggested complete utilization of grape pomace to recover bioactive compounds, dietary fiber, and oil. Zhang et al. [45] suggested the simultaneous extraction of dietary fiber and polyphenol and the application of remaining residue as fertilizer.

Jin et al. [46] developed an integrated process to fully utilize grape pomace that gives 71.9 g of oil, 322.8 g of polyphenols and 20.7 g of total ABE (acetone, butanol, and ethanol) from 1 kg of dry grape pomace. Benetto et al. [22] evaluated the overall environmental performance using a life cycle approach to produce grape marc pellets from grape marc for domestic heat applications. Muhlack et al. [39] reviewed the different extraction techniques such as pressurized liquid extraction, supercritical fluid extraction, PEF, HVED, etc., along with the conventional extraction process for extraction of the phenolic compounds from grape marc. Corbin et al. [47] illustrated the suitability of grape marc as a feedstock for bioethanol production with 270 L/t for white grape marc and 211 L/t for red grape marc upon fermentation. Muhlack et al. [39] summarized different thermal conversion process options for energy recovery from grape marc biomass such as combustion, gasification, pyrolysis, etc., with the illustration of the benefits and disadvantages of the process options along with the phenolic extraction.

Gunaseelan [48] used grape marc to produce methane with a yield of 420 CH_4/kg of volatile solids (VS) with a retention time of 20 days. Martinez et al. [43] demonstrated the methane production of 113 L/kg VS from the residual grape marc after polyphenol extraction. A study by Fabbri et al. [49] investigated the production of biogas and methane yields of 105.4–152.0 L/kg and 70.9-74.1 L/kg, respectively, for grape marc samples containing stalks, seeds, and skins. El Achkar et al. [50] estimated anaerobic digestion of grape marc for lab-scale with methane yields of 205 L/kg VS and larger pilot plant approximately 80% of lab-scale yield and found an increase in methane yield up to 280 L/kg VS after mechanical pretreatment. Díaz et al. [51] showed that the production of enzymes from grape marc using solid-state fermentation as an alternative option for producers for valorization. Javier et al. [52] evaluated the highest methane rate production of 47.97±4.29 mL/L h for 1.25 gVS/L under mesophilic conditions at laboratory scale for grape marc waste for the grape variety of Spain.

Several other applications of grape marc including compost [53], animal feed [54], biofertilizers [55], vermicomposting [53], soil conditioner [56], and removal of metal pollutants from industrial effluent are being produced using grape marc [23]. Grape marc entrapped within calcium alginate beads were investigated for the adsorption of micronutrients and sulfate from winery wastewaters [57]. Gómez-Brandón et al. [58] discussed recycling and revalorization strategies of grape marc such as composting, anaerobic digestion, vermicomposting, and co-composting with other organic materials such as municipal solid wastes. The addition of grape pomace as flour or various extracts into food formulas to increase the dietary ingestion of fiber and phenolic compounds provides an additional option of valorization and income opportunity for wine producers [59]. Papadaki and Mantzouridou [60] optimized for citric acid production using green olive processing wastewaters enriched with sugars from white grape pomace and Aspergillus niger B60 with yield (0.56 g/g) of the citric acid-containing maximum citric acid content of 85g/L. Gorans et al. [61] showed that phytosterols and squalene content in grape seed are 1.56 mg/g and 0.1 mg/g of oil. Along with that, grape seed oil contains 70% linoleic acid, with the yield of grapeseed oil 13% on w/w basis. Extracted Linolenic acid has been commonly used in the cosmetic and therapeutic sectors due to its skin revitalization properties.

19.3.3.2 Potential utilization of grape stalk

Grape stalk represents a rich source of polyphenols, mainly stilbenes and flavonoids. Diaz [51] found that total polyphenol and flavonoid contents were 38.2 ± 1.0 mg GAE/g DS (gallic acid equivalents/gm of dry stalk) and 37.6 ± 1.5 mg CATE/g DS (catechin equivalents/gm of dry stalk) for conventional solid–liquid extraction of the grape stalk. Microwave pretreatment showed an increase in the extraction yield of total phenolic content (TPC) and total flavonoid contents (TFC) of 19% and 23%, respectively. Diaz [51] also optimized grape stalks conversion into sugars-rich extracts via a hydrothermal process and found the most suitable temperature of 140°C to maximize the sugar yield (264 mg/g dry stalk). Amendola et al. [62] applied a two-step extraction with autohydrolysis pretreatment 180°C for 30 min followed by ethanol organosolv process at 180°C for 90 min isolating hemicelluloses, phenolic mixes, and lignin from red grape stalks. Organosolv alcohol with lignin contains phenols, which has potential application as a cancer prevention agent. Spigno et al. [29] explored the yields for hemicellulosic sugars, cancer prevention agent mixes, and cellulose for the red and white grape stalks and showed the impacts of the cultivar in lignocellulosic fractionation of the crude material.

Prozil et al. [63] evaluated the use of grape stalk for pellet production and they found that the specific energy consumption for pelletizing of grape stalks was approximately 25% lower when compared to that for softwood sawdust. Bastos-Arrieta et al. [64] demonstrated the utilization of polyphenols and reducing sugars present in extracts from grape stalk wastes as reducing agents and stabilizers in the silver nanoparticles production. Villaescusa et al. [65] investigated the usefulness of grape stalk for the removal of copper and nickel ions from metal-containing effluents. Teixeira et al. [66] tested different solid–liquid extraction conditions to obtain higher amounts of polyphenols and proposed the best conditions that include the use of a mixture of acetone/ethanol/water (1:1:1) at room temperature for 20 min, utilizing a prewash treatment with deionized water. Deiana et al. [67] presented the use of grape stalk to prepare activated carbon through the physical and chemical route and obtained BET areas between 1000 and 1500 m²/g, this is due to the catalytic effect of the minerals present in the grape stalk.

19.3.3.3 Potential utilization of wine lees
The winemaking produces a significant amount of wine lee residues and its management and disposal raise serious environmental concerns. Kontogiannopoulos et al. [34] showed 74.9% TA recovery from wine lees using cation exchange resin, by avoiding the waste calcium sulfate sludge of the conventional process. Rivas et al. [37] optimized the recovery of TA from distilled vinification lees up to 92.4% of the initial TA concentration. The residual lees were used as nutrients to produce LA by Lactobacillus pentosus CECT-4023 using hydrolysates of hemicellulosic vine shoot as a carbon source. Kopsahelis et al. [68] proposed the integrated refining of cheese whey and wine lees to produce commodity and specialty products, such as whey protein concentrate, antioxidants, ethanol, tartrate salts and microbial oil. Dimou et al. [9] demonstrated integrated biorefinery for the production of several value-added products s such as antioxidants, ethanol, and tartrate and the remaining stream as a nutrient for the production of poly (3-hydroxybutyrate) (PHB) production using the strain Cupriavidus necator DSM 7237.

Wine lees have been utilized for several applications such as the recovery phenolic compound extract using microwave-assisted extraction (MAE) that provides efficient extraction with a shorter time (17 min) than the conventional extraction for phenolic compounds (24 h) [69]. Naziri et al. [70] proposed the potential use of wine lees for the recovery of squalene using US-assisted extraction that is used for lowering cholesterol. Bustos et al. [71] used lees as a supplement of nutrients for LA production (105.5 g/L) using Lactobacillus rhamnosus. Wine lees have also found the application for the production of biogas through anaerobic co-digestion under mesophilic or thermophilic conditions showed similar yields (0.40 m³/kg COD fed) for both conditions [72]. The concentration of anthocyanins, a natural colorant with health-promoting properties is 10 times higher in wine lees than in grape skins. Red wine lees contain anthocyanins abundantly which are cyanidin, peonidin, delphinidin, petunidin, and malvidin along with that other phenolic compounds found in red wines are catechins, proanthocyanidins, and stilbenes. Diaz [51] analyzed TPC and TFC of wine lees extract. The mixture of 75:25(v/v) EtOH: H_2O showed the highest values with 254 mg gallic acid equivalents (GAE)/g dried extract (DE) and 146 mg catechin equivalents (CATE)/g DE respectively.

Jara-Palacios [73] summarized the extraction of antioxidant compounds using different extraction techniques, solvents, and pretreatment methods, focusing mainly on flavonols and anthocyanins from red grapes which are the most abundant phenolic compounds in wine lees. Tao et al. [74] investigated the extraction of total phenolics from wine lees by US ranged between 44 and 59 mg GAE/g dry matter (DM). Romero-Díez et al. [75] extracted wine lees utilizing various solvents, for example, water, methanol, ethanol, two hydroalcoholic blends as well as acetone and shows phenolic content in the range of 26 and 254 mg GAE/g DM, with the blend of 75:25 (v/v) EtOH: H_2O indicating the most elevated productivity. Delgado-Torre et al. [76] compared extracts from 18 different vineyard cultivars using different methods such as superheated liquid extraction (SHLE), microwave-assisted extraction (MAE) and US-assisted extraction (USAE) and they found optimized conditions for each extraction method as 80% (v/v) aqueous ethanol at pH 3, 180°C and 60 min for SHLE; 140 W and 5 min microwave irradiation for MAE; and 280 W, 50% duty cycle and 7.5 min extraction for USAE and reported better extraction efficiency using SHLE compared to the other two approaches [76].

19.3.3.4 Potential utilization of vine shoots

As vine shoots contents major composition of cellulose and hemicellulose, after hydrolysis these can be used for the production of biofuels and other platform chemicals and other sugar-based chemicals [4]. Dávila et al. [77] studied the suitability of coproduction of lignin and glucose from vine shoots using alkali delignification which can further be valorized to produce chemicals. Bustos et al. [71] carried out prehydrolysis to solubilize the hemicellulose and solid residues that contain mainly lignin and cellulose further it is subjected to simultaneous saccharification and fermentation (SSF) with *lactobaccillus rhamnosus* to produce LA. Sánchez-Gómez et al. [78] compared different extraction methods to extract polyphenols from vine shoots and studied its application in cosmetics, nutraceuticals, or pharmaceuticals due to the antifeedant and allelopathic activities of the extracts. Portilla et al. [79] employed bioconversion of hemicellulosic hydrolysates obtained from vine shoot to produce LA and biosurfactant using *Lactobacillus acidophilus* by complete conversion of glucose and xylitol from xylose using *Debaryomyces hansenii*.

Jiménez et al. [80] investigated the application of vine shoots for the preparation of cellulose pulp showing the 49.7% yield of cellulose pulp from vine shoot. The vine shoots that have been studied for the production of activated carbon demonstrates the BET surface area 528–1173 m^2/g and micropore volume 0.25–0.53 cm^3/g which can be used as adsorbent [81]. Along with these other applications such as a source prebiotic oligosaccharide from renewable sugars [82], sorbitol by glucose hydrogenation [4], LA, xylitol [37], ethanol, and biosurfactant was studied [83, 84].

Gullon et al. [17] analyzed five potential valorization schemes from vine shoots using the life cycle assessment and suggested most sustainable biorefining route of producing 2.8 kg of the antioxidant extract, 19.7 kg of hemicellulosic oligosaccharides, 16.6 kg of cellulose, 16 kg of lignin from 100 kg of vine shoots. Rodríguez-Pazo et al. [85] demonstrated the production of a combination of compounds such as LA, phenyllactic Acid and biosurfactants by using cultures L. plantarum and L. pentosus using hemicellulosic and cellulosic fractions of trimming vine shoots. This study shows the highest LA and phenyllactic Acid concentrations, 43.0 g/L, and 1.58 mM, that is obtained after 144 h by simultaneous saccharification and fermentation (SSF) using cocultures of Lactobacillus plantarum and Lactobacillus pentosus. Montalvo et al. [86] demonstrated the production of 876 ± 45 L CH_4/kg VS using combined waste wine lees and waste activated sludge aerobic treatment plant for the 30-day batch process. Pachón, et al. [87] analyzed two biorefinery scenarios using vine shoots as feedstock to co-produce chemicals from an environmental point of view for production of LA, and co-production of LA and furfural. This study showed that vine shoots offers environmental benefits for multiple impacts compared to conventional processes.

19.3.3.5 Biogas and other products from winery wastewater

Untreated wastewater created from wineries poses significant risks to the environment. Distinctive treatment alternatives are available for the management of wastewaters varying with cost, efficacy, and reliability. Makadia et al. [88] carried out anaerobic codigestion of winery wastewater and grape marc by producing methane for batch with the energy content of 5.04 MJ/kgVS. Mosse et al. [89] talked about treatment alternatives along with the mechanism of each process, advantages and disadvantages of their use to treat wastewater within the wine industry. Ros et al. [90] suggested waste-activated sludge from wastewater treatment and wine lees, to be cotreated using an anaerobic digestion process to produce biogas. Andreottola et al. [91] discussed the treatment of winery wastewater using several biological methods and presented existing status and advances in biological treatment of winery wastewater in the last decade, considering both lab, pilot, and full-scale studies. Moletta [93] carried out the winery wastewater treatment to produce biogas between 400 and 600 L/kg COD removed with 60% to 70% methane content using anaerobic digestion [92]. Andres et al. (1997) evaluated the technical feasibility of the recovery of TA up to 10 kg/m^3 of the winery wastewater using electrodialysis.

19.4 Proposed biorefinery scenario using zero-waste cascading valorization of wastes and residues

A biorefinery is an integrated facility that provides efficient, conversion of feedstocks through a combination of multiple processes, for example, physical, biochemical, and thermochemical into numerous products. The integrated approach is proposed and applied to winery waste and residues, for the optimal exploitation to produce value-added compounds. [94]. Biorefining aims to complete the utilization of winery waste by performing the different cascading processes with minimum energy and to maximize the overall value of the production chain. Several factors should be considered while assessing the practicality of such processes involved in the proposed approach. Therefore, several conversion technologies can be integrated, into the production of, value-added bioactives, chemicals along with the production of heat and electricity [8]. Literature provides various alternatives for the valorization of viticulture and viniculture waste/residue and its complete utilization will help to reduce the environmental impact significantly. Various studies have been carried for the extraction of polyphenolics from grape residue and waste such as the grape stalk, vine shoots, grape marc, wine lees. Zabaniotou et al. [3] proposed a closed-loop approach for cascade biorefinery to process winery waste using to produce multiproduct such as bio-oil, biochar, gas, hydrocolloid, and grape seed oil using hydrocolloid extraction followed by pyrolysis.

Based on the composition of different kinds of wastes and residues and the possible obtainable products using the advanced approaches and techniques several possible integral approaches can be feasible but the proposed approach considers a zero waste cascading approach for integral utilization of waste and residues. A proposed scenario for the complete utilization of the winery waste has been described in Fig. 19.1.

Typically, five stages may be defined in a proposed biorefinery:

1. Bioactive compounds extraction from raw biomass such as pigments, polyphenols, proteins, etc., from the winery waste and residues. The products obtained from this step have multiple applications in the nutraceutical and cosmetic sectors. Extraction of high-value compounds includes the recovery of bioactives with high economic interest by the partial detoxification of the waste ensuring removal of undesirable compounds for subsequent biological post-treatments. The interest of these value-added compounds has been increasing because of their different medical advantages including cancer prevention agents, cardiovascular, antihypertensive, and antiproliferative impacts.

2. Lignocellulosic residues that remained after the extraction are exposed to chemical treatment for the lignin extraction which can further be purified to produce some bioactive agents. The lignocellulosic part of the biomass residues contains also some interesting bio-active compounds that will be extracted from the biomass before being (bio)chemically modified to enhance their intrinsic biological properties or, even better, provide them with new ones. Lignin residues will be fractionated to building blocks that will be then modified to access new types of bioactive compounds. All these novel agents will be purified and various biological activities.

Currently, most lignocellulosic biorefineries using enzymes obtain lignin-rich streams by either (1) extracting the plant sugars to get the solid residue with high lignin content or (2) by pretreatments fractionating biomass to extract lignin before starch transformation [95]. Isolation of lignin from biomass can be accomplished by four main processes in industries such as the sulfite, soda, kraft, and organosolv processes. A few short-and long-haul viewpoints for lignin valorization to aromatics, cosmetics, polymers, and materials, are in-the-pipeline however it is expected to look at the limitations in current lignin valorization and create potential ways forward [96, 97].

1. Hydrolysis of the lignin portion can produce sweet-smelling phenolic mixes, for example, alcohols, aldehydes, ketones, or acids [76]. Dias et al. [98] talked about the

Potential value-added products from wineries residues 389

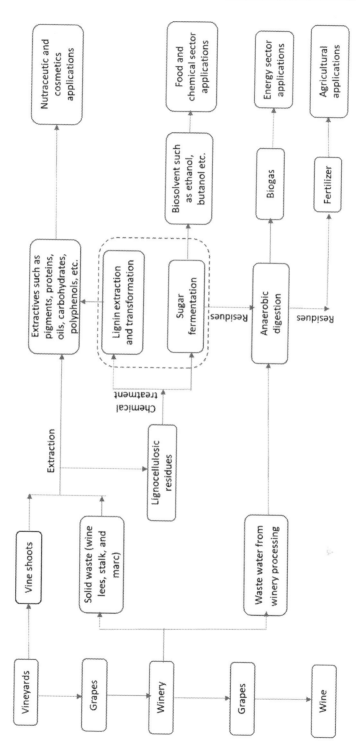

Fig. 19.1 A proposed cascading scenario for the utilization of winery waste.

future patterns for the lignin-based material, for example, retardants of fire, warm stabilizer, hydrophobicity operator, carbon fiber, aerogels, and cell reinforcements. Biomass residues after the removal of lignin contain sugars components such as cellulose and hemicelluloses which can be converted to bio solvents such as ethanol, butanol, etc. thermal pressure pretreatment and enzymatic hydrolysis followed by different types of microorganisms such as yeast and bacterial in a bioreactor. These bio solvents are broadly utilized in the food and chemical sector.

2. Residues in the process of separation of lignin as well as the conversion of sugars into bio solvents can further be utilized along with the winery wastewater to produce the biogas which can be used to deliver heat and electricity. Anaerobic digestion of these substrates can be combined with other substrates to produce higher biogas yield potential.
3. The remained residues after the biogas production can be utilized as fertilizer as an organic amendment as it contains nutrients and micronutrients.

Challenges with each of the described processes need to be addressed for its successful implementation. A biorefinery approach combining several processes could be attractive for the complete utilization of different winery wastes. Extraction and lignin valorization steps could provide high economic benefit by selling the recovered bioactive compounds. The third step in the proposed biorefinery produces bio solvents that can be used as a platform chemical, which can be used as precursors in different applications. The energy requirement for the different steps of the proposed biorefinery approach can be supplemented by producing biogas from anaerobic digestion in the fourth step. At the final stage, the digestate after biogas production process can be used for the recovery of the nutrients and organic matter for the agricultural sector. A solution would be to integrate the different processes to fully utilize the biomass using the biorefinery approach which will move towards economic, environmental, and social sustainability.

19.5 Conclusions and perspectives

The different types of winey wastes and residues should be considered for valorization using the biorefinery approach, as they meet different criteria, for example, the amount of waste as a raw material with high value-added components and potential of reducing environmental footprint. Proposed incorporated biorefinery could create some positive synergistic impacts such as (1) cost optimization by better exploiting the diverse types of waste (2) additional profits by generating multiple products in markets (3) minimizing waste generation, and (4) creation of renewable energies, synthetic compounds, and material (5) by avoiding the negative environmental impact. This approach could prompt general supportability by the execution of a zero-waste biorefinery approach for winey waste and residues.

At present, a large portion of research has been concentrating on the extraction of bio actives, since they can have a high market value. However, efficient valorization needs additional improvement in extraction, purification, and recovery methods to get a higher yield of bioactive phytochemicals for its targeted applications in cosmetics, pharmaceutical and food industries. Companies and different research organizations should put efforts and resources into new advances to diminish the effect of winery waste and buildups on nature and to set up new procedures that give additional income. The government, policymaker, and other actors should provide a suitable environment for the valorization of winery wastes. Along with that knowledge of the life cycle impact of valorization strategies and techno-economic viability are needed to provide the basis for policy decision making. Further research to improve resource recovery is needed to validate different biorefinery scenarios. As a range of applications, from winery waste and residues in the form of fuels, chemicals and materials offer significant opportunities for the winery sector.

The collaboration between the different partners and actors from research and industries will help to move on the sustainable path by reducing environmental impact by efficient valorization. Design and implementation of novel integrated biorefineries focusing on the potential recovery of winery waste and residues streams is the need of time that reduces GHG emissions by contributing to the climate change goals. Numerous nations have stepped up that encourages the adoption of environmental sustainability and waste valorization programs by improvement in resource efficiency which helps to satisfy the triple bottom line concept of combining financial viability with the requirement of social and environmental sustainability. The proposed biorefinery approach suggests treating different types of wastes and residues allowing the solution to an environmental problem and gaining economic benefit from the commercialization of the manufactured products. However, the continuous efforts from several actors of the production process need to be put in for the successful implementation of the valorization strategies considering different aspects of sustainability.

References

[1] R. Deutschmann, R.F.H. Dekker, From plant biomass to bio-based chemicals: latest developments in xylan research, Biotechnol. Adv. 30 (2012) 1627–1640.

[2] F.G. Fermoso, A. Serrano, B.A Fariñas, J.F. Bolaños, R. Borja, G.R. Gutiérrez, Valuable compound extraction, anaerobic digestion and composting: a leading biorefinery approach for agricultural wastes, J. Agric. Food Chem. 66 (2018) 8451–8468.

[3] A. Zabaniotou, P. Kamaterou, A. Pavlou, C. Panayiotou, Sustainable bioeconomy transitions: targeting value capture by integrating pyrolysis in a winery waste biorefinery, J. Clean. Prod. 172 (2018) 3387-3397.

[4] R.R. Diez, Development of an integrated and green biorefinery from winery waste: application to wine lees and grape stems Ph.D. Thesis, University of Valladolid, Valladolid, 2018, pp. 1–373.

[5] R. Devesa-Rey, X. Vecino, J.L. Varela-Alende, M.T. Barral, J. MCruz, A.B. Moldes, Valorization of winery waste vs. the costs of not recycling, Waste Manag. 31 (2011) 2327–2335.

[6] P. Masset, J.P. Weisskopf, Producing and consuming locally: Switzerland as a local market, in The Palgrave Handbook of Wine Industry Economics, Ugaglia A. A., Cardebat J.-M., Corsi A. (Eds.), Chap. 27, Springer Nature, Switzerland, 2019, pp. 507–522.

[7] E. Kalli, I. Lappa, P. Bouchagier, P.A. Tarantilis, E. Skotti, Novel application and industrial exploitation of winery by-products, Bioresour. Bioprocess. 5 (46) (2018) 1–21.

[8] M. Zacharof, Grape winery waste as feedstock for bioconversions: Applying the biorefinery concept, Waste Biomass Valorization 8 (4) (2017) 1011–1025.

[9] C. Dimou, N. Kopsahelis, A. Papadaki, S. Papanikolaou, I.K. Kookos, I. Mandala, A.A Koutinas, Wine lees valorization: biorefinery development including production of a generic fermentation feedstock employed for poly (3-hydroxybutyrate) synthesis, Food Res. Int. 73 (2015) 81–87.

[10] M. Oliveira, E. Duarte, Integrated approach to winery waste: waste generation and data consolidation, Front. Environ. Sci. Eng. 10 (1) (2016) 168–176.

[11] M. Bordiga, F. Travaglia, M. Locatelli, Valorisation of grape pomace: an approach that is increasingly reaching its maturity—a review, Int. J. Food Sci. Technol. 54 (2019) 933–942.

[12] A. Nayak, B. Bhushan, B.L. Rodriguez-Turienzo, Recovery of polyphenols onto porous carbons developed from exhausted grape pomace: a sustainable approach for the treatment of wine wastewaters, Water Res. 145 (2018) 741–756.

[13] A. Teixeira, N. Baenas, R. Dominguez-Perles, A. Barros, E. Rosa, D.A. Moreno, C.A. Viguera, Natural bioactive compounds from winery by-products as health promoters: a review, Int. J. Mol. Sci. 15 (9) (2014) 15638–15678.

[14] N. Uranic, T. Grabiel, Waste residues and coproducts, Briefing Note: Transport Environ. 2013 (2013) 1–7.

[15] International Sustainability & Carbon Certification ISCC, ISCC 201-1 Waste and Residues, Version 3, ISCC System GmbH, 2016, pp. 1–18.

[16] B. Lalevic, B.V. Sivcev, V.B. Raicevic, Z.R. Vasic, N. Petrovic, M. Milinkovic, Environmental impact of viticulture: Biofertilizer influence on pruning and wine waste, Bulg. J. Agric. Sci. 19 (5) (2013) 1027–1032.

[17] P. Gullón, B. Gullón, I. Dávila, J. Labidi, S. Gonzalez-Garcia, Comparative environmental life cycle assessment of integral revalorization of vine shoots from a biorefinery perspective, Sci. Total Environ. 624 (2018) 225–240.

[18] I. Dávila, O. Gordobil, J. Labidi, P. Gullón, Assessment of suitability of vine shoots for hemicellulosic oligosaccharides production through aqueous processing, Bioresour. Technol. 211 (2016) 636–644.

[19] D.R. Cinzia, Assessment of energy and material recovery from winery wastes Ph.D. Thesis, Ca' Foscari University of Venice, Italy, 2016, pp. 1–195.

[20] Wine Statistics, Federal office of Agriculture (FOAG) Switzerland. https://www.blw.admin.ch/blw/fr/home/nachhaltigeproduktion/pflanzlicheproduktion/weine-und-spirituosen/weinwirtschaftliche-statistik.html (Accessed 15 February 2020), (2018).

[21] A.P. Romani, M. Michelin, L. Domingues, J.A. Teixeira, Valorization of wastes from Agrofood and Pulp and Paper Industries within Biorefinery Concept Southwestern Europe Scenario, in Waste biorefinery : potential and perspectives, T. Bhaskar, A. Pandey, S.V. Mohan, D.J. Lee, S.K. Khana (Eds.), Chapter 16 2018, pp. 487–508.

[22] E. Benetto, C. Jury, G. Kneip, I. Vazquez-Rowe, V. Huck, F. Minette, Life cycle assessment of heat production from grape marc pellets, J. Clean. Prod. 87 (2015) 149–158.

[23] B. Max, J.M. Salgado, S. Cortes, Extraction of phenolic acids by alkaline hydrolysis from the solid residue obtained after prehydrolysis of trimming vine shoots, J. Agric. Food Chem. 58 (2010) 1909–1917.

[24] Q. Jin, A.P. Neilson, A.C. Stewart, S.F. O'Keefe, Y.T. Kim, M. McGuire, G. Wilder, H. Huang, Integrated approach for the valorization of red grape pomace: production of oil, polyphenols and acetone-butanol-ethanol, ACS Sustain. Chem. Eng. 6 (12) (2018) 16279–16286.

[25] I.S. Arvanitoyannis, D. Ladas, A. Mavromatis, Potential uses and applications of treated wine waste: a review, Int. J. Food Sci. Technol. 41 (2006) 475–487.

[26] J. Yu, M. Ahmedna, Invited review: functional components of grape pomace: their composition, biological properties and potential applications, Int. J. Food Sci. Technol. 48 (2013) 221–237.

[27] C. Beres, F.F. Simas-Tosin, I. Cabezudo, S.P. Freitas, M. Iacomini, C. Mellinger-Silva, L.M.C. Cabral, Antioxidant dietary fibre recovery from Brazilian Pinot noir grape pomace, Food Chem. 201 (2016) 145–152.

[28] E.T. Nerantzis, P. Tataridis, Integrated Enology- Utilization of winery by-products into high value-added products, e-J. Sci. Technol. 1 (3) (2006) 79–89.

[29] G. Spigno, L. Maggi, D. Amendola, M. Dragoni, D.M. De Faveri, Influence of cultivar on the lignocellulosic fractionation of grape stalks, Ind. Crops Prod. 46 (2013) 283–289.

[30] S.O. Prozil, D.V. Evtuguina, L.P.C. Lopes, Chemical composition of grape stalks of Vitis vinifera L. from red grape pomaces, Ind. Crops Prod. 35 (2012) 178–184.

[31] L. Ping, N. Brosse, L. Chrusciel, P. Navarrete, A. Pizzi, Extraction of condensed tannins from grape pomace for use as wood adhesives, Ind. Crops Prod. 33 (2011) 253–257.

[32] J. Domínguez, J.C. Sanchez-Hernandez, M. Lores, Vermicomposting of winemaking by-Products, in Handbook of Grape Processing By-Products, Handbook of Grape Processing By-Products. Academic Press, Ellsevier Inc. Chapter 3 2017, pp. 55–77. http://dx.doi.org/10.1016/B978-0-12-809870-7.00003-X.

[33] C.S.K. Lin, A.A. Koutinas, K. Stamatelatou, E.B. Mubofu, A.S. Matharu, N. Kopsahelis, L.A. Pfaltzgraff, J.H. Clark, S. Papanikolaou, T.H. Kwan, R. Luque, Current and future trends in food waste valorization for the production of chemicals, materials and fuels: a global perspective, Biofuels, Bioprod. Bioref. 8 (2014) 686–715.

[34] K.N. Kontogiannopoulos, S.I. Patsios, A.J. Karabelas, Tartaric acid recovery from winery lees using cation exchange resin: Optimization by response surface methodology, Sep. Purif. Technol. 165 (2016) 32–41.

[35] N. Zhang, A. Hoadley, J. Patel, S. Lim, C. Li, Sustainable options for the utilization of solid residues from wine production, Waste Manag. 60 (2017) 173–183.

[36] S. Takkellapati, T. Li, M.A. Gonzalez, An overview of biorefinery-derived platform chemicals from a cellulose and hemicellulose biorefinery, Clean Technol. Environ. Policy 20 (2018) 1615–1630.

[37] B. Rivas, A. Torrado, A.B. Moldes, J.M. Domianguez, Tartaric acid recovery from distilled lees and use of the residual solid as an economic nutrient for Lactobacillus, J. Agric. Food Chem. 54 (2006) 7904–7911.

[38] F.G. Braga, F.A.L. Silva, A. Alves, Recovery of winery by-products in the Douro Demarcated Region: production of calcium tartrate and grape pigments, Am. J. Enol. Viticult. 53 (1) (2002) 41–45.

[39] R.A. Muhlack, R. Potumarthi, D.W. Jeffery, Sustainable wineries through waste valorisation: a review of grape marc utilisation for value-added products, J. Clean. Prod. 172 (2018) 3387–3397.

[40] G. Spigno, L. Tramelli, D.M. De Faveri, Effects of extraction time, temperature and solvent on concentration and antioxidant activity of grape marc phenolics, J. Food Eng. 81 (2007) 200–208.

[41] R.R. Watson, Book Polyphenols in Plants Isolation, Purification and Extract Preparation Elsevier Inc. Chapter 10, 2018, pp. 151–164.

[42] A. De la Cerda-Carrasco, R. López-Solís, H. Nuñez-Kalasic, Á. Peña-Neira, E. Obreque-Slier, Phenolic composition and antioxidant capacity of pomaces from four grape varieties (Vitis vinifera L.), J. Sci. Food Agric. 95 (7) (2015) 1521–1527.

[43] G.A. Martinez, S. Rebecchi, D. Decorti, J.M.B. Domingos, A. Natolino, D.D. Rio, L. Bertin, C.D. Porto, F. Fava, Towards multi-purpose biorefinery platforms for the valorisation of red grape pomace: production of polyphenols, volatile fatty acids, polyhydroxyalkanoates and biogas, Green Chem 18 (2016) 261–270.

[44] C. Beres, G.N.S. Costa, I. Cabezudo, N.K. Silva-James, A.S.C. Teles, A.P.G. Cruz, C.M. Silva, R.V. Tonon, L.M.C. Cabral, S.P. Freitas, Towards integral utilization of grape pomace from winemaking process: a review, Waste Manag. 68 (2017) 581–594.

[45] L. Zhang, M.T. Zhu, T. Shi, C. Guo, Y.S. Huang, Y. Chen, M.Y. Xie, Recovery of dietary fiber and polyphenol from grape juice pomace and evaluation of their functional properties and polyphenol compositions, Food Funct 8 (2017) 341–351.

[46] Q. Jin, L. Yang, N. Poe, H. Huang, Review: Integrated processing of plant-derived waste to produce value-added products based on the biorefinery concept, Trends Food Sci. Technol. 74 (2018) 119–131.

[47] K.R. Corbin, Y.S.Y. Hsieh, N.S. Betts, C.S. Byrt, M. Henderson, J. Stork, S. DeBolt, G.B. Fincher, R.A. Burton, Grape marc as a source of carbohydrates for bioethanol: Chemical composition, pretreatment and saccharification, Bioresour. Technol. 193 (2015) 76–83.

[48] V.N. Gunaseelan, Biochemical methane potential of fruits and vegetable solid waste feedstocks, Biomass Bioenergy 26 (2004) 389–399.

[49] A. Fabbri, G. Bonifazi, S. Serranti, Micro-scale energy valorization of grape marcs in winery production plants, Waste Manag. 36 (2015) 156–165.

[50] J.H. El Achkar, T. Lendormi, Z. Hobaika, D. Salameh, N. Louka, R.G. Maroun, J.L. Lanoisellé, Anaerobic digestion of grape pomace: biochemical characterization of the fractions and methane production in batch and continuous digesters, Waste Manag. 50 (2016) 275–282.

[51] A.B. Díaz, I. Ory, I. Caro, A. Blandino, Enhance hydrolytic enzymes production by Aspergillus awamori on supplemented grape pomace, Food Bioproduct Process. 90 (1) (2012) 72–78.

[52] H. Javier, S.J. Ángel, G. Aida, G.M. Carmen, M.M. Ángeles, Revalorization of grape marc waste from liqueur wine: biomethanization, J. Chem. Technol. Biotechnol. 94 (2019) 1499–1508.

[53] R. Paradelo, A.B. Moldes, J.M. Dominguez, M.T. Barral, Reduction of water repellence of hydrophobic plant substrates using biosurfactant produced from hydrolyzed grape marc, J. Agric. Food Chem. 57 (2009) 4895–4899.

[54] A. Jayanegara, F. Leiber, M. Kreuzer, Meta-analysis of the relationship between dietary tannin level and methane formation in ruminants from in vivo and in vitro experiments, J. Anim. Physiol. Anim. Nutr. 96 (2012) 365–375.

[55] C. Nendel, S. Reuter, Kinetics of net nitrogen mineralisation from soil-applied grape residues, Nutr. Cycling Agroecosyst. 79 (2007) 233–241.

[56] M. Bustamante, R. Moral, C. Paredes, A. Pérez-Espinosa, J. Moreno-Caselles, M. Pérez-Murcia, Agrochemical characterisation of the solid by-products and residues from the winery and distillery industry, Waste Manag. 28 (2008) 372–380.

[57] M. Perez-Ameneiro, X. Vecino, L. Vega, R. Devesa-Rey, J. Cruz, A. Moldes, Elimination of micronutrients from winery wastewater using entrapped grape marc in alginate beads, CyTA-J. Food 12 (2014) 73–79.

[58] M. Gómez-Brandón, M. Lores, H. Insam, J. Domínguez, Strategies for recycling and valorization of grape marc, Critical Rev. Biotechnol. 39 (4) (2019) 1–14.

[59] C. Gaita, M.A. Poiana, Grape pomace as an innovative functional ingredient to design value-added food products: A review, J. Agroaliment. Processes Technol. 23 (4) (2017) 223–228.

[60] E. Papadaki, F.T. Mantzouridou, Citric acid production from the integration of Spanish-style green olive processing wastewaters with white grape pomace by Aspergillus niger, Bioresour. Technol. 280 (2019) 59–69.

[61] P. Górnas, M. Rudzinska, Seeds recovered from industry by-products of nine fruit species with a high potential utility as a source of unconventional oil for biodiesel and cosmetic and pharmaceutical sectors, Ind. Crops Prod. 83 (2016) 329–338.

[62] D. Amendola, D.M.D. Faveri, I. Egües, L. Serrano, J. Labidi, G. Spigno, Autohydrolysis and organosolv process for recovery of hemicelluloses, phenolic compounds and lignin from grape stalks, Biores. Technol. 107 (2012) 267–274.

[63] S. Prozil, D. Evtuguin, S. Lopes, L. Cruz-Lopes, A. Arshanitsa, V. Solodovnik, G. Telysheva, Evaluation of grape stalks as a feedstock for pellets production, 13th European Workshop on Lignocellulosics and Pulp, Instituto de Recursos Naturales y Agrobiología de Sevilla, Consejo Superior de Investigaciones Científicas (EWLP2014) 24-27 July, Seville Spain. ISBN: 978-84-616-9842-4, (2014) 683–686.

[64] J. Bastos-Arrieta, A. Florido, C. Pérez-Ràfols, N. Serrano, N. Fiol, J. Poch, I. Villaescusa, Green synthesis of Ag nanoparticles using grape stalk waste extract for the modification of screen-printed electrodes, Nanomaterials 8 (11) (2018) 1–14.

[65] I. Villaescusa, N. Fiol, M. Martınez, N. Miralles, J. Poch, J. Serarols, Removal of copper and nickel ions from aqueous solutions by grape stalks wastes, Water Res. 38 (4) (2004) 992–1002.

[66] N. Teixeira, N. Mateus, V. Freitas, J. Oliveira, Wine industry by-product: full polyphenolic characterization of grape stalks, Food Chem. 268 (2018) 110–117.

[67] A.C. Deiana, M.F. Sardella, H. Silva, A. Amaya, N. Tancredi, Use of grape stalk, a waste of the viticulture industry, to obtain activated carbon, J. Haz. Mater. 172 (2009) 13–19.

[68] N. Kopsahelis, C. Dimou, A. Papadaki, E. Xenopoulos, M. Kyraleou, S. Kallithraka, Y. Kotseridis, S. Papanikolaou, A.A. Koutinas, Refining of wine lees and cheese whey for the production of microbial oil, polyphenol-rich extracts, and value-added co-products, J. Chem. Technol. Biotechnol. 93 (2018) 257–268.

[69] J.A. Pérez-Serradilla, M.D.L. Castro, Microwave-assisted extraction of phenolic compounds from wine lees and spray-drying of the extract, Food Chem. 124 (2014) 1652–1659.

[70] E. Naziri, N. Nenadis, F.T. Mantzouridou, M.Z. Tsimidou, Valorization of the major agrifood industrial by-products and waste from Central Macedonia (Greece) for the recovery of compounds for food applications, Food Res. Int. 65 (2014) 350–358.

[71] G. Bustos, A.B. Moldes, J.M. Cruz, J.M. Dominguez, Production of lactic acid from vine-trimming wastes and viticulture lees using a simultaneous saccharification fermentation method, J. Sci Food Agric. 85 (2005) 466–472.

[72] C. Da Ros, C. Cavinato, P. Pavan, D. Bolzonella, Winery waste recycling through anaerobic co-digestion with waste activated sludge, Waste Manag. 34 (11) (2014) 2028–2035.

[73] M.J. Jara-Palacios, Review Wine Lees as a source of antioxidant compounds, Antioxidants 8 (45) (2019) 1–10.

[74] Y. Tao, D. Wu, Q.A. Zhang, D.W. Sun, Ultrasound-assisted extraction of phenolics from wine lees: Modeling, optimization, and stability of extracts during storage, Ultrason. Sonochem. 21 (2014) 706–715.

[75] R. Romero-Díez, S. Rodríguez-Rojo, M.J. Cocero, C.M.M. Duarte, A.A. Matias, M.R. Bronze, Phenolic characterization of aging wine lees: Correlation with antioxidant activities, Food Chem 259 (2018) 188–195.

[76] M.P. Delgado-Torre, C. Ferreiro-Vera, F. Priego-Capote, P.M. Pérez-Juan, M.D.L. Castro, Comparison of accelerated methods for the extraction of phenolic compounds from different vine-shoot cultivars, J. Agric. Food Chem. 60 (2012) 3051–3060.

[77] I. Dávila, P. Gullón, M.A. Andrés, J. Labidi, Coproduction of lignin and glucose from vine shoots by eco-friendly strategies: Toward the development of an integrated biorefinery, Biores. Technol. 244 (2017) 328–337.

[78] R.S. Sánchez-Gómez, R.S. Vioque, O.S. Méridas, M.M. Bejerano, G.L. Alonso, M.R. Salinas, A. Zalacain, A potential use of vine-shoot wastes: the antioxidant, antifeedant and phytotoxic activities of their aqueous extracts, Ind. Crops Products 97 (2017) 120–127.

[79] O.M. Portilla, B. Rivas, A. Torrado, A.B. Moldes, J.M. Dominguez, Revalorisation of vine trimming wastes using Lactobacillus acidophilus and Debaryomyces hansenii, J Sci. Food Agric. 88 (2008) 2298–2308.

[80] L. Jiménez, V. Angulo, E. Ramos, M.J. De la Torre, J.L. Ferrer, Comparison of various pulping processes for producing pulp from vine shoots, Ind. Crops Prod. 23 (2) (2006) 122–130.

[81] J.M.V. Nabais, C. Laginhas, P.J.M. Carrott, M.M.L. RibeiroCarrott, Thermal conversion of a novel biomass agricultural residue (vine shoots) into activated carbon using activation with CO_2, J. Analyt. Appl. Pyrol. 87 (1) (2010) 8–13.

[82] I. Dávila, B. Gullón, J.L. Alonsoc, J. Labidi, P. Gullón, Vine shoots as new source for the manufacture of prebiotic oligosaccharides, Carbohyd. Pol. 207 (2019) 34–43.

[83] S. Cortes-Camargo, N. Perez-Rodriguez, R.P. Souza Oliveira, B.E. Barragan Huerta, J.M. Dominguez, Production of biosurfactants from vine-trimming shoots using the halotolerant strain Bacillus tequilensis ZSB10, Ind. Crops Prod. 79 (2016) 258–266.

[84] G. Bustos, N.D. Torre, A.B. Moldes, J.M. Cruz, J.M. Dominguez, Revalorization of hemicellulosic trimming vine shoots hydrolyzates through continuous production of lactic acid and biosurfactants by L. pentosus, J. Food Eng. 78 (2007) 405–412.

[85] N. Rodríguez-Pazo, J.M. Salgado, S. Cortés-Diéguez, J.M. Domínguez, Biotechnological production of phenyllactic acid and biosurfactants from trimming vine shoot hydrolyzates by microbial coculture fermentation, Appl. Biochem. Biotechnol. 169 (2013) 2175–2188.

[86] S. Montalvo, J. Martinez, A. Castillo, C. Huiliñir, R. Borja, V. García, R. Salazar, Sustainable energy for a winery through biogas production and its utilization: A Chilean case study, Sustain. Energy Technol. Assess. 37 (2020) 1–9.

[87] E.R. Pachón, P. Mandade, E. Gnansounou, Conversion of vine shoots into bioethanol and chemicals: Prospective LCA of biorefinery concept, Biores. Technol. 303 (122946) (2020) 1–9.

[88] T.H. Makadia, E. Shahsavari, E.M. Adetutu, P.J. Sheppard, A.S. Ball, Effect of anaerobic co-digestion of grape marc and winery wastewater on energy production, Aust. J. Crop Sci. 10 (1) (2016) 57–61.

[89] K.P.M. Mosse, A.F. Patti, E.W. Christen, T.R. Cavagnaro, Review: Winery wastewater quality and treatment options in Australia, Aust. J. Grape Wine Res. 17 (2011) 111–122.

[90] C.D. Ros, C. Cavinato, P. Pavan, D. Bolzonella, Mesophilic and thermophilic anaerobic co-digestion of winery wastewater sludge and wine lees: An integrated approach for sustainable wine production, J. Environ. Manag. 203 (2) (2017) 745–752.

[91] G. Andreottola, P. Foladori, G. Ziglio, Biological treatment of winery wastewater: an overview, Water Sci. Technol. 60 (5) (2009) 1117–1125.

[92] R. Moletta, Winery and distillery wastewater treatment by anaerobic digestion, Water Sci. Technol. 51 (1) (2005) 137–144.

[93] L. Andres, F. Riera, R. Alvarez, Recovery and concentration by electrodialysis of tartaric acid from fruit juice industries wastewater, J. Chem. Technol. Biotechnol. 70 (1997) 247–252.

[94] M. Lucarini, A. Durazzo, A. Romani, M. Campo, G. Lombardi-Boccia, F. Cecchini, Bio-Based compounds from grape seeds: a biorefinery approach, Molecules 23 (2018) 1–12.

[95] A.J. Ragauskas, G.T. Beckham, M.J. Biddy, R. Chandra, F. Chen, M.F. Davis, B.H. Davison, R.A. Dixon, P. Gilna, M. Keller, P. Langan, A.K. Naskar, J.N. Saddler, T.J. Tschaplinski, G.A. Tuskan, C.E. Wyman, Lignin valorization: Improving lignin processing in the biorefinery, Science 344 (2014) 709–719.

[96] Z. Strassberger, S. Tanase, G. Rothenberg, The pros and cons of lignin valorisation in an integrated biorefinery, RSC Adv. 4 (2014) 25310–25318.

[97] M. Kleinert, T. Barth, Phenols from lignin, Chem. Eng. Technol. 31 (5) (2008) 736–745.

[98] O.A.T. Dias, D.R. Negrão, D.F.C. Gonçalves, I. Cesarino, A.L. Leão, Recent approaches and future trends for lignin-based materials, J. Mol. Cryst. Liq. Cryst. 655 (1) (2017) 204–223.

[99] F.J. Barba, S. Brianceau, M. Turk, N. Boussetta, E. Vorobiev, Effect of Alternative Physical Treatments (Ultrasounds, Pulsed Electric Fields, and High-Voltage Electrical Discharges) on Selective Recovery of Bio-compounds from Fermented Grape Pomace, Food Bioprocess. Technol. 8 (2015) 1139–1148.

[100] C.D. Porto, A. Natolino, M. Scalet, Ultrasound-assisted extraction of proanthocyanidins from vine shoots of Vitis vinifera, J. Wine Res. 29 (4) (2018) 290–301.

[101] Y. Jiang, J. Simonsen, Y. Zhao, Compression-molded biocomposite boards from red and white wine grape pomaces, J. Appl. Polym. Sci. 119 (5) (2011) 2834–2846.

[102] A.B.B. Bender, M.M. Luvielmo, B.B. Loureiro, C.S. Speroni, A.A. Boligon, L.P. Silva, N.G. Penna, Obtaining and characterizing grape skin flour and its use in extruded snacks, Braz. J. Food Technol. 19 (2016) 1–9.

[103] L. Luo, Y. Cui, S. Zhang, L. Li, Y. Li, P. Zhou, B. Sun, Preparative separation of grape skin polyphenols by high-speed counter-current chromatography, Food Chem. 212 (2016) 712–721.

[104] F. Pasini, F. Chinnici, M.F. Caboni, V. Verardo, Recovery of Oligomeric Proanthocyanidins and Other Phenolic Compounds with Established Bioactivity from Grape Seed By-Products, Molecules 24 (4) (2019) 1–12.

[105] Y.H. Wei, C.L. Chou, J.S. Chang, Rhamnolipid production by indigenous Pseudomonas aeruginosa J4 originating from petrochemical wastewater, Biochem. Eng. J. 27 (2) (2005) 146–154.

[106] A. Shrikhande, Wine byproducts with health benefits, Food Res. Int. 33 (2000) 469–474.

[107] S. Surini, K. Nursatyani, D. Ramadon, Gel formulation containing microcapsules of Grape seed oil (Vitis vinifera L.) for skin moisturizer, J. Young Pharmacists 10 (2018) 41–47.

[108] Y. Zheng, C. Lee, C. Yu, Y.S. Cheng, C.W. Simmons, R. Zhang, B.M. Jenkins, J.S. VanderGheynst, Ensilage and Bioconversion of Grape Pomace into Fuel Ethanol, J. Agric. Food Chem. 60 (2012) 11128–11134.

[109] Y. Zhijing, A. Shavandi, R. Harrison, A.E.A. Bekhit, Characterization of Phenolic Compounds in Wine Lees, Antioxidants 7 (48) (2018) 1–13.

[110] F. Smagge, J. Mourgues, J. Escudier, T. Conte, J. Molinier, C. Malmary, Recovery of calcium tartrate and calcium malate in effluents from grape sugar production by electrodialysis, Bioresour. Technol. 39 (1992) 85–189.

Index

Page numbers followed by "*f*" and "*t*" indicate, figures and tables respectively.

A

Accidental adversaries, 10
Agricultural residues
  advantages of, 288
  and bioethanol, 210
  mass composition of, 290
Agricultural systems, 117
Agrivoltaics, 189
Agrivoltaic systems, 204
Agro-Environmental Sugarcane Industry Protocol, 318–319
Air quality, in Sao Paulo (Brazil), 333
Air separation unit (ASU), 244
Amazon rainforests, 100–101
Ammonia production, 244
Amorphous thin-film silicon (a-Si, TF-Si), 203
Anaerobic digestion (AD)
  in biomethane production, 284
  of organic materials, 231
Animal diversity, in Sao Paulo (Brazil), 335–336
Antifragility, 114–115
Ash content, 350–351
Aspen Energy Analyzer V8.8, 313
Aspen Plus, 25–26

B

Bark residues, 346
Benin
  in 2019, popupation of, 280*f*
  cases of, 300
  cellulose, hemicellulose and lignin available in, 291*t*
  energy consumption by, 270*t*
  feedstock available in, 272*t*
  total agricultural residues in, 282*f*
Bioactive compounds, 388
Biobutanol, 283
Biodiversity, in Sao Paulo (Brazil), 335
Bioethanol, 261, 283
  plant
    cash flow analysis for 300 MT/day, 219*t*
    economic analysis for 300 MT/day, 219*t*
    material input costs for 300 MT/day, 219*t*
  production, rice straw in, 216
  production from agricultural residues, advantage of, 210–216
  resource assessment for, 210
Biofuels, 147, 153
  in India, social aspects of, 224
  matrix, 301–302 *See also* (West Africa)
  in West Africa, 263
  policies, 209
  second-generation, 228
  strategies, in West Africa, 261, 269–271, 290
  supply, relative cost function of, 261
  technologies, system analysis of, 209
Biogas, 231
  carbon dioxide emissions from, 236
  composition of, 231
  cost of upgrading, 236
  in developed country, purification of, 234
  digesters, cost of, 235–236
  production with carbon capture, 231
  systems, in developed country, 234*f*
  upgrading techniques, energy consumption and operational expenditure for, 239*t*
  usage in developing country, 234
Biogenic CO_2, 231
Bio-geochemical cycles, 236
Biomass
  availability, 210
  feedstock, 281
  gasification, 285
Biomass integrated gasification combined cycle (BIGCC), 285, 287
Biomethane, 231, 284
  burning of, 236
  market, global, 232
  natural gas and, 236
  from organic wastes, 232
  production
    from anaerobic digestion, 284
    from biomass gasification, 285
  systems analysis, dimensions of, 233
    economics, 235
    environment, 236
    market, 237
    policy, 237
    social aspects, 237
    technology, 233

Biomethane (*continued*)
　use, case study of Ireland for, 238
　　background, 238
　　policy influences technology, 239
　　role of technology and economics in, 239
　use in Ireland, 238
　　background, 238
　　policy influences technology, 239
　　role of technology and economics in, 239
Biomethanol, 283–284
Biorefineries, semantic sustainability characterization of, 311–312
　case study, 316
　　air quality, 333
　　animal diversity, 335–336
　　biodiversity, 335
　　context, 318
　　economic specific indicators, 319
　　energy affordability, 326–327
　　energy dependence, 326
　　fossil fuel conservation, 327
　　greenhouse gas emissions, 336
　　land productivity, 331–332
　　landscape and cultural heritage conservation, 328–329
　　land use management, 331
　　net energy balance, 327
　　net nonrenewable energy indicator, 336–337
　　process design, 316
　　rural development and workforce, 329
　　rural livelihood improvement, 329–330
　　social affordability, 321–323
　　social well-being and prosperity, 323
　　soil quality, 332
　　stakeholders' participation, 323
　　technological innovation and development of skilled workforce, 330–331
　　transparency, 323
　　water conservation, 327–328
　　water quality, 333–334
　problematic of, 313
　　criteria for selecting the indicators, 314
　　economic indicators, 314
　　environmental indicators, 315
　　improvement of the energy efficiency, 313
　　logic-based model, 313
　　model, background of, 313
　　rule-base, 315
　　social indicators, 314
Biorefinery, 380
　establishment and biomass density, 210–216
　process of, 222
　surplus electricity product, 327
Bio-Synthetic Natural Gas (Bio-SNG), 286–287

Brazil, in Corruption Perception Index ranking, 323
Brazilian Regulatory Electricity Agency (ANEEL), 326–327
Briquettes and pellets, 348
Brundtland report, 314
Burkina Faso
　in 2019, popupation of, 280*f*
　cellulose, hemicellulose and lignin available in, 291*t*
　energy consumption by, 270*t*
　total agricultural residues in, 282*f*
Business as usual (BAU), 289
Butanol, 283

C

Cadmium telluride (CdTe), 203
Canal-based solar plants, 189
Capital expenditure (CAPEX), 235
Carbon and hydrogen, 351
Carbon dioxide capturing technologies, 239–240
Carbon footprint, 37
Carbon tax, 239–240
Cash flow analysis, 21
Cellulosic ethanol, 164, 283
Cereals production, in single day, 243
Child labor, in Brazil, 323–324
Coal Integrated Gasification and Combined Cycle, 287
CO_2 emission
　from Haber-Bosch process, 246–247
　per capita (2019), average, 231
Commission for Additional Sources of Energy (CASE), 175–180
Continuously stirred tank reactors (CSTR), 234
Conversion cost, economic viability of, 320
Copper indium gallium diselenide (CIGS), 203
Corn dry grind ethanol process, 91, 154
Corn dry milling technology, 147–149
Corn ethanol plants, in the United States, 235
Council for Scientific and Industrial Research, 175–180
Crop production, in single day, fertilizers used in, 243
Crop yields, 86
Crude oil price volatility, 320
Crystalline silicon solar panels, primary components of, 203

D

Demand reduction programs, 189–191
Demand-side policies, 362
Digestate, 237
Direct capital costs, 21
Direct electrochemical nitrogen reduction, 248
Dispatchability, defined, 196
Dow Jones Sustainability World Index, 128–129
Dry animal manure, 345

E

Ecoinvent database 3.33, 222
Ecological footprint, 34
Ecological risk assessment (ERA), 63
Ecology, 135–136
Economic affordability, of farm gate feedstock price, 319
Economic indicators and biorefineries, 314
Economic profitability, 140
Economics, of Haber-Bosch process, 244–246
Economic viability, of electricity produced, 321
Ecotoxicity, 332–333
Electricity for transportation, in West Africa, 285
Electric Reliability Council of Texas (ERCOT), 250
Electrochemical H_2 production and H-B process, 248
Electrochemical nitrogen reduction (ENR), 247, 248
Energy affordability, 326–327
Energy dependence, 326
Energy requirements, of Haber-Bosch process, 244
Energy storage technologies, 198
"Enlightened anthropocentrism,", 133
Environmental assessment, of second-generation biofuels, 228
Environmental emissions, 355–358
Environmental impact assessment, 199, 222
    goal and scope, 222
    solar, 201
    wind, 199
Environmental impacts, 33
    methods used for, 33–37
Environmental incentives, 137
Environmental indicators, 315
Environmental policy, 138
Environmental risk assessment, 53
Environment-centered sustainability, 142f
Enzymatic hydrolysis and fermentation, 153
Ethanol, production of, 227
Ethanol recovery technologies, 153
Ethanol yield, 210
Ethylene vinyl acetate (EVA), 203
European Commission, 128–129
    and biofuel, 269
Eutrophication, 334
Event tree, 62
Exclusion and consistency rules, 132

F

FAOSTAT, role of, 288–289
Feedstock
    for advanced biofuels, in West Africa, 281
    availability, 364
    in case of the recommended strategy, 272
    logistic cost, economic affordability of, 319–320
    pressure on, 266 *See also* (West Africa)

Fertilizer synthesis and Haber-Bosch process, 244
First and Second Generation (1G2G) ethanol plant, 313
First-generation biofuels, 259–260
First-generation solar panel resource requirements, 202–203
Fixed capital costs, 19–20
Fixed dome model, 234
Floating drum model, 234
Flovoltaics, 189
Food and Agriculture Organization (FAO), 289
    business as usual, 289
    stratified societies, 289
    toward sustainability, 289
Food and Agriculture Organization Corporate Statistical Database, 371
Food security, in Sao Paulo (Brazil), 325–326
Food waste generation, in Ireland, 238
Fossil fuel
    conservation, 327
    energy, 196
Fragility, 114
Fruits production, in single day, 243
Fuel cells, 284
Fuelwood, 345
Futurol, 283

G

Gambia
    in 2019, popupation of, 280f
    cellulose, hemicellulose and lignin available in, 291t
    energy consumption by, 270t
    total agricultural residues in, 282f
Gate fee, 237
General equilibrium models (GEM), 259–260
Ghana
    in 2019, popupation of, 280f
    cellulose, hemicellulose and lignin available in, 291t
    energy consumption by, 270t
    feedstock available in, 272t
    total agricultural residues in, 282f
Gini coefficient, 321–323
Global ecological damages, 137
Global solid waste production, in 2016, 231–232
Global vegetables production, in single day, 243
Good governance, 105–106
Governance, 105–106
Grape processing, 378
Grape stalks, 378
Green harvesting residues (GHR), composition of, 316t
Greenhouse gas (GHG) emissions, 336, 371
Green Product or System", 128–129
Greywater, 34

Gross national product (GNP), 10
Guinea
  in 2019, popupation of, 280f
  cellulose, hemicellulose and lignin available in, 291t
  energy consumption by, 270t
  feedstock available in, 272t
  total agricultural residues in, 282f

H

Haber-Bosch (H-B) process, 243, 244
  CO_2 emissions, 246–247
  economics of, 244–246
  electrochemical H_2 production, 248
  energy requirements of, 244
H2A Distributed Hydrogen Production Model, 246
Hazard analysis and critical control points (HACCP), 59–61
Heat exchanger network (HEN), 313
Heating value (calorific value), 351
Heat Recovery Steam Generator, 287
Heavy metals, 352
Heirarchist (H) impact assessment model, 222–223
Henry Hub natural gas price, 244–246
Human resource barriers, 187–188
Hybrid methods, 45f
Hydrogen, 284
  synthesis, 247
Hydrothermal carbonization, 348

I

Increasing severity factors, 152–153
India
  annual agricultural residue consumption in, 212t
  annual agricultural residue generation in, 211t
  annual agricultural residue surplus in, 211t
  annual surplus biomass density, 214t
  biofuel technologies in, 209
  biomass surplus and biomass density in, 217f
  environmental impact assessment, 222
  policy and social aspects of biofuels in, 224
  resource assessment for bioethanol, 210
  state wise cellulosic ethanol potential of, 215t
  techno-economic analysis, 216
Indium, usage of, 203
Industrial mega solar parks, 188
Inherent randomness, 23
Input-output matrix, 44
Input-output model of an economy, 43–44
Intended Nationally Determined Contribution (INDC), 175
Intergovernmental Panel on Climate Change (IPCC), 199t, 199
Internal rate of return (IRR), 22, 216–218

International Energy Agency, 283
International Renewable Energy Agency, 259–260
Ireland
  biomethane use in, 238
    background, 238
    policy influences technology, 239
    role of technology and economics in, 239
Ivory Coast
  in 2019, popupation of, 280f
  cellulose, hemicellulose and lignin available in, 291t
  energy consumption by, 270t
  feedstock available in, 272t
  total agricultural residues in, 282f

J

Jet cooking and liquefaction, 147–149
Job creation and 1G2G ethanol value chain, 329

K

Kharif season, 210

L

Land and nutrient resource, 161–162
Land productivity, in Brazil, 331–332
Land resources, 76
Landscape and cultural heritage conservation, in Sao Paulo (Brazil), 328–329
Land use management, in Brazil, 331
Liberia
  in 2019, popupation of, 280f
  cellulose, hemicellulose and lignin available in, 270t
  energy consumption by, 272t
  feedstock available in, 282t
  total agricultural residues in, 282f
Life cycle assessment/analysis (LCA), 37–38
  definition and scoping, 39–40
  four steps, 39f
  inventory, 39f
  inventory databases, 41t
  inventory steps, 40f
  environmental impacts, 355
  impact assessment, 46
  software V1.10.3, 222
  step, 42
  interpretation, 46–47
  inventory, 222
  inventory and emissions, 160–161
Lignocellulosic bioethanol plant, 286
Lignocellulosic biomass, 281
  amount of, 281
  categories, 281
Lignocellulosic residues, 388

Index

Linear event-based thinking, 3f
Logic-Based Model (LBM), 311–312
Long Fuse Big Bang processes, 123–124

M

Mali
    in 2019, population of, 280f
    cellulose, hemicellulose and lignin available in, 291t
    energy consumption by, 270t
    feedstock available in, 272t
    total agricultural residues in, 282f
Maximum purchasing price (MPP), 319
Measurement error, 23
Message maps, 64–65
Methane, 237
Methyl Terbutyl Ether, 147
Microalgae upgrading, 239
Milena-type gasifier, 286–287
Minimum selling price (MSP), 314
Model uncertainty, 23
Moisture content, 348–349
Montreal protocol, 101
Multifeedstock plants, in West Africa, 285
    biomass integrated gasification combined cycle, 287
    lignocellulosic bioethanol plant, 286
    synthetic natural gas, 286–287
Municipal solid waste, 347

N

National Biofuel Coordination Committee, 224
National biofuels policy 2018, 209
National policy on biofuels (NPB), 224
National Renewable Energy Laboratory (NREL), 197
National solar mission (NSM), 180–182
Natural gas reformation, 244
Natural variation, 23
Negative emission technology (NET), 231
Net energy balance, of biorefinery supply chain, 327
Net nonrenewable energy indicator, 336–337
Net present value, 18
Nexus barriers, 188
Nigeria
    in 2019, popupation of, 280f
    cellulose, hemicellulose and lignin available in, 291t
    energy consumption by, 270t
    feedstock available in, 272t
    total agricultural residues in, 282f
Nitrogen, for plant growth, 243
Nuclear fission, 4–5

O

Open burning, adverse effects of, 231–232
Operational costs, 21
Operation and maintenance (O&M) costs, 246
Organization Environmental Footprint Sector Rules (OEFSR), 129

P

Paris Climate Accords, 205–206
Particle size and particle density, 349–350
Payback period, 21
Peak power balancing, 191
Photovoltaic (PV) solar cells, 196
Pigouvian tax, 100–101
Plant-based biomass, 375–376
Plant diversity, in Sao Paulo (Brazil), 335
Policy
    for biomethane production, 239–240
    for renewable energy production, 237
    and social aspects of biofuels, in India, 224
Policy, governance and financial barriers, 187
Policy making
    and governance, 99
    frameworks, 103
    models, 101–102
Polycrystalline silicon panels, 202–203
Potassium, 90
Primary sources, 344
Processed solid biofuels, 347
Product Environmental Footprint (PEF), 129
Product price volatility, economic sensitivity to, 320
Pump-Stored Hydro (PSH), 198
Punjab, stubble burning in, 224–227
Purchase agreements (PPAs), 182
    PPA contracts, 191–192

Q

Quantitative tools, 15
Quartz, 203

R

Rabi season, 210
Real-life processes, 46
Real option
    analysis, 24
    strategy, 24–25
Reinforcing and balancing loops, 5–12
Renewable Electricity Futures Study (RE-Futures Study), 197
Renewable energy, advantages of, 195
Renewable Energy Directive and Fuel Quality Directive (RED-FQD) wastes, 373
Renewable Energy Directive of the European Union of 2018 (REDII), 269
Renewable energy sources, 343

Renewable Fuel Standards (RFS), 100–101
Renewable portfolio standards (RPS), 205, 206f
Residual risk, 55
Residue-to-product ratio (RPR), 288
Resilience, 115–117
Resilience thinking, 113
Resilience thinking in systems analysis, 123
Resource Barriers, 187
Resource environmental profile analysis, 37–38
Resources, 75
Rhizobium, 86–87
Ribeirão Preto municipality, 321–323
Rice husk, 346
Rice straw
    in bioethanol production, 216
    ethanol
        life cycle impacts of, 223t
        process, 223t
    stubble burning, 224–227
Risk analysis method, 56
    ISO framework, 57f
Risk assessment process, 53
Risk management, 56–57
Robustness, 114
Robust strategy, determination of, 271t
Rule base, 142
Rule beating, 10
Rural livelihood, 329–330

S

Sao Paulo (Brazil)
    air quality in, 333
    animal diversity in, 335–336
    biorefinery plant in, 318
    Gini coefficient in, 321–323
    land productivity in, 331–332
    landscape and cultural heritage conservation in, 328–329
    mechanical harvesting of sugarcane in, 323–324
    plant diversity in, 335
    poor people in, 321–323
    social indicators, 314–315
    soil quality in, 332
    sugarcane and ethanol production industries in, 323–324
        safety and health of the employees, 324–325
        sugar-ethanol chain index in, 324–325
    tropospheric ozone in, 333
    water quality in, 333–334
Saw dust, 346
Schockley-Quisser efficiency limit, 204
Second-generation
    biofuels, 228
    ethanol, 150–152
    solar panel resource requirements, 203

Selenium wafers, usage of, 196
Senate Bill 100 (SB100), 205–206
Senegal
    cellulose, hemicellulose and lignin available in, 291t
    energy consumption by, 270t
    feedstock available in, 272t
    total agricultural residues in, 282f
Sierra Leona
    cellulose, hemicellulose and lignin available in, 291t
    energy consumption by, 270t
    feedstock available in, 272t
    total agricultural residues in, 282f
Silica sand, 203
Silicon-based (C-si) solar panel, 205
Silicon PV cells, 196
Six-step approach structures, 133
Slurry viscosity, 150
Social affordability, 321–323
Social impact analysis, 205
Social indicators
    and bioenergy production, 314–315
    evaluation of, 314–315
Social well-being and prosperity, 323
Socio-ecological systems, 123
Soil quality, in Sao Paulo (Brazil), 332
Solar energy, 175–180
    end of life, 204
    generation in the United States, 196
    and GHG emissions, 201
    history of, 196
    land use, 204
    resource sustainability, 202
        concentrating solar energy resource requirements, 204
        first-generation solar panel resource requirements, 202
        second-generation solar panel resource requirements, 203
        third-generation solar panel resource requirements, 204
    usage of, 196
    water use, 204
Solar panels, 189
Solar tracking systems, 186–187
Solid biofuel, 348
    bulk density of, 348–349
    mechanical durability of, 350
Solid biofuel policies, 361
Solid biofuels, 343, 344, 352, 353, 355–358
Stakeholders' participation, 323
Stratified societies, 289 *See also* Food and Agriculture Organization (FAO)
Stubble burning, 224–227
Subjective judgment, 23–24

Index

Sugarcane
    composition of, 316t
    usage of, 210–216
    industries, in Sao Paulo (Brazil), 323–324
Sulphur content, 351–352
SuperPro Designer, 25–26
SuperPro V 8.5, 218
Sustainability, 113, 127, 131
Sustainability indicators, 131
"Sustainability Value Added (SVA), 130
Sustainable bioeconomy, 14–15
Synoptic rationality model of policy making, 102
Synthetic natural gas (SNG), 260, 286–287
    advantage of, 263–265
    higher yields of, 269
    share of, 265
Synthetic nitrogenous fertilizers, in agriculture, 243
SysRes Model, 120
System archetypes, 7f, 8f, 11f
Systematic error:, 23
System dynamics, 4–5
Systems analysis, 2
Systems thinking, 2

T

Techno-economic analysis, 18, 216 *See also* India
Techno-economic assessment, 17
Technology readiness level (TRL), 232
Tellurium, global supply of, 203
Terrestrial acidification, 332–333
Thin film (TF) solar cells, 203
Third-generation solar panel resource requirements, 204
Togo
    in 2019, popupation of, 280f
    cases of, 300
    cellulose, hemicellulose and lignin available in, 291t
    energy consumption by, 270t
    feedstock available in, 272t
    total agricultural residues in, 282f
Torrefied biomass and hydrochar, 347
Toward sustainability (TS), 289
TRACI (tools for reduction and assessments of chemicals and other environmental impacts), 236
Traditional investment strategies, 24
Transformed risk, 55
Tropospheric ozone, in Sao Paulo (Brazil), 333
Tubers production, in single day, 243

U

UN Brundlandt Commission, 113
Uncertainty and sensitivity analysis, 22–24
United Nations Environmental Programme Green-Economy, 128–129
United States, solar and wind energy in, 195
    environmental impact assessment, 199
        solar, 201
        wind, 199
    perspectives, 207
    policy, governance, and social impact analysis, 205
    renewable electricity futures study, 197
    resource sustainability analysis, 201
        wind energy land and water use, 201
        wind energy resource sustainability, 201
        wind turbine end of life, 202
    solar energy resource sustainability, 202
        concentrating solar energy resource requirements, 204
        first-generation solar panel resource requirements, 202
        second-generation solar panel resource requirements, 203
        solar energy end of life, 204
        solar energy land use, 204
        solar energy water use, 204
        third-generation solar panel resource requirements, 204
    storage, 198
    technical feasibility analysis, 196
United States Renewable Electricity Production Tax Credit, 362
Unnat Jyoti by Affordable LEDs for all (UJALA) scheme, 189–191
Unprocessed solid biofuels, 345

V

Value added, 131
Viticulture waste, 373–374
Volatile matter, 350
Vulnerability, 114

W

Wafer-based solar cells, 202–203
Waste Framework Directive (WFD), 373
Water conservation, 327–328
Water footprints, 36t
Water quality, in Sao Paulo (Brazil), 333–334
Water scrubbing, 239
Weak sustainability, 127
West Africa
    biofuels for transportation in, 279
        biobutanol, 283
        bioethanol, 283
        biomass integrated gasification combined cycle, 287
        biomethane, 284

West Africa (continued)
   biomethane production from anaerobic digestion, 284
   biomethane production from biomass gasification, 285
   biomethanol, 283–284
   electricity, 285
   feedstock for advanced biofuels, 281
   hydrogen, 284
   lignocellulosic bioethanol plant, 286
   multifeedstock plants, 285
   synthetic natural gas, 286–287
evaluation of the available feedstock, 288
influences of the methodology, 287–288
matrix of biofuels, 263
   bioenergy consumption, 274
   common results for the countries, 263
   country specific results, 265
   pressure on the feedstock, 272
   recommended biofuels strategy, 269
optimal biofuel strategies, 261, 290
   scenario 1, 291–293
   scenario 2, 293
   scenario 3, 293
   scenario 4, 293
   scenario 5, 293–294
   scenario 6, 294
   scenario 7, 295
   scenario 8, 295
   scenario 9, 295–298
   scenario 10, 298
   scenario 11, 298
   scenario 12, 298
   scenario 13, 298–300
   scenario 14, 300
   scenario 15, 300
   scenario 16, 300
robustness analyses, 301
   matrix of biofuels, 301–302
Wet milling technology, 150
Wind energy, land and water use, 201–202
Windmill
   for irrigation, 195
   in the United States, first, 195
Wind plants installation, in United States, 195–196
Wind power
   for pumping water and grinding grain, 195
   technologies, 199
      environmental impact of, 199–201
      life cycle of, 199
Wind turbine
   composition of, 201
   end of life, 202
Wine production, 374–375
Winery wastes, 377
Winery wastes and residues, 374–376, 378–379
Wood and agricultural industries, 346
Wood chips, 346
Wood residues, moisture content of, 288
World Business Council, 130

Y

Yara/BASF ammonia plant, 246

Z

Zaid season, 210
Zero carbon sources, 205–206
Zero-order estimates, 18

Printed in the United States
by Baker & Taylor Publisher Services